Catalytic Reforming and Hydrogen Production: From the Past to the Future

Catalytic Reforming and Hydrogen Production: From the Past to the Future

Guest Editors

Georgios Bampos
Paraskevi Panagiotopoulou
Eleni A. Kyriakidou

Basel • Beijing • Wuhan • Barcelona • Belgrade • Novi Sad • Cluj • Manchester

Guest Editors

Georgios Bampos
Department of Chemical
Engineering
University of Patras
Patras
Greece

Paraskevi Panagiotopoulou
School of Chemical and
Environmental Engineering
Technical University of Crete
Chania
Greece

Eleni A. Kyriakidou
School of Engineering and
Applied Sciences
University at Buffalo
Buffalo, NY
United States

Editorial Office
MDPI AG
Grosspeteranlage 5
4052 Basel, Switzerland

This is a reprint of the Special Issue, published open access by the journal *Catalysts* (ISSN 2073-4344), freely accessible at: https://www.mdpi.com/journal/catalysts/special_issues/QEF21TP8VH.

For citation purposes, cite each article independently as indicated on the article page online and as indicated below:

Lastname, A.A.; Lastname, B.B. Article Title. *Journal Name* **Year**, *Volume Number*, Page Range.

ISBN 978-3-7258-3920-9 (Hbk)
ISBN 978-3-7258-3919-3 (PDF)
https://doi.org/10.3390/books978-3-7258-3919-3

Contents

Editorial

Catalytic Reforming and Hydrogen Production: From the Past to the Future

Georgios Bampos [1,*], Paraskevi Panagiotopoulou [2] and Eleni A. Kyriakidou [3]

[1] Department of Chemical Engineering, University of Patras, GR-26504 Patras, Greece
[2] School of Environmental Engineering, Technical University of Crete, GR-73100 Chania, Greece; ppanagiotopoulou@isc.tuc.gr
[3] Department of Chemical and Biological Engineering, University at Buffalo, The State University of New York, Buffalo, NY 14260, USA; elenikyr@buffalo.edu
* Correspondence: geoba@chemeng.upatras.gr

Received: 26 March 2025
Accepted: 28 March 2025
Published: 31 March 2025

Citation: Bampos, G.; Panagiotopoulou, P.; Kyriakidou, E.A. Catalytic Reforming and Hydrogen Production: From the Past to the Future. *Catalysts* **2025**, *15*, 332. https://doi.org/10.3390/catal15040332

1. Introduction

Continuously increasing energy demands and the intense environmental pollution caused by the increasing global population and modern lifestyles have driven research interest toward finding alternative sustainable energy sources [1]. These renewable energy sources can be divided into primary and secondary energy sources depending on whether the produced energy comes directly from a natural source, such as solar or wind energy, or requires conversion into another form [2]. One of the most characteristic secondary energy sources is hydrogen (H_2), typically produced via water electrolysis or the reforming of biomass [3–6]. Because of hydrogen's high energy content, the reforming processes used for various hydrocarbons obtained from non-renewable sources, such as natural gas, have been extensively investigated over recent decades [7–10]. The significant development of high-yield H_2-fueled low-temperature fuel cells has contributed toward this research direction [11]. The development of innovative catalytic systems for the aforementioned hydrogen-related processes and their optimization has attracted the interest of the research community. This Special Issue includes eight research articles, and two review papers related to catalytic reforming processes for H_2 production, highlighting the significance of this secondary energy source of H_2 in a fossil-fuel-free future. The detailed research articles introduce potential readers to new insights and perspectives concerning catalytic processes for hydrogen production, whereas the reviews summarize recent developments in catalytic systems for reforming methanol and the synthesis and catalytic properties of cobalt manganese oxide spinels typically applied in processes such as the production of chemicals and fuels using the Fischer–Tropsch process.

2. Overview of Published Articles

The articles published in this Special Issue focus on advanced catalytic processes for H_2 production, including the dry reforming of methane (DRM), methanol steam reforming (MSR), methanol aqueous-phase reforming (APR), and photocatalytic processes.

DRM is among the most promising technologies, utilizing both CH_4 and CO_2, i.e., the two primary greenhouse gases, for syngas ($CO+H_2$) production. One of the most significant challenges hampering DRM applications is the severe carbon formation on the catalytic surface, thus resulting in rapid catalytic deactivation. Henni et al. (contribution 8) studied DRM over Ni-Ag catalytic systems and found that the addition of Ag stabilizes the Ni phase, thus reducing carbon accumulation on the catalytic surface. A catalytic system with a Ni/Ag ratio equal to 1 exhibited the highest performance for DRM reaction for temperatures up to 650 °C. In similar work related to the development of active catalytic systems for DRM, Elnour et al. (contribution 6) investigated the addition of Ga in Ni-Ga/(Mg, Al)O_x catalytic systems. The catalyst with a Ga/Ni ratio equal to 0.3 exhibited the highest catalytic activity and stability, suppressing carbon formation and, thus, facilitating CO_2 utilization. Trishch et al. (contribution 4) focused on the study of the conservatively perturbed equilibrium (CPE) phenomenon, which can result in an increase in the produced H_2 and CO concentrations up to 39% and 49%, respectively, while reducing the required reactor length to achieve equilibrium.

Regarding methanol reforming, some of the included research focuses on the APR process. Sousa et al. (contribution 5) examined the performance of APR over Pt/Al_2O_3 catalytic systems. It was observed that an increase in pressure and methanol concentration reduced catalytic activity, whereas the gradual deactivation of the applied catalyst was attributed to the sintering and leaching of Pt crystallites as well as the phase change of the alumina support to boehmite. Nguyen et al. (contribution 7) studied the effect of the size of Pt nanoparticles and the nature of carbon support (various commercial carbon supports, such as Vulcan XC72) on APR activity and stability. Platinum nanoparticles with an average diameter of 1.5 nm, which were well dispersed on Ketjenblack carbon, were found to exhibit a high hydrogen site time yield (8.92 min^{-1} at 220 °C) along with excellent stability under high-temperature treatment conditions and multiple recycling cycles.

Photocatalytic hydrogen production is one of the most promising technologies for the production of green energy through the utilization of solar radiation. Chang et al. (contribution 1) studied the effect of the calcination-induced oxidation method for the synthesis of cubic Cu_2O/CuO nanomaterials on photocatalytic H_2 production. The researchers found that the formation of a Cu_2O/CuO heterostructure improved light absorption while increasing the separation capacity between electrons and holes, thus enhancing H_2 production (11.888 $\mu\text{mol h}^{-1}\text{g}^{-1}$). In another study, Chaudhary et al. (contribution 2) studied the development and photocatalytic performance of La-doped MoS_2 (La-MoS_2) for the hydrogen evolution reaction (HER). Doping with La resulted in a decrease in the band gap of the material from 1.80 eV (MoS_2) to 1.68 eV (La-MoS_2), thus enhancing H_2 production (1670 $\mu\text{mol g}^{-1}$).

A unique contribution to this Special Issue is the work of Valecillos et al. (contribution 3) investigating bio-oil reformation via a combined steam and CO_2 reforming (CSDR) process that uses a Rh/ZDC catalyst, which initially exhibited high activity (yield of syngas equal to 77% and H_2/CO ratio equal to 1.2). However, the deactivation of the catalyst during the reaction mainly affected methane conversion, while regeneration through carbon black combustion was not able to fully restore activity due to structural changes in the support (CeO_2-ZrO_2) and the formation of amorphous carbon black. The results indicated that Rh/ZDC is an efficient catalyst for the conversion of bio-oil to syngas, although its limited stability during regeneration is a critical obstacle for long-term applications.

The publication of the two review papers in this Special Issue significantly increases its scientific impact. These reviews provide thorough investigations into recent developments in catalytic systems for MSR (contribution 10) and Co- and Mn-based spinels applied in various applications, such as volatile organic compound (VOC) degradation and Fischer–Tropsch processes (contributor 9).

Béres et al.'s (contribution 9) review focuses on cobalt manganese oxide spinels ($Co_xMn_{3-x}O_4$, $0 < x < 3$), typically applied in industrial applications, i.e., the CO oxidation process, the removal of VOCs, and Fischer–Tropsch synthesis. The distribution of Co/Mn ions in tetrahedral and octahedral sites of the spinel framework was found to directly affect the catalytic properties, whereas the nature of the structure phases and the synthesis conditions were strongly related to their chemical and redox properties.

Zhang et al. (contribution 10), studying MSR, found that Cu-based catalysts, such as $Cu/ZnO/Al_2O_3$, exhibited high activity but reduced stability due to Cu particle agglomeration. On the contrary, noble-metal-based catalysts (e.g., Pd/ZnO or $Pt/In_2O_3/CeO_2$) demonstrated lower CO production and higher resistance to poisoning, whereas novel single-atom alloy catalysts (e.g., Pd-Cu system) appeared to be promising materials for MSR.

3. Conclusions

The research and reviews published in this Special Issue entitled "Catalytic Reforming and Hydrogen Production: From the Past to the Future" highlight significant advances in the development of innovative catalytic systems and provide an in-depth understanding of the insights into the various mechanisms of reaction.

Improvements in the catalytic stability of DRM catalysts, with a focus on the addition of Ag and Ga in Ni-based systems to reduce the carbon accumulation on catalytic surfaces and the utilization of the CPE phenomenon to optimize H_2 and CO production, were among the main challenges investigated in DRM-related work. The significance of Pt nanoparticle size and the type of support for catalytic stability in methanol reforming was underlined; moreover, the review highlighted the need for developing resilient Cu-based and Pd-Cu single-atom alloy catalytic systems to increase the conversion of methanol to H_2 while reducing byproducts.

Extremely intriguing were the results presented in the articles investigating the photocatalytic production of H_2. Cu_2O/CuO heterostructures and $La-MoS_2$-doped materials improve the separation capacity of electron–hole pairs and enhance photocatalytic performance toward H_2. Finally, despite the initial enhanced catalytic activity of Rh-based catalysts for the production H_2 via bio-oil reforming processes, the irreversible catalytic deactivation caused by structural changes in the support, as well as intense carbon accumulation on the surface, hampers its potential for larger-scale applications.

The papers included in this Special Issue contribute significantly to promoting innovative solutions for sustainable H_2 production. Novel catalytic systems with advanced properties that consider resistance to carbon deposition or catalytic poisoning are thoroughly investigated, and important issues requiring immediate attention such as catalytic deactivation are underscored. Future research efforts for H_2 production technologies should focus on (a) the development of nanostructured, multi-metallic, hybrid catalytic systems that may be fine-tuned with respect to their surface structure and chemistry, characterized by high activity, stability, and low cost; and (b) the detailed investigation of reaction kinetics and mechanisms.

Funding: This research received no external funding.

Acknowledgments: We are thankful to all of the authors for submitting their impressive work to this Special Issue and to the reviewers for their time and effort in reviewing the manuscripts.

Conflicts of Interest: The authors declare no conflicts of interest.

List of Contributors

1. Chang, C.-J.; Kang, C.-W.; Pundi, A. Effect of Calcination-Induced Oxidation on the Photocatalytic H_2 Production Performance of Cubic Cu_2O/CuO Composite Photocatalysts. *Catalysts* **2024**, *14*, 499.
2. Chaudhary, A.; Khan, R.A.; Almadhhi, S.S.; Alsulmi, A.; Ahmad, K.; Oh, T.H. Hydrothermal Synthesis of $La-MoS_2$ and Its Catalytic Activity for Improved Hydrogen Evolution Reaction. *Catalysts* **2024**, *14*, 893.

3. Nguyen, X.T.; Kitching, E.; Slater, T.; Pitzalis, E.; Filippi, J.; Oberhauser, W.; Evangelisti, C. Aqueous Phase Reforming by Platinum Catalysts: Effect of Particle Size and Carbon Support. *Catalysts* **2024**, *14*, 798.

4. Trishch, V.R.; Vilboi, M.O.; Yablonsky, G.S.; Kovaliuk, D.O. Hydrogen and CO Over-Equilibria in Catalytic Reactions of Methane Reforming. *Catalysts* **2024**, *14*, 773.

5. Sousa, J.; Lakhtaria, P.; Ribeirinha, P.; Huhtinen, W.; Tallgren, J.; Mendes, A. Kinetic Characterization of Pt/Al$_2$O$_3$ Catalyst for Hydrogen Production via Methanol Aqueous-Phase Reforming. *Catalysts* **2024**, *14*, 741.

6. Elnour, A.Y.; Abasaeed, A.E.; Fakeeha, A.H.; Ibrahim, A.A.; Alreshaidan, S.B.; Al-Fatesh, A.S. Dry Reforming of Methane (DRM) over Hydrotalcite-Based Ni-Ga/(Mg, Al)O$_x$ Catalysts: Tailoring Ga Content for Improved Stability. *Catalysts* **2024**, *14*, 721.

7. Valecillos, J.; Landa, L.; Elordi, G.; Remiro, A.; Bilbao, J.; Gayubo, A.G. Are Rh Catalysts a Suitable Choice for Bio-Oil Reforming? The Case of a Commercial Rh Catalyst in the Combined H$_2$O and CO$_2$ Reforming of Bio-Oil. *Catalysts* **2024**, *14*, 571.

8. Henni, H.; Benrabaa, R.; Roussel, P.; Löfberg, A. Ni-Ag Catalysts for Hydrogen Production through Dry Reforming of Methane: Characterization and Performance Evaluation. *Catalysts* **2024**, *14*, 400.

9. Béres, K.A.; Homonnay, Z.; Kótai, L. Review on Synthesis and Catalytic Properties of Cobalt Manganese Oxide Spinels (Co$_x$Mn$_{3-x}$O$_4$, $0 < x < 3$). *Catalysts* **2025**, *15*, 82.

10. Zhang, M.; Liu, D.; Wang, Y.; Zhao, L.; Xu, G.; Yu, Y.; He, H. Recent Advances in Methanol Steam Reforming Catalysts for Hydrogen Production. *Catalysts* **2025**, *15*, 36.

References

1. Sharma, R.; Shahbaz, M.; Kautish, P.; Vo, X.V. Does energy consumption reinforce environmental pollution? Evidence from emerging Asian economies. *J. Environ. Manag.* **2021**, *297*, 113272. [CrossRef]

2. Ang, T.-Z.; Salem, M.; Kamarol, M.; Das, H.S.; Nazari, M.A.; Prabaharan, N. A comprehensive study of renewable energy sources: Classifications, challenges and suggestions. *Energy Strateg. Rev.* **2022**, *43*, 100939. [CrossRef]

3. Zhu, M.; Ai, X.; Fang, J.; Wu, K.; Zheng, L.; Wei, L.; Wen, J. Optimal integration of electrolysis, gasification and reforming for stable hydrogen production. *Energy Convers. Manag.* **2023**, *292*, 117400. [CrossRef]

4. Salvi, B.L.; Subramanian, K.A. Sustainable development of road transportation sector using hydrogen energy system. *Renew. Sustain. Energy Rev.* **2015**, *51*, 1132–1155. [CrossRef]

5. Sezer, N.; Bayhan, S.; Fesli, U.; Sanfilippo, A. A comprehensive review of the state-of-the-art of proton exchange membrane water electrolysis. *Mater. Sci. Energy Technol.* **2025**, *8*, 44–65. [CrossRef]

6. Ahmadipour, M.; Ridha, H.M.; Ali, Z.; Zhining, Z.; Ahmadipour, M.; Othman, M.M.; Ramachandaramurthy, V.K. A comprehensive review on biomass energy system optimization approaches: Challenges and issues. *Int. J. Hydrogen Energy* **2025**, *106*, 1167–1183. [CrossRef]

7. Agún, B.; Abánades, A. Comprehensive review on dry reforming of methane: Challenges and potential for greenhouse gas mitigation. *Int. J. Hydrogen Energy* **2025**, *103*, 395–414. [CrossRef]

8. Liew, W.M.; Ainirazali, N. Cutting-edge innovations in bio-alcohol reforming: Pioneering pathways to high-purity hydrogen: A review. *Energy Convers. Manag.* **2025**, *326*, 119463. [CrossRef]

9. Fayaz, H.; Saidur, R.; Razali, N.; Anuar, F.S.; Saleman, A.R.; Islam, M.R. An overview of hydrogen as a vehicle fuel. *Renew. Sustain. Energy Rev.* **2012**, *16*, 5511–5528. [CrossRef]

10. Yentekakis, I.V.; Panagiotopoulou, P.; Artemakis, G. A review of recent efforts to promote dry reforming of methane (DRM) to syngas production via bimetallic catalyst formulations. *Appl. Catal. B Environ.* **2021**, *296*, 120210. [CrossRef]

11. Wood, D.A. Critical review of development challenges for expanding hydrogen-fuelled energy systems. *Fuel* **2025**, *387*, 134394. [CrossRef]

Review

Review on Synthesis and Catalytic Properties of Cobalt Manganese Oxide Spinels ($Co_xMn_{3-x}O_4$, $0 < x < 3$)

Kende Attila Béres [1,2], Zoltán Homonnay [1,2] and László Kótai [1,*]

1   HUN-REN Research Centre for Natural Sciences, Institute of Materials and Environmental Chemistry, Magyar Tudósok krt. 2., H–1117 Budapest, Hungary; beres.kende.attila@ttk.hu (K.A.B.); homonnay.zoltan@ttk.elte.hu (Z.H.)

2   Department of Analytical Chemistry, Institute of Chemistry, Faculty of Natural Sciences, ELTE Eötvös Loránd University, Pázmány Péter s. 1/A, H–1117 Budapest, Hungary

*   Correspondence: kotai.laszlo@ttk.hu

Abstract: The cobalt manganese oxides, especially the spinels and related (multiphase) materials described with the formula $Co_xMn_{3-x}O_4$ ($0 < x < 3$), are widely used catalysts in a range of processes in significant industrial and environmental areas. The great diversity in the phase relations, composition, and metal ion valences, together with ion and vacancy site distribution variations, results in great variety and activity as catalysts in various industrially important redox processes such as the removal of CO or volatile organic substances (VOCs) from the air and oxidative destruction of pollutants such as dyes and pharmaceuticals from wastewater using peroxides. These mixed oxides can gain application in the selective oxidation of organic molecules like 5-hydroxyfurfural or aromatic alcohols such as vanillyl alcohol or in the production of fuels and other valuable chemicals (alcohols, esters) with the Fischer–Tropsch method. In this review, we summarize these redox-based reactions in light of the chemical and phase composition of the catalysts with the formula $Co_xMn_{3-x}O_4$ with $0 < x < 3$.

Keywords: cobalt manganese oxide spinels; Fischer–Tropsch synthesis; carbon monoxide oxidation; VOCs removal; advanced oxidation processes; degradation of dyes and pharmaceuticals; oxygen evolution and reduction reactions; 2,5-diformylfuran; vanillin; oxidation of alcohols

Academic Editors: Paraskevi Panagiotopoulou, Eleni A. Kyriakidou and Georgios Bampos

Received: 31 December 2024
Revised: 11 January 2025
Accepted: 14 January 2025
Published: 16 January 2025

Citation: Béres, K.A.; Homonnay, Z.; Kótai, L. Review on Synthesis and Catalytic Properties of Cobalt Manganese Oxide Spinels ($Co_xMn_{3-x}O_4$, $0 < x < 3$). *Catalysts* **2025**, *15*, 82. https://doi.org/10.3390/catal15010082

1. Introduction

The enormous importance of cobalt manganese oxides, especially that of the cobalt manganese oxide spinels and related (multiphase) materials described with the formula $Co_xMn_{3-x}O_4$, is evident from both the fundamental and industrial application points of view. The great diversity in the phase composition, the distribution of metal ion and vacancy sites, and the metal ion valences results in various chemical, electrochemical, and catalytic properties. The development of electrodes containing cobalt manganese oxides for lithium [1,2] or sodium batteries [3] and supercapacitors [4,5] is in the focus of current interest.

Cobalt and manganese oxide-containing materials are widely used in various engineering fields, including chemical, environmental, and energy technologies [6]. These belong to the science and engineering of batteries [7], supercapacitors, sensors [8,9], protective coatings on metals or solid oxide fuel cells [10,11], the removal of mercury and other poisonous materials [12] from wastewater, etc. The main electrochemical characteristics of cobalt manganese oxides have been reviewed in detail [1,13–17]; however, their catalytic performance, with a special focus on how it depends on the synthesis conditions, has not yet been comprehensively summarized.

The cobalt manganese oxide-based materials, especially the $Co_xMn_{3-x}O_4$ spinels, the composition of which may vary from chemically pure Mn_3O_4 ($x = 0$) to Co_3O_4 ($x = 3$), are frequently used catalysts in processes of the chemical and environmental industries such as the removal of volatile organic materials (VOCs) [18–21], ammonia-SCR reactions with nitrogen oxides [22], selective oxidation of aromatic and non-aromatic alcohols [20,23], and oxidation or reduction of carbon monoxide [24], including the Fischer–Tropsch synthesis [25–30]. The electrochemical oxygen reduction [31], advanced oxidation processes (AOPs) with peroxides [32–35] and ozone [36], green hydrogen production [37], as well as solid oxide fuel cell processes also belong to the cobalt manganese oxide spinel-catalyzed processes [30]. The most important compounds in these processes, such as $CoMn_2O_4$, $Co_{1.5}Mn_{1.5}O_4$, and Co_2MnO_4, as well as the solid solutions of $Co_xMn_{3-x}O_4$ spinels ($0 < x < 3$) and various composites with inhomogeneous phase distribution have also been studied. Several important properties of these compounds may be attributed to structural stress at phase boundaries.

The composition and phase relations of $Co_xMn_{3-x}O_4$ components in the cobalt manganese oxide-based catalysts depend on the synthesis conditions, precursors, and catalytic reaction conditions, including the oxidative/reductive nature of the atmosphere and reactants. The re-distribution of the same ions between the tetrahedral (A) and octahedral (B) lattice sites [38] has an elementary effect on the band gap and the catalytic properties. Some examples are given in Table 1 about distributions of valence states and metal ions between spinel sites A and B depending on the synthesis method.

Table 1. Summary of the different synthesis modes applied for the different compositions of Co-Mn oxides catalysts.

References	Refs.	Synthesis Method	Composition
Wickham and Croft et al.	[39]	Solid phase reaction, heating of oxides at 900 °C	$Co^{2+}_xMn^{2+}_{1-x}[Mn^{3+}_2]O_4$ for $0 < x < 1$
Aoki et al.	[38]	Wet method, heating of nitrates at 1200 °C	$Co^{2+}_x[Mn^{2+}_{x-1}Mn^{3+}_{3-x}]O_4$ for $1.8 < x < 3$
Boucher et al.	[40]	Solid phase reaction, heating of oxides at 900 °C	$Co^{2+}[Co^{2+}Mn^{4+}]O_4$
Yamamoto et al.	[41]	Wet method, heating of hydroxides at 1000 °C	$Mn_{0.25}Co_{0.75}[Mn_{0.75}Co_{1.25}]O_4$
Jabry et al.	[42]	Wet method, heating of oxalates at 1350 °C	$Co^{2+}[Mn^{3+}_{2-2u}Mn^{4+}_uCo^{2+}_u]O_4$ when $u = x-1$ and $1 < x < 2$

Cobalt manganese oxide spinels are generally metal-deficient [43–45] due to the possible presence of high metal ion valences (e.g., Mn^{IV}). For example, a change of in the $CoMn_2O_4$ structure from $Co^{II,III}[Mn^{II–IV}_2O_4]$ to $Mn^{II–IV}[Co^{II,III}Mn^{II–IV}O_4]$ (metal exchange and electron hopping) [46,47] decreases the bandgap to one-third of its original value [48]. In this paper, instead of the real $Co_xMn_{3-x}O_4 + \delta$ ($\delta \leq 0$) (or $Co_{x-\square A}Mn_{3-x-\square B}O_4$ with $0 < x < 3$, $\square_A$ and $\square_B$ are the tetrahedral and octahedral vacancies, respectively) formulas, these cobalt manganese oxide spinels are given with their simplified formula as $Co_xMn_{3-x}O_4$ ($0 < x < 3$).

Thus, in this review, we summarize the influence of the preparation conditions and the phase and chemical composition on the catalytic properties of $Co_xMn_{3-x}O_4$ spinels with $0 < x < 3$.

2. Main Preparation Routes and Properties of $Co_xMn_{3-x}O_4$ $(0 < x < 3)$ Spinels

Co_3O_4 and Mn_3O_4 are cubic and distorted (Jahn–Teller effect) tetragonal spinels, with Co^{II} or Mn^{II} ions in the tetrahedral (A) and Co^{III} and Mn^{III} ions in the octahedral (B) sites, respectively. The cubic Mn_3O_4 exists only above 1473 K [49]. The existence of mixed spinels derived from Co_3O_4-and Mn_3O_4 in the Co-Mn-O phase diagram was studied first by Aoki et al. [38] with high-temperature X-ray diffraction and resistivity measurement methods for the full range of $Co_xMn_{3-x}O_4$ $(0 \leq x \leq 3)$ solid solutions [38]. They found thermal hysteresis and a tetragonal-to-cubic phase transition. The hysteresis loops were observed at the temperatures where cubic-to-tetragonal transitions were found, and this temperature of the cubic spinel phase formation strongly depended on the cobalt content, especially over 50 $w\%$. The increase in cobalt concentration of the spinel phase is carried out by increasing the manganese content on the B sites of the spinel lattice and, accordingly, decreasing the lattice constants [50], and a similar effect was observed by Bulavchenko et al. [51].

Thermogravimetric studies on the Co–Mn–O system between 500 and 2000 K also showed cobalt-rich spinel (cubic) and manganese-rich tetragonal phases in the entire composition of the Co_3O_4-Mn_3O_4 subsystem [50], as was also found by Golikov et al. [52] $(0 \leq x \leq 3, \Delta x = 0.5, T = 1000–1623$ K) (Figure 1). The same results were found for the phase relations of the Co-Mn-O system with XRD between 1070 and 1570 K in air by Chaly et al. [53]. Golikov et al. [54,55] used high-temperature XRD (675–1623 K) for Co-Mn-O samples synthesized in a solid-state reaction and constructed a semiquantitative $T–x$ phase diagram in air [56].

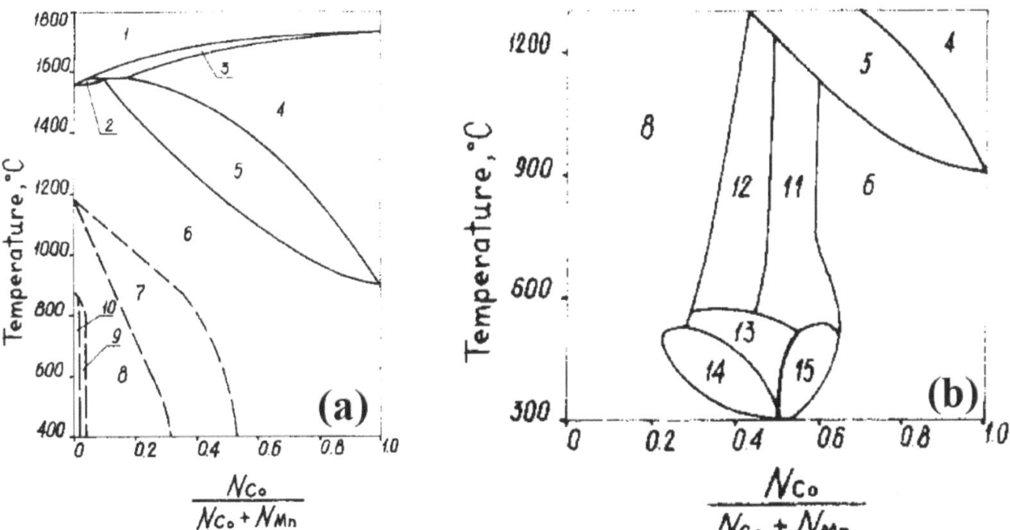

Figure 1. Phase diagram of the Co-Mn-O system air: (1) Liquid; (2) Liquid + Spinel; (3) Liquid + $Co_NMn_{1-N}O$; (4) $Co_NMn_{1-N}O$; (5) Spinel + $Co_NMn_{1-N}O$; (6) Spinel; (7) Spinel + Tetragonal Spinel; (8) Tetragonal Spinel; (9) Tetragonal Spinel + $(Mn,Co)_2O_3$; (10) $(Mn,Co)_2O_3$ **(a)** and (11) Tetragonal Spinel; (12) Tetragonal Spinels; (13; 14; 15) Tetragonal Spinels + $MnCoO_3$ **(b)**. Reproduced from ref. [52].

Vidales et al. studied the $Co_xMn_{3-x}O_4$ $(1 \leq x \leq 3)$ solid solutions prepared from the appropriate divalent metal chlorides and butylamine with heat treatments between 473 and 1273 K using the PXRD method. They found two different kinds of spinel-type solid solutions: a cubic (*Fd3m*) spinel-type when $x \leq 1.4$, a tetragonal ($I4_1/amd$) single phase when $2.0 \leq x \leq 3.0$ and also the same tetragonal single phase when $1.4 \leq x \leq 2.0$ after quenching from 1000 °C [57]. The phases formed at a given temperature and x-value are summarized in Table 2 [57].

Table 2. The identified phases and distortion parameters c/a' for $Co_xMn_{3-x}O_4$ ($1.4 \leq x \leq 2.6$) solid solutions between 200 and 1000 °C (reproduced from ref. [57]).

x	1.4	1.6	2.0		2.6	
T [°C]	Phase Identified	Phase Identified	Phase Identified	c/a'	Phase Identified	c/a'
200	C	T + R	T	1.125	T	1.154
300	C	T + R	T	1.131	T	1.150
400	C	T	T	1.138	T	1.152
500	C	T + C	T	1.140	T + B(8%)	1.153
600	C	T + C	T	1.143	T + B(88%)	1.148
700	C	T + C	T	1.145	T + B(88%)	1.148
800	C	T + C	T	1.146	T + B(82%)	1.152
900	C	T + C	T	1.144	T	1.155
1000	C	T + C	T	1.139	T	1.155

The main preparation methods of cobalt manganese oxides can be divided into several groups depending on the synthesis methods of intermediates that are calcined to form the spinel phases. Solid phase syntheses may start from Mn_3O_4-Co_3O_4 mixtures, oxidation of metals or lower valence oxides/compounds, and even by reduction of cobalt(III) permanganate complexes with their own reducing ligands such as ammonia. Intermediate compounds such as hydroxides, carbonates, oxalates, basic salts, and others may be prepared with sol-gel and other methods, with or without solvothermal or hydrothermal treatment, before the controlled heat treatment.

The structure of $CoMn_2O_4$ is shown in Figure 2. When the cobalt to manganese ratio is changed ($Co_xMn_{3-x}O_4$ ($0 \leq x \leq 3$)), the cell parameters are also changed, but the structure of the spinel phase is not altered drastically, only the exchange of the cobalt to manganese or Mn to Co takes place. The electronic structure of the $Co_xMn_{3-x}O_4$ ($0 \leq x \leq 3$) spinels, however, is sensitively dependent on the cobalt to manganese ratio. The bandgap, which is in a strong relationship with the presence of the high-spin or low-spin state, as well as with the structural distortions at the tetrahedral and octahedral sites, is one of the most important parameters having a significant effect on the catalytic properties [58]. For example, Yuval et al. [48] studied the electronic structure of this spinel with $x = 1$ in normal and inverse structures. The bandgap was reduced from 1.194 to 0.327 eV by exchanging half of the manganese cations at the octahedral sites with cobalt cations from the tetrahedral sites ($CoMn_2O_4 \rightarrow [Mn(CoMn)]O_4$) [48]. The electron configuration determination, however, is not easy because of the averaged oxidation states of Mn due to the electron hopping between the Mn^{IV} and Mn^{III} [59]. Additionally, the synthesis routes have enormous effects on the crystal and electron structure [48]; therefore, this paper focuses on these details.

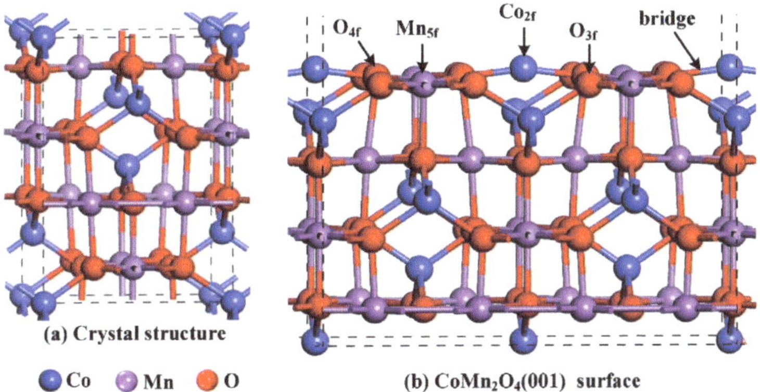

Figure 2. (a)The crystal structure of $CoMn_2O_4$. (b) $CoMn_2O_4(001)$ surface with five different adsorption sites. Mn_{5f} site denotes 5-fold coordinated Mn. Co_{2f} site denotes 2-fold coordinated Co. O_{3f} and O_{4f} denote 3-fold and 4-fold coordinated O, respectively. The bridge sites denote the centers between two adjacent surface atoms. Reproduced from ref. [60].

2.1. Solid Phase Syntheses

In ref. [47], a mixture of cobalt and manganese oxides, Mn_3O_4 and Co_3O_4, was dispersed in 10% poly(vinyl alcohol) (PVA) solution and spray-dried with subsequent heating to 873 K for an hour and then sintered at 1473 K for 8 h in a furnace. With the use of air as an oxidant, the $Co_xMn_{3-x}O_4$ spinels were prepared by heating a mixture of MnO and Co_3O_4 [61]. As a manganese source, aqueous manganese(II) nitrate solution was used to impregnate solid Co_3O_4 powder at 35 °C for 2 h with subsequent heating at 200 °C for 3 h [25]. Cobalt(II) nitrate and manganese(II) acetate solutions were used to impregnate TiO_2 support with subsequent calcination [26]. Cobalt(II) and manganese(II) acetate solutions were adsorbed on a carbon carrier, whereas cobalt(II) and manganese(II) nitrate salts were used to impregnate carbon nanotubes with drying at 60 °C, with heating at 500 °C for 2 h [62] or pyrolyzing at 360 °C for 2 h [30]. Drop casting of cobalt(II) and manganese(II) nitrate solutions on Pt/Ti9 fiber carrier was used with heating at 300 °C for 12 h [63]. Cobalt-impregnated zeolites were reacted with 2-methylimidazole, and after wetting with manganese(II) nitrate and heating at 500 °C for 2 h, $Co_xMn_{3-x}O_4$ spinels were obtained [64].

The oxidation of Co-Mn alloys on ferritic stainless-steel surfaces in an oxygen atmosphere at 800 °C in 4 h [65] or the oxidation of bulk alloys with various compositions up to 23.0% Mn content between 1273–1474 K and $10–10^5$ Pa oxygen pressure were performed. The $(Co,Mn)_3O_4$ phases were identified as an intermittent layer on a (Co,Mn)O core in pure form and as a co-phase in the outer part together with (Co,Mn)O [66].

The decreasing of the oxidation numbers of metals in complex compounds (Co^{III} and Mn^{VII}) containing reducing ligands (coordinated ammonia) and permanganate anion in a solid-phase thermal decomposition process is a relatively new possibility [44,45,67–69]. The Mn to Co ratio can be controlled with the selection of the anions, e.g., Mn to Co ratios 1:1, 1:2, and 1:3 were made available with the use of $[Co(NH_3)_4CO_3]MnO_4.H_2O$ and $[Co(NH_3)_6]Cl_2MnO_4$ precursors, $[Co(NH_3)_5Cl](MnO_4)_2$ and $[Co(NH_3)_6][KCl_2(MnO_4)_2]$ precursors, and $[Co(NH_3)_6](MnO_4)_3$ precursor, respectively [44,45,68,69]. The thermal decomposition process can be controlled using a low-boiling point oxidation-resistant organic solvent like toluene, whose boiling point is close to the thermal decomposition temperature of these compounds. The evaporation heat of the boiling solvents absorbs the decomposition reaction heat. Thus, amorphous intermediates are formed, which can be further heated under controlled conditions to prepare nanosized (2–50 nm) mixed cobalt manganese oxide spinels (Figure 3).

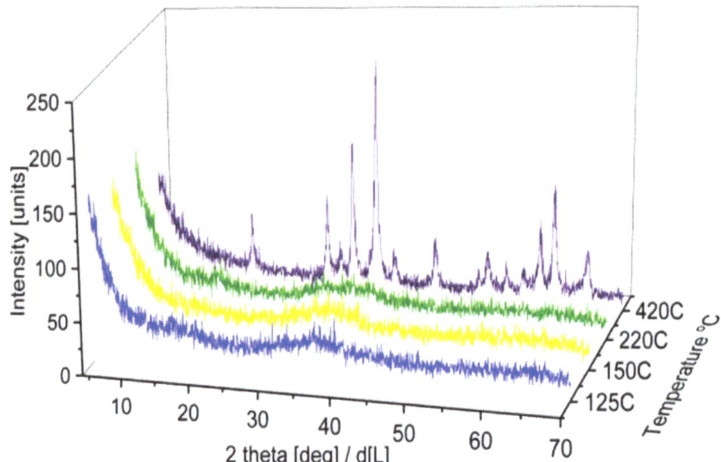

Figure 3. The crystallite sizes of $CoMn_2O_4$ prepared from $[Co(NH_3)_5Cl](MnO_4)_2$ after different heat treatments (reproduced from ref. [44]).

The aqueous leaching of manganese-free cobalt-containing intermediates such as water-soluble $[Co(NH_3)_5Cl]Cl_2$ formed in the first step of the solid phase thermal decomposition resulted in a phase that was more Mn-rich, as expected based on the initial Co to Mn stoichiometry 1:2 of the starting compound. It gives a new possibility for tuning the Co to Mn ratio in the $Co_xMn_{3-x}O_4$ compounds. The distribution of cations between T-4 (tetrahedral) and OC-6 (octahedral) (A and B, respectively) sites depends on the precursors and synthesis conditions [44,45,68,69].

2.2. Sol-Gel Synthesis from Divalent Metal Salts and Alkali Hydroxides, Ammonia, or Ammonia Precursors

The sol-gel syntheses were performed through the formation of cobalt and manganese hydroxides or other insoluble salts, such as carbonates and oxalates, using some soluble cobalt and manganese salts (chloride, nitrate, acetate, sulfate), with the use of alkaline materials like NaOH, KOH, ammonia, or ammonia precursors such as urea, alkali carbonates, hydrogen carbonates, or oxalates [70].

Using alkali hydroxides such as NaOH or KOH is the simplest way to prepare hydroxides because neither cobalt(II) nor manganese(II) hydroxides are soluble in the excess alkali hydroxide. Tessonier and coworkers synthesized Co-Mn oxide spinels as catalysts for carbon MWNT synthesis from the nitrates of manganese(II) and cobalt(II) using sodium hydroxide at pH = 10 [71]. The precipitate was dried in air at 180 °C for 5 h and calcined at 400 °C for 4 h to decompose it into the corresponding spinel-type mixed oxide [71]. The $Co_xMn_{3-x}O_4$ spinels were also prepared from these nitrate salts with aqueous NaOH at 90 °C for 2 h with subsequent calcination at 250 °C for 2 h [72]. The use of potassium hydroxide at 60 °C resulted in the $x = 1$ spinel after calcination at 300 °C for 3 h [73].

Cobalt(II) and manganese(II) acetate solutions mixed in various ratios were reacted with aqueous ammonia solution to adjust the pH to 10, and after stirring for 2 h, the hydroxides were aged overnight at room temperature. The dried precipitate was calcined at 400 °C for 4 h [74]. In the presence of glycerol, the ammonia-derived gel was combusted and calcined at 1073 °C for 16 h in order to obtain spinels [75]. Using the same acetate salts but hydrothermal conditions for a Co to Mn ratio 1:2 at 140 °C for 20 h [31] or at 150 °C for 3 h [76] resulted in the formation of $MnCo_2O_4$. Hydrothermal treatment in aqueous ethylene glycol in the presence of ammonia (pH = 10) at 180 °C for 78 h, with subsequent calcination on glass at 550 °C for 2 h, resulted in $Co_xMn_{3-x}O_4$ spinels [77]. The nitrate salts are also good precursors in the ammoniacal neutralization process. With the use of aqueous ammonia at 60 °C and after 24 h, the precipitate was calcined at 500 °C for 6 h [78], or in the presence of reduced graphene oxide (rGO) or activated carbon with subsequent calcination at 350 °C for 4 h and at 500 °C for 5 h [29] resulted in spinels [20].

The hydrothermal conditions ensure hydrolysis of urea; thus, the cobalt(II) and manganese(II) chloride salts could be transformed into spinels with aqueous urea as ammonia precursor by treatment at 200 °C for 12 h [79] or at 120 °C for 6 h [80], with subsequent calcination at 400 °C for 5 h on a platinized titanium mesh [79] or at 400 °C for 2 h in the pure form [80]. Cobalt(II) nitrate and manganese(II) acetate in a molar ratio of 4:5 at 90 °C turned into a precipitate with urea that was heated at 500 °C and resulted in cobalt manganese spinel oxide [81]. Hydrothermal treatment of a 1:1 mixture of the appropriate nitrate salts with urea at 175–190 °C using microwave heating to initiate the ignition also resulted in spinels [24]. Spinels formed when these nitrate salts and urea were subjected to hydrothermal treatment at 130 °C for 12 h in the presence of NH_4F, with subsequent calcination at 350 °C for 4 h on IrO_2 [82].

2.3. Sol-Gel Synthesis from Divalent Metal Salts and Alkali or Ammonium Carbonates/Oxalates

The basic cobalt(II) and manganese(II) carbonates were precipitated from the appropriate chloride salts with sodium carbonate at pH = 9 with stirring at 30–40 °C for 1 h [70], which were later calcined into the appropriate $Co_xMn_{3-x}O_4$ spinels. The nitrate salts mixed in various ratios were reacted with sodium carbonate at 30 °C, aged for 2 h, then calcined at 330 °C for 3 h [25], whereas the appropriate acetate salts with an Mn to Co ratio of 1:2 in ethylene glycol under solvothermal conditions at 160 °C [83] and heat treatment at 450 °C for 4 h resulted in $MnCo_2O_4$. A mixture of cobalt(II) nitrate and manganese(II) acetate was reacted in a 5:1 molar ratio with ammonium carbonate at 40 °C for 6 h, followed by calcination at 400 °C for 4 h [84]. The appropriate chloride salts, in the presence of PAA ($M_w = 2.5 \times 10^4$), were also reacted with ammonium carbonate, and the precipitate was heated at 400 °C in air for 3 h [85]. Over the hydroxide carbonate precipitating agents, their mixture and $NaHCO_3$ can also be used. The cobalt(II) and manganese(II) nitrates were reacted with a mixture of sodium carbonate and sodium hydroxide (pH = 10) at room temperature and subsequently calcined at different temperatures up to 1100 °C [86]. Sodium hydrogen carbonate was used to prepare $Co_{1.5}Mn_{1.5}O_4$ from nitrate salts in aqueous solution with 24 h stirring and subsequent calcination at 600 °C for 2 h [87].

Cobalt(II) and manganese(II) oxalate mixtures were precipitated from the appropriate acetate salts in ethanol at 80 °C; then, the precipitates were heated at 400 °C for 3 h [88]. A hydrothermal method was also developed to prepare the $Co_xMn_{3-x}O_4$ spinels from cobalt and manganese oxalate mixtures ($x = 1.5$) in aqueous ethylene glycol at 140 °C, with subsequent heating at 450 °C for 2 h [22]. The presence of polyvinylpyrrolidone at 170 or 160 °C for 1 or 6 h with 500 °C heat temperature for 4 h resulted in $x = 1.0$ spinels [89]. Ultrasonic impregnation of oxalates in a 1:2 molar ratio ($x = 1$) in aqueous solution at 100 °C on honeycomb ceramics with subsequent heating at 500 °C for 4 h also resulted in spinels [18]. The trituration of cobalt(II) and manganese(II) oxalate mixtures with 25% aqueous ammonia with subsequent heating at 400 °C for 2 h gave a spinel that was possible to load ultrasonically on activated carbon [90].

2.4. Solution Phase Redox, Hydro/Solvothermal Decomposition, Pyrolysis, and Other Reactions for the Synthesis of $Co_xMn_{3-x}O_4$ Spinels

Under hydrothermal or solvothermal conditions, the manganese(II) and cobalt(II) salts can decompose/transform into precipitates, which may be calcined into spinel compounds. Similarly, the oxidation of low-valent or reduction of high-valent metals in their salts, to be built into the spinel structure, can give intermediates that may be calcined into cobalt manganese oxide spinel compounds.

Cobalt(II) and manganese(II) nitrate (tetra and hexahydrate, respectively, were dissolved in ethanol in various ratios and heated in a Teflon autoclave at 140 °C for 24 h [91,92] or refluxed in the presence of silica for several hours [93]. The use of a glyceroliso-propanol solvent mixture at 180 °C resulted in a precipitate, which was transformed into spinels at 500 °C [94]. The increase in the hydrothermal treatment temperature (190 °C, 12 h) led to a decrease in the calcination temperature to 350 °C [23].

As reported by another group [95], divalent cobalt and manganese nitrates were dissolved in water in the presence of graphite and heated in a closed tube at 270 °C for 6 h. The main product was nanosized $(Co,Mn)(Co,Mn)_2O_4$ oxide with mixed metal ion site (A or B) distributions.

The solvothermal treatments using different recipes, i.e., (1) cobalt(II) and manganese(II) nitrates in a 1:1 ratio with poly(1,4-phenylene ether-ether sulfone), dimethylacetamide, and nitric acid [96], (2) the appropriate acetate salts using aqueous ethanol at 150 °C in the presence of carbon nanotubes [97], and (3) a mixture of cobalt(II) chloride and manganese(II) acetate, oleyl amine, and stearic acid in xylene solution at 120 °C for 3 h in the presence of reduced graphene oxide [98] all led to $Co_xMn_{3-x}O_4$ spinels [96]. The solvothermal treatment of manganese(II) oleate and cobalt(II) stearate in 1-octadecene at 120 °C for 1 h in a vacuum, in the presence of a $SrTiO_3$ carrier, with subsequent heating at 350 °C in N_2 for 0.45 h was also reported [9].

The solvothermal synthesis of spinels from cobalt(II) chloride and tetrabutylammonium permanganate (1:1 molar ratio) in isopropanol with 3 h reflux and drying at 120 °C was reported in [99]. Co-precipitation from cobalt(II) nitrate with potassium permanganate in the presence of maleic acid [100] or without maleic acid but at 70 °C for 70 min [101] with subsequent calcination at 500 °C for 4 h [100] or 350 °C for 2 h [101], respectively, were also tested. Instead of permanganate, hydrogen peroxide can also be used. Cobalt(II) nitrate and manganese(II) chloride were mixed in the presence of $NaNO_3$ and NH_4F and oxidized with hydrogen peroxide; then, after adding NaOH, the precipitate was heated at 300 °C [102] or at 750 °C for 4 h to gain the spinel structure [103].

The reaction of manganese(II) acetate and $K_3[Co(CN)_6]$ in the presence of poly (vinylpyrrolidone) [19,104] or aqueous citrate solution with stirring for 24 h [105] resulted in manganese(II) [hexacyanocobaltate(III)]-containing precursors which were transformed at 450 °C for 2 h [19,104] or at 350 °C for 2 h [105] into $Co_xMn_{3-x}O_4$ spinels.

Pulsed-spray evaporation and chemical vapor deposition was performed with cobalt(II) acetylacetonate and manganese(II) 2,2,6,6-tetramethyl-3,5-heptanedionate at 210 °C, with 4 Hz, 2.5 ms, heating in N_2/O_2 atm at 230 °C and then at 400 °C at 32 mbar pressure by the authors of ref. [106]. Spray pyrolysis was used for the tetraethylene glycol solutions of the appropriate metal(II) nitrates to ethanol with subsequent sintering at 600 and 800 °C for 2 h each, as reported in [107]. In another report [108], spray pyrolysis of aqueous cobalt(II) and manganese(II) acetates on a glass coated with fluorine-doped Sn-oxide was performed at 150 °C for 1 h.

Electrophoretic deposition of spinels was conducted from the appropriate nitrate salts in aqueous nitric acid solution, with heating at 500 °C for 20 min in Ar-H_2 followed by calcination in air at 690 °C for 2 h [109]. A $Co_{1.5}Mn_{1.5}O_4$ spinel was prepared with polarization by asymmetric AC from cobalt nitrate and manganese sulfate solutions in a 1:1 ratio, at pH = 3.5–4.0, with I_c/I_a = 1:1.25 at 40 °C for 60 min [109].

Electrospinning of cobalt(II) and manganese(II) acetate solutions with polyvinylpyrrolidone in DMF (25 kV, 1.2 mL/h), then heating at 600 °C for 3 h was performed to prepare a spinel in ref. [110]. Coaxial electrospinning (DC = 13 kV, 10 μL/min) and heating at 900 °C in N_2 for 100 min and then at 250 °C in the air for 30 min on carbon nanofiber carriers were also reported [25]. In ref. [111], the appropriate acetate salts were mixed with polyvinylpyrrolidone in EtOH, stirred for 12 h, then again for 12 h with acetic acid, and electrospinning was performed at 18 kV, 2 mL/h, with heating at 400 °C for 4 h.

The synthesis possibilities of $Co_xMn_{3-x}O_4$ spinels $(0 < x < 3)$ are summarized schematically in Figure 4.

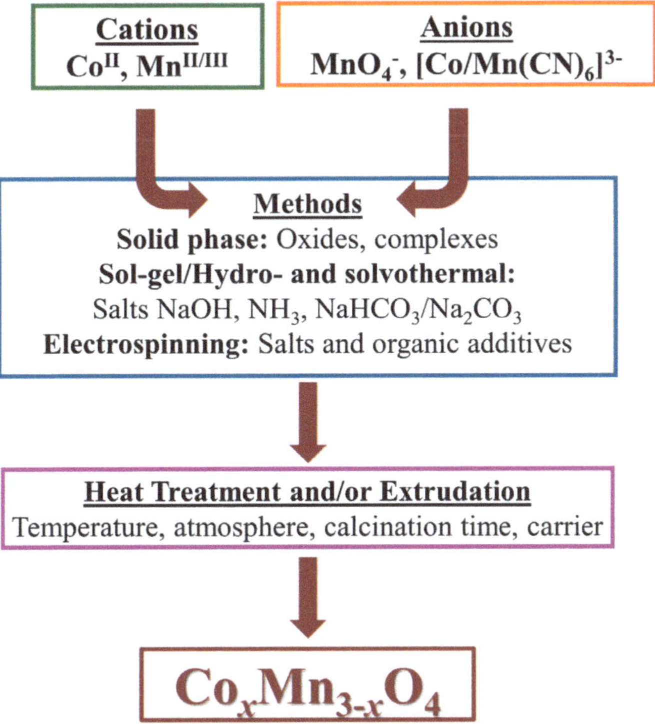

Figure 4. A schematic summary of different starting materials, additives in different synthesis routes, and heat treatments that resulted in $Co_xMn_{3-x}O_4$ $(0 < x < 3)$ spinels.

3. The $Co_xMn_{3-x}O_4$ $(0 < x < 3)$ Spinels as Catalysts

The different $Co_xMn_{3-x}O_4$ spinels with $0 < x < 3$ prepared on different reaction routes were described as versatile catalysts in various industrially important reactions such as the NH_3-SCR reaction [22] and removal of volatile organic contaminants and carbon monoxide [94] from air [18–21]. They serve as a good catalyst to enhance propellant combustion [112]. These spinels catalyze oxygen evolution (OER) [113] and oxygen reduction (ORR) [98] reactions and advanced oxidation processes performed with peroxides [101]. The preparation of fuel and high-value chemicals from synthesis gas by the Fischer–Tropsch process is also catalyzed by these cobalt manganese oxide spinels [71].

3.1. $Co_xMn_{3-x}O_4$ Catalysts in NH_3-SCR Reaction

The catalytic activity of $Co_xMn_{3-x}O_4$ ($0 < x < 3$) spinels prepared from cobalt(II) and manganese(II) oxalates in hydrothermal reactions in ethylene glycol was comprehensively tested in the NH_3-SCR reaction under different circumstances [22]. The $Co_xMn_{3-x}O_4$ spinel with $x = 1.5$ was prepared in a hydrothermal reaction of cobalt(II) and manganese(II) oxalates in aq. ethylene glycol at 140 °C and showed excellent N_2 selectivity (99% at 50 °C) [22]. Cai et al. [81] tested the influence of different metallic mesh carriers (Fe, Ti, Cu, and Ni) on the catalytic effect of $Co_xMn_{3-x}O_4$ spinels prepared from cobalt(II) nitrate, manganese(II) acetate and urea as the ammonia precursor under hydrothermal conditions. The spinel with $x = 1.33$ was tested on ceramic carriers and iron mesh between 90 and 360 °C. The iron mesh showed a stable NH_3 absorption for 60 h operation and kept the NO conversion over 90% with >98% N_2 selectivity (Table 3). Furthermore, adding 10 v% of H_2O to the system did not change the catalytic effect, whereas 60% conversion was obtained when 100 ppm SO_2 was injected.

Table 3. The catalytic activity of $Co_xMn_{3-x}O_4$ ($0 < x < 3$) spinels prepared on various reaction routes in NH_3-SCR reaction.

$Co_xMn_{3-x}O_4$	Precursors, Preparation Conditions	Catalyst Properties and Reaction Conditions	Refs.	Remarks
$x = 0.75$	$K_3[Co(CN)_6]$ and Mn^{II} acetate, in 1:3 ratio aq. sol. with poly(vinylpyrrolidone), calcination at 450 °C for 2 h	[NO] = 500 ppm T_{50} = 85 °C, Conv.: 98.5% (150 °C), N_2O: ~100 ppm (150 °C), rate: $8.37 \cdot 10^{-9}$ mol·s^{-1}, O_2:H_2O 3:8 v%, GHSV: 38,000 h^{-1}, flow rate 210 mL/min SO_2 (200 ppm) the NO conv.: 95% 8 v% H_2O the NO conv.: 99% Both, the NO conv.: 90% [NH_3] = 500 ppm $T_{50} \approx 250$ °C, Conv.: >99% (350 °C), Selectivity of $N_2O \approx 10$% O_2:H_2O 3:8 v%, GHSV: 38,000 h^{-1}, flow rate 210 mL/min	[19]	S_{BET} = 77.1 m^2 g^{-1}, V_p = 0.318 cm^3 g^{-1}
$x = 1.33$	Hydrothermal, Co^{II} nitrate and Mn^{II} acetate, 4:5 molar ratio, urea aq. sol, 90 °C, calcination at 500 °C	N_2 selectivity: 100% between 90 and 200 °C, NO_x selectivity: 100% between 130 and 270 °C, GHSV: 10,000 h^{-1},	[81]	S_{BET} = 66.1 m^2 g^{-1}, d_p = 9.5 nm,
$x = 1.5$	Hydrothermal, Co^{II} and Mn^{II} oxalates, 1:1 molar ratio, aq. ethylene glycol, 140 °C, calcination 450 °C for 2 h	N_2 selectivity: 99% to 86% at 50 to 200 °C, resp.; NO_x selectivity: 93%/75 °C and 95%/100–200 °C, GHSV: 32,000 h^{-1}, E_a 23.5 kJ/mol, flow rate 100 mL/min	[22]	S_{BET} = 103 m^2 g^{-1}, d_p = 5.8 nm, V_p = 0.44 cm^3 g^{-1}

Zhang et al. [19] studied the influence of morphology, stability, and the presence of inhibitors like H_2O and/or SO_2 on the catalytic activity of $Co_xMn_{3-x}O_4$ with $x = 0.75$ prepared by thermal decomposition of manganese(II) [hexacyanocobaltate(III)] at 450 °C for 2 h. The results and influence of different reaction conditions are summarized in Table 3.

3.2. Removal of Volatile Organic Compounds (VOCs) and Some Harmful Inorganic Gases

The volatile organic and inorganic contaminants such as aliphatic and aromatic hydrocarbons, organic compounds (esters, alcohols), carbon monoxide, and dinitrogen oxide are harmful constituents, the removal of which from the environment is more than an important task, especially via finding active catalysts to decrease the temperature of treatment and increase the efficiency of contaminant removal. Recently, the catalytic decomposition of different VOCs has been one of the hottest topics and one of the main possibilities in the fight against global warming [114]. There are a lot of mixed catalysts containing $Co_xMn_{3-x}O_4$

spinels to increase the efficiency in VOC removal [115], and now we summarize the results of removal of typical VOCs in the presence of $Co_xMn_{3-x}O_4$ spinels $(0 < x < 3)$.

3.2.1. Removal of Aromatic Hydrocarbons and Esters

The removal of ethyl acetate and toluene from the atmospheric environment in the presence of $Co_xMn_{3-x}O_4$ spinels $(0 < x < 3)$ is summarized in Table 4.

Table 4. The catalytic activity of $Co_xMn_{3-x}O_4$ spinels $(0 < x < 3)$ in the oxidation of toluene and ethyl acetate in air.

$Co_xMn_{3-x}O_4$	Precursors and Preparation Conditions	Catalyst Efficiency and Reaction Conditions	Refs.	Remarks
$x = 0.27$	Co^{II} and Mn^{II} nitrates in 1:10 ratio, 2 h at 500 °C, zeolitic 2-methylimidazole frameworks (ZIFs)	1000 ppm of toluene, 20% O_2 T_{50} = 236.6 °C and T_{90} = 242.0 °C, E_a = 136 kJ/mol, GHSV: 120,000 h^{-1} flow rate 50 mL/min	[64]	S_{BET} = 112 m^2 g^{-1}, d_p = 2.0 nm, V_p = 0.56 cm^3 g^{-1}
$x = 0.5$	Co^{II} and Mn^{II} nitrates in 1:5 ratio, 2 h at 500 °C, zeolitic 2-methylimidazole frameworks (ZIFs)	1000 ppm of toluene, 20% O_2 T_{50} = 241.7 °C and T_{90} = 245.6 °C, E_a = 186 kJ/mol, GHSV: 120,000 h^{-1} flow rate 50 mL/min	[64]	S_{BET} = 74 m^2 g^{-1}, d_p = 2.1 nm, V_p = 0.39 cm^3 g^{-1}
$x = 1.0$	Ultrasonic impregnation, Co^{II} and Mn^{II} oxalates, 1:2 molar ratio, aq. sol., 100 °C, 500 °C for 4 h, on ceramics	500 ppm of toluene or ethyl acetate, toluene conv.: 100% at 220 °C, ethyl acetate conv.: 100% at 180 °C, GHSV: 45,000 h^{-1} flow rate 75 mL/min	[18]	No data
$x = 1.0$	Hydrothermal, Co^{II} and Mn^{II} oxalates in 1:2 ratio, poly(vinyl pyrrolidone), ethylene glycol., 1 h at 170 °C, 150 °C (6 h), 500 °C for 4 h, treatment with 0.05 M HNO_3	1000 ppm of toluene, 21% O_2 T_{50} = 227 °C and T_{90} = 238 °C, E_a = 134 kJ/mol, GHSV: 90,000 h^{-1} flow rate 100 mL/min	[89]	S_{BET} = 55.9 m^2 g^{-1}, d_p = 12.24 nm, V_p = 0.19 cm^3 g^{-1}
$x = 1.0$	Co^{II} and Mn^{II} acetate, 1:2 molar ratio, ethanol, oxalic acid, 80 °C, calcination at 400 °C for 3 h	500 ppm of toluene and 20% O_2 T_{50} = 202 °C and T_{90} = 210 °C, E_a = 35.5 kJ/mol, GHSV: 22,500 h^{-1} T_{90} = 227 °C, GHSV: 45,000 h^{-1}	[88]	S_{BET} = 124.4 m^2 g^{-1}, d_p = 3.7 nm, V_p = 0.24 cm^3 g^{-1}
$x = 1.5$	Ultrasonic impregnation, Co^{II} and Mn^{II} oxalates, 1:2 molar ratio, aq. sol., 100 °C, calcination at 500 °C for 4 h, on ceramics	500 ppm of toluene or ethyl acetate, toluene conv.: 100% at 220 °C, ethyl acetate conv.: 100% at 180 °C, GHSV: 45,000 h^{-1} flow rate 75 mL/min	[18]	No data
$x = 1.5$	Co^{II} and Mn^{II} nitrates in 1:1 ratio, coatings, 2 h, 500 °C, on zeolitic 2-methylimidazole frameworks (ZIFs)	1000 ppm of toluene, 20% O_2 T_{50} = 243.8 °C and T_{90} = 249.7 °C, E_a = 194 kJ/mol, GHSV: 120,000 h^{-1} flow rate 50 mL/min	[64]	S_{BET} = 54 m^2 g^{-1}, d_p = 2.1 nm, V_p = 0.56 cm^3 g^{-1}
$x = 2.0$	Ultrasonic impregnation, Co^{II} and Mn^{II} oxalates, 1:2 molar ratio, aq. soln., 100 °C, 500 °C for 4 h on ceramics	500 ppm of toluene or ethyl acetate, toluene conv.: 100% at 220 °C, ethyl acetate conv.: 100% at 180 °C, GHSV: 45,000 h^{-1} flow rate 75 mL/min; 500 ppm of toluene or ethyl acetate, toluene conv.: 100% at 230 °C, GHSV: 60,000 h^{-1} flow rate 75 mL/min; 500 ppm of toluene or ethyl acetate, toluene conv.: 100% at 250 °C, GHSV: 78,000 h^{-1} flow rate 75 mL/min	[18]	No data
$x = 2.0$	Ultrasonic impregnation, Co^{II} and Mn^{II} oxalates, 1:2 molar ratio, aq. soln., 100 °C, calcination 500 °C for 4 h, on ceramics	500 ppm–500 ppm mixture of toluene and ethyl acetate: T_{20} = 182 °C, T_{50} = 204 °C, and T_{90} = 217 °C, 1000 ppm–1000 ppm mixture of toluene and ethyl acetate: T_{90} = 230 °C, GHSV: 45,000 h^{-1} flow rate 75 mL/min		

Table 4. *Cont.*

$Co_xMn_{3-x}O_4$	Precursors and Preparation Conditions	Catalyst Efficiency and Reaction Conditions	Refs.	Remarks
$x = 2.0$	Co^{II} nitrate, Mn^{II} chloride, 2:1 molar ratio aq. sol. and Mn^{II} oxidation with H_2O_2, NaOH, 300 °C	400 ppm of toluene, 20% O_2 $T_{50} = 239$ °C and $T_{90} = 259$ °C, $E_a = 68.7$ kJ/mol, GHSV: 12,000 h^{-1} flow rate 160 mL/min 700 ppm of toluene, 20% O_2 $T_{50} = 250$ °C and $T_{90} = 271$ °C, $E_a = 68.7$ kJ/mol, GHSV: 12,000 h^{-1} flow rate 160 mL/min 1000 ppm of toluene, 20% O_2 $T_{50} = 256$ °C and $T_{90} = 272$ °C, $E_a = 68.7$ kJ/mol, GHSV: 12,000 h^{-1}	[102]	No data
$x = 2.0$	Co^{II} nitrate and Mn^{II} chloride, 2:1 ratio, aq. H_2O_2, NaOH, and NH_4F solution, 12 h, 750 °C, 4 h	toluene (5000 ppm) conversion > 90% in ~10 min; selectivity of H_2: 1.4% in ~30 min $v(H_2) = 5.58\ 10^{-4}$ mol·L^{-1}·s^{-1}, at 400 °C, flow rate 75 mL/min	[103]	$S_{BET} = 28.34$ m^2 g^{-1}, $d_P = 17.02$ nm, $V_P = 0.121$ cm^3 g$^-$
$x = 2.4$	Ultrasonic impregnation, Co^{II} and Mn^{II} oxalates, 1:2 molar ratio, aq. sol., 100 °C, 500 °C for 4 h, on ceramics	500 ppm of toluene or ethyl acetate, toluene conv.: 100% at 220 °C, ethyl acetate conv.: 100% at 180 °C, GHSV: 45,000 h^{-1} flow rate 75 mL/min	[18]	No data

The catalytic activity of the $Co_xMn_{3-x}O_4$ spinels ($x = 1, 1.5, 2$, and 2.4) prepared by ultrasonic impregnation of Co^{II} and Mn^{II} oxalates at 100 °C with subsequent heating at 500 °C for 4 h on ceramics were comprehensively tested by Zhao et al. [18] in the oxidation of toluene and ethyl acetate. The relation of the catalytic activities (A) in the toluene oxidation for different x values was $A(x = 2.0) > A(x = 1.5) > A(x = 1.0) \approx A(x) = A(2.4)$ (Table 4). The best performance was found in the case of the $Co_xMn_{3-x}O_4$ ($x = 2$), when the $T_{20} = 182$ °C, $T_{50} = 204$ °C, and $T_{90} = 217$ °C. The best $Co_xMn_{3-x}O_4$ ($x = 2$) catalyst showed 7 cycles/2500 min stability; the T_{90} for toluene decreased to 225 °C, whereas for ethyl acetate, it increased to 192 °C. Even 2.0 vol% water did not disturb the stability of this catalyst, but the number of vacancies greatly affects their activity [18].

The effect of the vacancies on the catalytic activity of $CoMn_2O_4$ prepared in the hydrothermal reaction of Co^{II} and Mn^{II} oxalates in ethylene glycol in the presence of poly(vinylpyrrolidone) (PVP) followed by calcination at 500 °C for 4 h with subsequent generation of vacancies in the spinel structure with acidic leaching (HNO_3 solutions, $c = 0.00, 0.01, 0.05, 0.1$, and 0.5 M) was studied for toluene degradation by Wang et al. [89]. The best efficiency was obtained when the sample was treated with 0.05 M nitric acid (Table 4). The stability of the sample at $c = 0.05$ was found to be excellent since, applying it in a 5 v% H_2O humid condition, the conversion ratio was at least 80%. The effect of the calcination temperature on the catalytic efficiency of $MnCo_2O_4$ prepared from nitrate salts with NaOH and oxidation by H_2O_2 with subsequent calcination at 300 °C was tested for toluene removal by Wang et al. [102]. They applied 400, 700, and 1000 ppm toluene concentration, and in all the cases, the T_{50} and T_{90} decreased in the following order of calcination temperature: 750 °C > 600 °C > 450 °C > 300 °C. The stability of the best-performing spinel oxide (heat-treated at 300 °C) was 140 h when the toluene concentration was 1000 ppm was excellent since, at 290 and 270 °C, the conversation rates were $99 \pm 0.4\%$ and $76 \pm 0.5\%$, respectively. Under the same circumstances, but with water vapor ($RH = 85\%$) content, these numbers did not change; they were 99 and 77%, respectively, even after 120 min.

Dong et al. [88] tested the catalytic conversion of toluene with the spinel oxide ($x = 1$) prepared from the acetate salts in ethanol with co-precipitation. The stability of the spinel was excellent at 220 °C since after 700 min, the conversion rate dropped only to 98%, even in the presence of 2.0 v% water vapor. The increasing GHSV has a slight effect since 45,000 h^{-1} decreased the conversion rate only to 90% at 227 °C. The effect of the presence of oxygen was also tested. Without 20% O_2, the conversion rate reached 80% (for 14 min); however, it dropped to 16% after 33 min. They concluded that the surface lattice oxygen took part in the catalytic reaction, and the toluene oxidation over this catalyst occurs through benzyl alcohol and benzoate. Dong et al. also studied the effect of Pt-doping (0.28 w%) [116], which resulted in a great efficacy increase (the Pt-CoMn$_2$O$_4$ reached 99% conversion at 160 °C, and T_{50} and T_{90} were 134 and 150 °C, respectively). In contrast, at the same temperature, the undoped sample reached only 7% conversion. Similarly, Chen et al. [117] claimed that if Cu was added to the Co-Mn spinels, the toluene oxidation was found to be promoted, the T_{90} was around 210 °C, and water formation was observed only for limited periods (under 1600 min).

Han et al. [64] prepared Co$_x$Mn$_{3-x}$O$_4$ (0 < x < 3) spinels with Mn substitution at the octahedral sites of the cubic Co$_3$O$_4$ spinel-like structure. The metal nitrates were reacted on a zeolitic imidazole framework, with calcination at 500 °C for 2 h. The best T_{90} value was obtained with the x = 0.27 catalyst (242.0 ± 1.5 °C). The catalyst was stable; the toluene conversion was around 94% even after 55 h.

Wang et al. [103] studied the catalytic activity of the regular CoMn$_2$O$_4$ spinel with excess oxygen content in the low-temperature pyrolysis of toluene. The best catalyst was obtained by co-precipitation from nitrate salts with NaOH in the presence of NH$_4$F and hydrogen peroxide, with calcination at 750 °C for 4 h. Their isotherm heating studies showed that the best toluene concentration at 400 °C was 5000 ppm, below that, a lower amount of H_2 formed, whereas, above that, the toluene conversion became low. Above 500 °C, large amounts of methane and benzene formed.

3.2.2. Removal of Aliphatic Saturated and Unsaturated Hydrocarbons, Alcohols, and Inorganic Gases in the Presence of Co$_x$Mn$_{3-x}$O$_4$ Spinels (0 < x < 3)

The results about the catalytic activity of Co$_x$Mn$_{3-x}$O$_4$ (0 < x < 3) spinels in the oxidative removal of carbon monoxide, N$_2$O, aliphatic hydrocarbons (methane, propane, propene, and acetylene), and alcohols such as ethanol and n-butanol are summarized in Table 5.

Table 5. The catalytic activity of Co$_x$Mn$_{3-x}$O$_4$ in the oxidation of CO, N$_2$O, aliphatic hydrocarbons, and alcohols in air.

Co$_x$Mn$_{3-x}$O$_4$	Precursors and Preparation Conditions	Catalyst Efficiency and Reaction Conditions	Refs.	Remarks
x = 0.5	Co^{II} and Mn^{II} nitrate in 1:5 ratio, aq. NaOH sol. (pH = 11), 400 °C for 4 h	1 v% CO, T_{50} = 183 °C, E_a = 49 kJ/mol CO conv. rate: 0.2376 cm^3·g^{-1}·s^{-1} at 250 °C, contact time 0.12 s	[118]	S_{BET} = 60 m^2 g^{-1},
x = 1	Hydrothermal, pH = 12, (KOH), Co^{II} and Mn^{II} nitrate, 1:1 molar ratio, aq. sol., 300 °C for 3 h	0.2 V% propane and 2.0 V% O_2, T_{50} = 180 °C and T_{90} = 228 °C, GHSV: 18,000 h^{-1}, E_a = 72.2 kJ/mol, K = 0.389 μmol/g s	[119]	S_{BET} = 90.8 m^2 g^{-1}, d_p = 13.1 nm, V_p = 0.46 cm^3 g^{-1}
x = 1.5	Hydrothermal, Co^{II} and Mn^{II} nitrate, 1:1 molar ratio, aq. urea, 175–190 °C, microwave heating upon ignition temp.	100% conversion of CO to CO$_2$ at 200 °C in 9 min, E_a = 33 kJ/mol	[24]	No data
x = 2	Hydrothermal, Co^{II} and Mn^{II} nitrate, 1:1 molar ratio, aq. urea, 175–190 °C, microwave heating upon ignition temp.	100% conversion of CO to CO$_2$ at 180 °C in 10 min, E_a = 31 kJ/mol		

Table 5. *Cont.*

$Co_xMn_{3-x}O_4$	Precursors and Preparation Conditions	Catalyst Efficiency and Reaction Conditions	Refs.	Remarks
$x = 2.0$	Solvothermal, Co^{II} and Mn^{II} nitrate, 2:1 ratio, Glycerol-i-PrOH, 180 °C, calcination at 500 °C	1-1 $v\%$ of CO and O_2, $T_{20} = 60$ °C, $T_{50} = 90$ °C, $T_{90} = 110$ °C $E_a = 30.1$ kJ·mol^{-1}, TOF = 9.52·10^{-4} s^{-1}), GHSV: 36,000 h^{-1} flow rate 60 mL/min 1-1 $v\%$ of CO and O_2, and 2 $v\%$ H_2O $T_{20} = 90$ °C, $T_{50} = 130$ °C, $T_{90} = 170$ °C TOF = 1.75·10^{-4} s^{-1}, GHSV: 36,000 h^{-1} flow rate 60 mL/min	[94]	$S_{BET} = 203.5$ m^2 g^{-1}, $d_p = 3.59$ nm
$x = 2$	Na_2CO_3-NaOH mixt., Mn^{II} and Co^{II} nitrate, 1:2 ratio, $T = 25$ °C pH = 10, 2 h, 1100 °C for 2 h (inert atmosphere), 500 °C for 4 h (air)	EtOH (1 ppm) $T_{50} = 165$ °C, $T_{90} \approx 250$ °C flow rate: 2500 mL·h^{-1} N_2O (1000 ppm) T = 350–450 °C $k = 4.30·10^{11}$ mol$_{N2O}$·s^{-1}·m^{-2}·Pa^{-1} flow rate: 100 mL·h^{-1}	[86]	$S_{BET} = 43.6$ m^2 g^{-1},
$x = 2.0$	Co^{II} and Mn^{II} acetate, 2:1 molar ratio, aq. NH_3, 60 °C, 24 h, then 500 °C for 6 h	400 ppm n-butanol, $T_{50} = 120$ °C and $T_{80} = 250$ °C GHSV: 21,000 h^{-1}	[78]	$S_{BET} = 86$ m^2 g^{-1}, $d_p = 12.1$ nm, $V_p = 0.27$ cm^3 g^{-1}
$x = 2.5$	Co^{II} nitrate and Mn^{II} acetate, 5:1 ratio, aq. $(NH_4)_2CO_3$, 40 °C for 6 h, then 400 °C for 4 h	2 $v\%$ CH_4–6 $v\%$ O_2, $T_{50} = 306$ °C, $T_{50} = 382$ °C, $T_{90} = 473$ °C flow rate 100 mL/min	[84]	$S_{BET} = 108$ m^2 g^{-1}, $d_p = 7.48$ nm, $V_p = 0.263$ cm^3 g^{-1}
$x = 2.66$	Pulsed-spray evaporation chemical vapor deposition, Co^{II} acetylacetonate, Mn^{II} 2,2,6,6-tetramethyl-3,5-heptanedionate in 2:1 ratio, 4 Hz, 2.5 ms, 210 °C, N_2/O_2 atm. at 230 °C, 400 °C at 32 mbar.	1 $v\%$ C_2H_2–20 $v\%$ O_2, $T_{50} = 282$ °C, $T_{90} = 294$ °C CO_2 selectivity: $T_{50} = 282$ °C, $T_{90} = 295$ °C $E_a = 131.86$ kJ·mol^{-1}, flow rate 15 mL/min 1 $v\%$ C_3H_6–20 $v\%$ O_2, $T_{50} = 321$ °C, $T_{90} = 356$ °C CO_2 selectivity: $T_{50} = 323$ °C, $T_{90} = 357$ °C $E_a = 114.59$ kJ·mol^{-1}, flow rate 15 mL/min	[106]	No data

Wright et al. [120] did some preliminary tests with CO oxidation in the presence of amorphous and crystalline $Co_xMn_{3-x}O_4$ catalysts when the oxidation reaction started above 110 °C. In the presence of $Co_xMn_{3-x}O_4$ catalysts prepared by microwave heating from the appropriate divalent metal nitrates in the presence of urea under hydrothermal conditions with $x = 2$ and $x = 1$, a complete conversion of CO to CO_2 in 9 and 10 min at 200 and 180 °C was obtained, respectively, as reported in ref. [24] (Table 5). Bulavchenko et al. [118] studied the effect of calcination temperatures on the catalytic activity of $Co_{0.5}Mn_{2.5}O_4$ (prepared from Co^{II} and Mn^{II} nitrates, using aq. NaOH at pH = 11, with calcination at 400 °C for 4 h) in CO oxidation, and 400 °C was found to be the best choice when T_{50} for CO oxidation was 183 °C and the E_a value was 49 kJ/mol. The increase in calcination temperature to 600 and 800 °C resulted in the rise of T_{50} values to 250 °C and 461 °C, and the activation energies were also increased up to 56 and 67 kJ/mol, respectively [118]. The oxidation of carbon monoxide was studied in detail by Xu et al. [94] with $Co_xMn_{3-x}O_4$ catalysts that had different compositions and structures. The catalysts were prepared on a solvothermal route from Co^{II} and Mn^{III} nitrates in a glycerol-i-PrOH mixture at 180 °C, with subsequent calcination at 500 °C. The best performance, moisture tolerance ($v\% = 2$), and long-term durability (>30 h) were obtained with $x = 2$ with cubic structure when at 90 °C the conversion was 54.95%, whereas, in the presence of water, it was reached at around 130 °C [94]. All the catalytic results obtained by Xu et al. [94] are summarized in Figure 5.

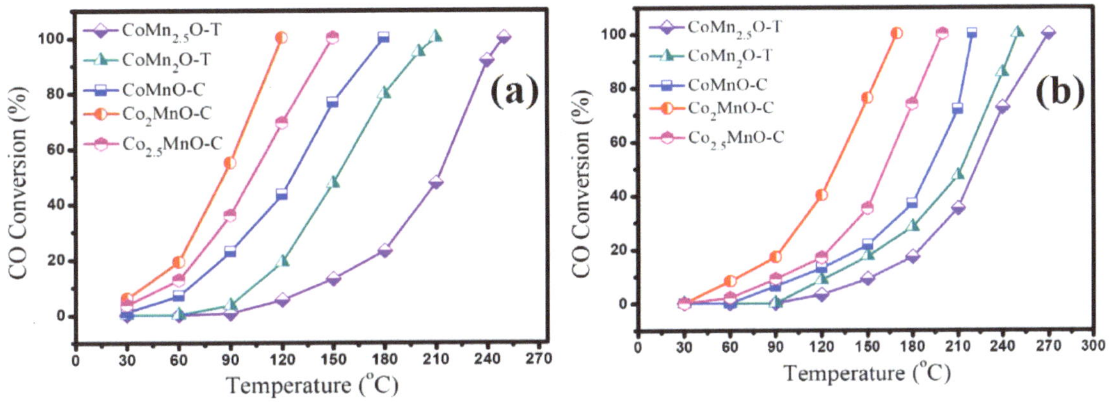

Figure 5. The CO conversion as a function of reaction temperature (**a**) and CO conversion as a function of reaction temperature in the presence of $v\%$ = 2 H$_2$O (**b**) for different Co$_x$Mn$_{3-x}$O$_4$ studied by Xu et al. [94]. Reproduced from ref. [94].

Rotko et al. [84] studied different Co$_x$Mn$_{3-x}$O$_4$ (x = 0.5, 1.5, and 2.5) catalysts in the oxidative removal of methane. The catalysts were prepared in a hydrothermal reaction of the metal nitrates with KOH at pH = 12, with subsequent heating at 300 °C for 3 h. The best catalytic performance was obtained with the lowest Mn-containing sample (Co$_{2.5}$Mn$_{0.5}$O$_4$) when the T_{10}, $T_{50,}$ and T_{90} values were found to be much lower than those with the other catalysts, even the Pd-Pt-containing ones (Table 5). The increase in the number of vacancies in Co$_{1.5}$Mn$_{1.5}$O$_4$ is advantageous for the complete oxidation of propane. Thus, the amorphous nature of the spinel oxide resulted in lower T_{50} and T_{90} values (169 °C and 207 °C) than those for the crystalline one (180 °C and 228 °C) with the same composition [119].

Tian et al. [106] prepared some low-cobalt containing Co$_x$Mn$_{3-x}$O$_4$ ($x \leq 0.34$) catalysts with the pulsed-spray evaporation-driven chemical vapor deposition (PSE-CVD) technique from manganese(II) acetylacetonate and cobalt(II) 2,2,6,6-tetramethylheptandedionate and tested for the oxidation of unsaturated hydrocarbons such as propylene and acetylene. The highest Mn-containing spinel (x = 0.34) had the highest electrical resistivity (16.50 $\Omega\cdot$cm) and was found to be stable up to 690 °C. The conversion and the selectivity for CO$_2$ formation are summarized in Table 5, and the results were found to be close to those obtained for an expensive Au/Al$_2$O$_3$ catalyst [106].

Kovanada et al. also studied the oxidation of N$_2$O and EtOH with different mixed oxide catalysts [86] made from nitrate salts at pH = 10 in 2 h at room temperature with subsequent calcination and extrudation at 1100 °C for 2 and at 500 °C for 4 h, respectively. The MnCo$_2$O$_4$ catalyst resulted in a midfield of conversion of EtOH and N$_2$O between 330 and 450 °C (Table 5). The increase in the catalytic efficiency of cobalt manganese spinels was obtained in the presence of other cations, such as Al and Mg, as was observed by Linde et al., who found that Al or transition metal doped (Cu, Ni, Fe) Co-Mn spinels had better catalytic efficiency in the oxidation of N$_2$O than the undoped cobalt manganese oxide spinels [121]. Mitran et al. studied MnCo$_2$O$_4$ catalysts prepared at different pH (adjusted by aq. NH$_3$ solution) from acetate salts at 60 °C for 24 h and with calcination at 500 °C for 6 h in the oxidation of n-butanol at 400 ppm concentration [78]. The highest n-butanol conversion was reached at 250 °C; other products were also identified as acetaldehyde and propionaldehyde [78].

3.3. Co$_x$Mn$_{3-x}$O$_4$ (0 < x < 3) Catalyst in the Oxidative or Reductive Transformations of Organic Compounds

The Co$_x$Mn$_{3-x}$O$_4$ (0 < x < 3) spinels were tested in numerous organic reactions on a preparative scale, for example, in oxidation reactions of alcohols or amines into valuable substances. For example, the spinel with x = 1.5 (S_{BET} = 137 m^2 g^{-1}, d_p = 3.8 nm, V_p = 0.12 cm^3 g^{-1}) prepared from cobalt and manganese nitrate in the presence of poly(1,4-phenylene ether-ether sulfone), dimethylacetamide, and nitric acid [96] catalyzed the oxidation of benzylamine in toluene with 76% conversion at 110 °C in 11 h with 99% selectivity to N-benzylidenebenzylamine. The authors concluded that no active metal leaching occurred during the reaction [96].

3.3.1. Transformations of 5-Hydroxymethylfurfural

The oxidative conversion of 5-hydroxymethylfurfural has many possible reaction pathways (Figure 6).

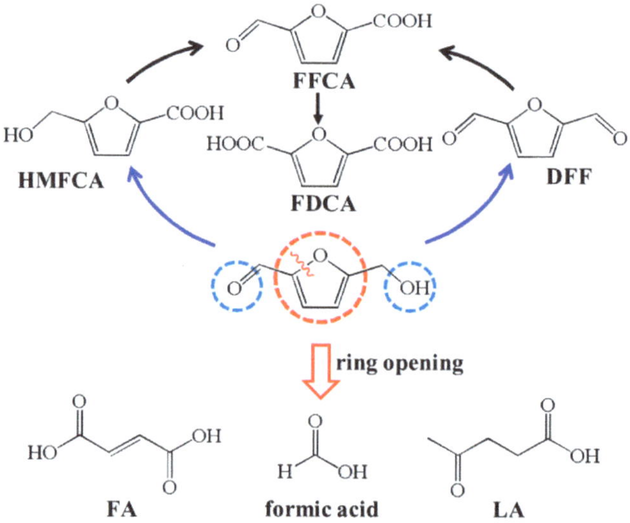

Figure 6. The possible oxidation products of 5-hydroxymethylfurfural. Reproduced from ref. [23].

The oxidation of 5-hydroxymethylfurfural (HFM) into 2,5-diformylfurane (DFF) is the most important reaction route, which can easily be controlled with the use of Co$_x$Mn$_{3-x}$O$_4$ spinels prepared under hydrothermal conditions (glycerol-isopropanol, 190 °C for 12 h) with subsequent calcination at 350 °C from the divalent metal nitrates [23]. Ding et al. studied these Co$_x$Mn$_{3-x}$O$_4$ catalysts (x = 1; 1.2, 1.5, and 2), and first, the temperature dependence was tested with the sample x = 1.2. The increase in temperature from 90 to 160 °C increased the conversion from 29.6 to 84.9%, but the selectivity was changed in the opposite direction, decreasing from 100% (100 °C) to 24.0% (160 °C). The catalytic activity of the different Co$_x$Mn$_{3-x}$O$_4$ (x = 1; 1.2, 1.5, and 2) spinel catalysts were tested at 100 °C, and the results are summarized in Table 6 [23].

Table 6. The result of the 5-hydroxymethylfurfural transformations catalyzed by $Co_xMn_{3-x}O_4$ ($x = 1$–2) spinels.

$Co_xMn_{3-x}O_4$	Precursors, Preparation Conditions	Catalyst Efficiency and Reaction Conditions	Refs.	Remarks
$x = 1$	Hydrothermal, Co^{II} and Mn^{II} nitrate, 1:2 ratio, glycerol and water in IPA, 190 °C for 12 h, 350 °C for 3 h	HFM conversion: 34.2% DFF selectivity: >99% $P(O_2) = 0.75$ MPa, $T = 100$ °C, in DMF	[23]	$S_{BET} = 82.6$ m^2 g^{-1}, d_p distribution = 3–10 nm
$x = 1.2$	Hydrothermal, Co^{II} and Mn^{II} nitrate, 2:3 molar ratio, glycerol and water in IPA, 190 °C for 12 h, 350 °C for 3 h	HFM conversion: 41.6% DFF selectivity: >99% $P(O_2) = 0.75$ MPa, $T = 100$ °C, in DMF		
$x = 1.5$	Hydrothermal, Co^{II} and Mn^{II} nitrate, 1:1 ratio, glycerol and water in IPA, 190 °C for 12 h, 350 °C for 3 h	HFM conversion: 33.7% DFF selectivity: >99% $P(O_2) = 0.75$ MPa, $T = 100$ °C, in DMF		
$x = 2$	Hydrothermal, Co^{II} and Mn^{II} nitrate, 2:1 ratio, glycerol and water in IPA, 190 °C for 12 h, 350 °C for 3 h	HFM conversion: 36.0% DFF selectivity: 92.1% $P(O_2) = 0.75$ MPa, $T = 100$ °C, in DMF		
$x = 2$	Mn^{II} and Co^{II} acetate, 1:2 ratio, 6 h, calcination at 425 °C for 8 h, Impregnation on Ru NPs	HFM conversion: > 90% BHMF selectivity: 98.5% $P(H_2) = 8.2$ MPa, $T = 100$ °C, $t = 4$ h, in MeOH HFM conversion: 98.7% BHMTHF selectivity: 97.3% $P(H_2) = 8.2$ MPa, $T = 100$°C, $t = 16$ h, in MeOH	[122]	$S_{BET} = 135.4$ m^2 g^{-1}, $d_p = 9.1$ nm, $V_p = 0.3082$ cm^3 g^{-1}

The conversion to selectivity ratio changes with the amount of the catalyst; the best ratio was 100:40 (Figure 7) [23], and a controversial effect was found in the case of reaction time as well; the best duration of the reaction was 2 h. The catalyst was dried at 70 °C for 10 h and calcined at 350 °C. It was found to be reusable for 6 cycles with the same efficiency (the recovery process was always applied between subsequent cycles) [23].

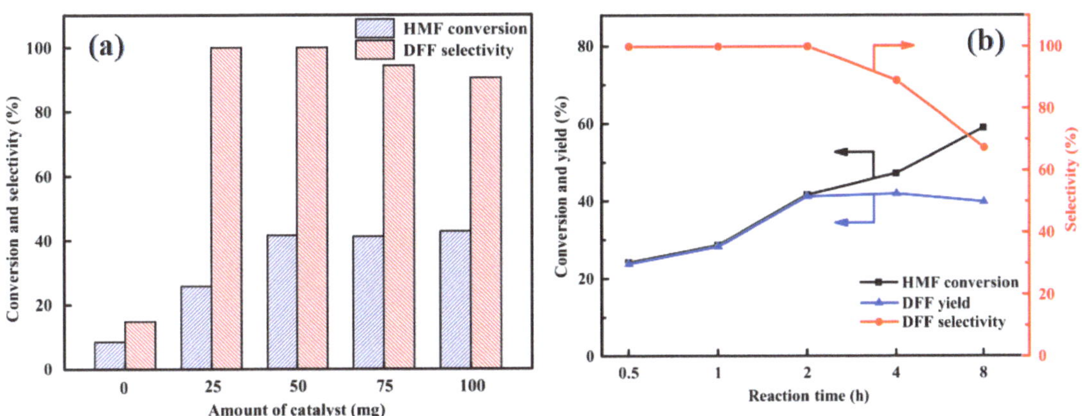

Figure 7. The conversion of 5-hydroxymethylfurfural into 2,5-diformylfuran in the presence of CoMn$_2$O$_4$ catalyst. Influence of the catalyst amount for selective aerobic oxidation of HFM (**a**) and Time course study of HFM oxidation with CoMn$_2$O$_4$ catalyst (**b**). Reproduced from ref. [23].

Hydrogenation of HMF in the presence of CoMn$_2$O$_4$ doped with Ru NPs resulted in 2,5-bis(hydroxymethyl)furan (BHMF) and 2,5-bis-(hydroxymethyl)tetrahydrofuran (BHMTHF) in a hydrogenation reaction [122].

3.3.2. Organic Oxidation Reactions of Aromatic Alcohol Derivatives Catalyzed by $Co_xMn_{3-x}O_4$ Spinels ($0 < x < 3$)

Jha et al. studied the conversion of vanillyl alcohol into vanillin and tested the effect of the Co/Mn ratio and the conditions of the reactions, including the solvent, temperature, reaction time, catalyst concentration, and partial O_2 pressure [83]. The catalyst was prepared from Co^{II} and Mn^{III} acetates in ethylene glycol and aqueous Na_2CO_3 under solvothermal conditions (160 °C) followed by calcination at 450 °C for 4 h. The best performance was obtained with acetonitrile at 140 °C and 21 bar pressure with the $MnCo_2O_4$ catalyst in 2 h (Table 7).

Table 7. Oxidation of vanillyl alcohol with $MnCo_2O_4$ samples prepared in hydrothermal/solvothermal reactions in ethylene glycol from acetate salts with sodium carbonate or ammonia.

$Co_xMn_{3-x}O_4$	Precursors and Preparation Conditions	Catalyst Efficiency and Reaction Conditions	Refs.	Remarks
$x = 2.0$	Solvothermal, Co^{II} and Mn^{II} acetate, in ethylene glycol and aq. Na_2CO_3, 160 °C, 450 °C 4 h	Vanillyl alcohol conversion: 64% Selectivity: vanillin: 83%, vanillic acid: 6%, other: 11% acetonitrile, $T = 140$ °C, 21 bar, 2 h	[83]	$S_{BET} = 105$ $m^2\,g^{-1}$
$x = 2.0$	Hydrothermal, Co^{II} and Mn^{II} acetate, 2:1 molar ratio, ethylene glycol and aq. NH_3, 160 °C 2 h, 350 °C 4 h, with 1% rGO carrier	Vanillyl alcohol conversion: 74% Selectivity: vanillin:83%, Vanillic acid: 8%, Quinone: 8% acetonitrile, 140 °C, 21 bar, 2 h	[20]	$S_{BET} = 111$ $m^2\,g^{-1}$, d_p distribution = 3–20 nm, $V_p = 0.393$ $cm^3\,g^{-1}$

Changing the Co to Mn ratio in the spinels had no important effect on the conversion. The increase in the reaction time from 1 to 4 h increased the conversion from 43 to 80%, with a decrease in the selectivity for vanillin from 79 to 72% [83]. The reaction time was chosen to be $t = 2$ h; then, the increase in temperature from 100 to 160 °C increased the conversion from 41 to 76%, with decreasing vanillin selectivity from 57 to 83% up to 140 °C, followed by a drop to 74%. The pressure of air had only a minuscule effect on the conversion and selectivity, but above 21 bar, it started to drop [83]. Regarding the recycling of the catalyst, simple washing with MeOH and drying at 100 °C was not found to be enough; calcination at 450 °C for 1 h was needed to obtain a similar catalytic performance as the fresh sample, and there was no leaching of Co and Mn observed [83].

A composite catalyst prepared in a hydrothermal reaction from acetate salts with aqueous NH_3 in ethylene glycol at 160 °C for 2 h, followed by calcination at 350 °C for 4 h with 1% rGO-containing $MnCo_2O_4$ (rGO = graphene oxide) was also tested in the oxidation of different aromatic alcohols in the liquid phase (acetonitrile), including vanillyl alcohol [20]. The best results (74% conversion, the selectivity for vanillin, vanillic acid, and quinone being 83, 8, and 10%, respectively) at 140 °C for 2 h in acetonitrile were obtained in the presence of 1% rGO. Increasing the reaction time decreased the selectivity for the vanillin formation, and 2-metoxybenzoquinone occurred. The 1% rGO containing $MnCo_2O_4$ catalyst was found to be stable and was heat treated at 350 °C for 2 h after each cycle, obtaining the same results as with the fresh catalyst [20].

Other different aromatic alcohols were also used by Jha et al., and the results are summarized in Table 7 [20]. Their test showed that 4-hydroxy-3-methoxy-α-methyl benzyl alcohol (a secondary alcohol) showed the highest conversion (92%) and the lowest selectivity to ketone (57%) compared to primary alcohols. In contrast, the para-positioned alcohols were found to be less reactive than the alcohols with o- or m-positions [20].

3.4. $Co_xMn_{3-x}O_4$ ($0 < x < 3$) Catalysts in Oxygen Evolution Reactions (OERs) and Oxygen Reduction Reactions (ORRs)

The oxygen evolution reactions (OERs) are hot topics in environmental chemistry and are key reactions for sustainable chemical energy storage and production of different chemicals and fuels [113]. Recently, there has been significant progress in generating O_2 via chemical reactions, such as oxidation of water via photolysis, electrolysis, or electrocatalytic oxygen evolution from oxides and oxoacids. Similarly, oxygen reduction reactions (ORRs) are evolving as an exciting area in developing methods for the production of H_2O and/or H_2O_2 in an electrochemical cell [123]. These processes (OERs and ORRs) can be catalyzed with $Co_xMn_{3-x}O_4$ ($0 < x < 3$) spinels [31,98,104,124,125]. A general scheme of OERs is shown in Figure 8.

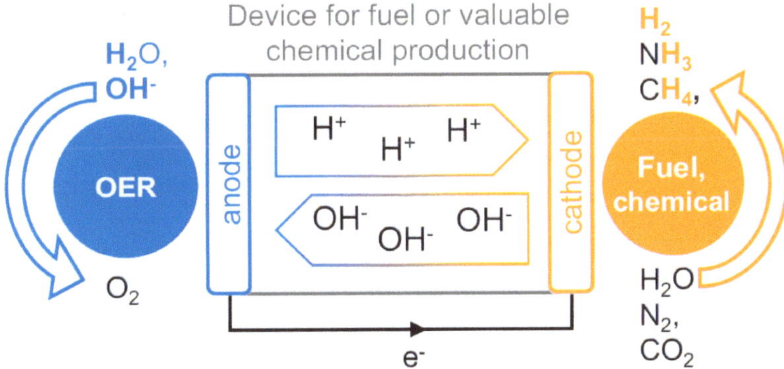

Figure 8. Uses of OERs to produce different chemicals and fuels. Reproduced from ref. [113].

Rios et al. studied the catalytic activity of $Co_xMn_{3-x}O_4$ (x = 0.25, 0.5, 075 and 1) coatings made on glass/SnO_2-F electrodes (spray pyrolysis of aq. acetate salts at 150 °C for 1 h) in OERs. They found that the Tafel slopes were almost independent of the Mn to Co ratio. However, the current density (J) and the overpotential (η) were highly dependent on it (Table 8), and the best Mn to Co ratio was 1:11 [108].

Xie et al. prepared a porous hollow sphere $MnCo_2O_4$ catalyst (solvothermal reaction of cobalt(II) nitrate and manganese(II) acetate in an iPrOH/glycerol mixture at 140 °C, with subsequent heating at 350 °C for 2 h) and studied it comprehensively under different conditions in OERs [91]. The catalyst was found to be stable over 12 h at J = 10 mA·cm^{-2}. The Tafel slope (A) related to the activation energy of the reaction was found to be lower than that for the Co_3O_4. The double-layer capacitance (C_{dl}) and the charge transfer resistance (R_{CT}) were 0.75 mF·cm^{-2} and 252 Ω, respectively [91]. Lee et al. [63] prepared a $MnCo_2O_4$ catalyst with the same composition but with a drop-casting method from Co^{II} and Mn^{III} nitrate solutions (300 °C for 12 h) on a Pt/Ti fiber carrier that had excellent applicability in the acidic electrolyte (0.05 M H_2SO_4). After 15,000 CV cycles, the η = 420 mV was found at current density J = 10 mA·cm^{-2}. The stability at 100 mA·cm^{-2} was 190 h, whereas, at J = 400 mA·cm^{-2}, the catalyst was completely dissolved after 40 h OER. The Ni-doping improved all the OER parameters of the spinel catalyst even at 5% Ni content [63].

Table 8. The catalytic activity of $Co_xMn_{3-x}O_4$ ($0 < x < 3$) spinels in oxygen evolution reactions (OERs).

$Co_xMn_{3-x}O_4$	Precursors and Preparation Conditions	Catalyst Results and Conditions	Refs.	Remarks
$x = 0.75$	Hydrothermal, $K_3[Co(CN)_6]$ and Mn^{II} acetate, in 2:3 ratio aqueous solution with polyvinylpyrrolidone, 450 °C, 2 h	$\eta = 710$ mV when $J = 1$ mA·cm^{-2} in phosphate buffer (pH = 7.0)	[104]	$S_{BET} = 61.4$ m^2 g^{-1},
$x = 1$	Solvothermal, Co^{II} and Mn^{II} acetate, 1:2 molar ratio, ethanol, 150 °C, 12 h, on CNTs	$\eta = 560$ mV when $J = 10$ mA·cm^{-2} disk $A = 288.19$ mV·dec^{-1}, disk rot. 1600 rpm, in 0.1 M KOH electrolyte	[97]	no data
$x = 1$	Solvothermal, Co^{II} chloride and Mn^{II} acetate, oleylamine, stearic acid in xylene sol. 120 °C, 3 h, on a reduced graphene oxide carrier	$\eta = 1490$ mV when $J = 10$ mA·cm^{-2} $A = 56$ mV·dec^{-1} disk rot. 1600 rpm, in 0.1 M KOH electrolyte	[98]	Crystallite size distribution: 1.7–3 nm, $S_{BET} = 412$ m^2 g^{-1}, $d_p = 3.7$ nm
$x = 1$	Electrospinning, Co^{II} and Mn^{II} acetate, 1:2 ratio, mixed with polyvinylpyrrolidone in DMF, 25 kV, 1.2 mL·h^{-1}, 600 °C 3 h	Mass activity: 19.0 A·g^{-1} Spec. surface activity: 119 mA·cm^{-2} $V = 1.55$ (V vs. RHE) in 0.1 M KOH electrolyte	[110]	$S_{BET} = 16.0$ m^2 g^{-1}, $d_p = 19.4$ nm, $V_p = 0.078$ cm^3 g^{-1}
$x = 1.2$	$K_3[Co(CN)_6]$ and Mn^{II} acetate, 1:1.5 ratio, aq. citrate solution, 24 h, 350 °C, 2 h	$\eta = 455$ mV when $J = 10$ mA·cm^{-2} $\eta = 582$ mV when $J = 50$ mA·cm^{-2} $A = 194.8$ mV·dec^{-1}, $R_{CT} = 129.4$ Ω $C_{dl} = 129.8$ μF·cm^{-2} in 0.5 M H_2SO_4 electrolyte	[105]	No data
$x = 1.2$	Solvothermal, Mn^{II} oleate and Co^{II} stearate, 1.5:1 ratio, 1-octadecene, 120 °C (vac.) 1 h, 305 °C (N_2 atm.) 0.5 h on $SrTiO_3$ carrier	$A = 218$ mV·dec^{-1}, in 0.1 M phosphate electrolyte (pH = 6.9)	[9]	No data
$x = 1.5$	$K_3[Co(CN)_6]$ and Mn^{II} acetate, 1:1 ratio, aq. citrate solution, 24 h, 350 °C, 2 h	$\eta = 415$ mV when $J = 10$ mA·cm^{-2} $\eta = 552$ mV when $J = 50$ mA·cm^{-2} $A = 195.4$ mV·dec^{-1}, $R_{CT} = 88.1$ Ω $C_{dl} = 188.9$ μF·cm^{-2} in 0.5 M H_2SO_4 electrolyte	[105]	No data
$x = 1.5$	Solvothermal, Co^{II} chloride and n-Bu$_4$NMnO$_4$, 1:1 molar ratio, 2-propanol, 3 h reflux, 120 °C	$\eta = 400$ mV when $J = 10$ mA·cm^{-2} disk $A = 57$ mV·dec^{-1}, disk rot. 1600 rpm, 0.63 cm^2 BET·cm^{-2} disk in 0.1 M KOH electrolyte	[99]	$S_{BET} = 223$ m^2 g^{-1}, TUNNEL
		$\eta = 400$ mV when $J = 10$ mA·cm^{-2} disk $A = 52$ mV·dec^{-1}, disk rot. 1600 rpm, 0.63 cm^2 BET·cm^{-2} disk in 0.1 M KOH electrolyte		$S_{BET} = 35$ m^2 g^{-1}, LINEAR
		$\eta = 420$ mV when $J = 10$ mA·cm^{-2} disk $A = 65$ mV·dec^{-1}, disk rot. 1600 rpm, 0.63 cm^2 BET·cm^{-2} disk in 0.1 M KOH electrolyte		$S_{BET} = 283$ m^2 g^{-1}, SPINEL
$x = 2$	Solvothermal, Co^{II} nitrate and Mn^{II} acetate, 2:1 molar ratio, 2-propanol and glycerol, 140 °C; 350 °C for 2 h	$\eta = 305$ mV when $J = 10$ mA·cm^{-2} $\eta = 336$ mV when $J = 50$ mA·cm^{-2} $\eta = 355$ mV when $J = 100$ mA·cm^{-2} $A = 98.5$ mV·dec^{-1}, $R_{CT} = 252$ Ω $C_{dl} = 0.75$ mF·cm^{-2} in 1 M KOH electrolyte	[91]	$S_{BET} = 78.5$ m^2 g^{-1}, $d_p = 5.57$ nm,
		$A = 27.6$ mV·dec^{-1}, $J = 10$ mA·cm^{-2} $R_{CT} = 1472.9$ Ω, $C_{dl} = 0.11$ mF·cm^{-2} in neutral electrolyte		

Table 8. *Cont.*

$Co_xMn_{3-x}O_4$	Precursors and Preparation Conditions	Catalyst Results and Conditions	Refs.	Remarks
$x = 2$	Drop casting method, Co^{II} and Mn^{II} nitrate solution, 300 °C, 12 h on Pt/Ti fiber carrier	$\eta = 371$ mV when $J = 10$ mA·cm^{-2} $\eta = 460$ mV when $J = 100$ mA·cm^{-2} $A = 163$ mV·dec^{-1}, in 0.05 M H_2SO_4 electrolyte	[63]	$S_{BET} = 103$ m^2 g^{-1}, $d_p = 5.8$ nm, $V_p = 0.44$ cm^3 g^{-1}
$x = 2$	Hydrothermal, Co^{II} and Mn^{II} acetate, 2:1 molar ratio, ethanol, 150 °C, 40 °C for 12 h on CNTs	$\eta = 450$ mV when $J = 10$ mA·cm^{-2} $A = 176.86$ mV·dec^{-1}, disk rot. 1600 rpm, in 0.1 M KOH electrolyte	[97]	no data
$x = 2$	$K_3[Co(CN)_6]$ and Mn^{II} acetate, 2:1 ratio, aq. citrate solution, 24 h, 350 °C for 2 h	$\eta = 489$ mV when $J = 10$ mA·cm^{-2} $\eta = 651$ mV when $J = 50$ mA·cm^{-2} $A = 235.6$ mV·dec^{-1}, $R_{CT} = 154.2$ Ω $C_{dl} = 101.3$ µF·cm^{-2} in 0.5 M H_2SO_4 electrolyte	[105]	No data
$x = 2$	Hydrothermal, Co^{II} and Mn^{II} nitrate, 2:1 ratio, aq. urea and NH_4F solution, 130 °C for 12 h, 350 °C for 4 h on IrO_2 carrier	$\eta = 210$ mV when $J = 10$ mA·cm^{-2} $A = 69.5$ mV·dec^{-1}, $C_{dl} = 12.8$ µF·cm^{-2} in 0.1 M $HClO_4$ electrolyte	[82]	$S_{BET} = 108.8$ m^2 g^{-1}
$x = 2$	Electrospinning, Co^{II} and Mn^{II} acetate, 1:2 ratio, mixed with polyvinylpyrrolidone in DMF, 25 kV, 1.2 mL·h^{-1}, 600 °C 3 h	Mass activity: 17.4 A·g^{-1}, specific surface activity: 90 mA·cm^{-2}, $V = 1.55$ V, in 0.1 M KOH electrolyte	[110]	$S_{BET} = 19.3$ m^2 g^{-1}, $d_p = 21.1$ nm, $V_p = 0.102$ cm^3 g^{-1}
$x = 2$	Spray pyrolysis, Co^{II} and Mn^{II} acetate, 2:1 ratio, aq. solution, sprayed on glass coating of fluorine-doped Sn-oxide, 150 °C for 1 h	$\eta = 145$ mV when $J = 3$ mA·cm^{-2} $A = 69$ mV·dec^{-1}, $C_{dl} = 60$ µF·cm^{-2} at T = 20 °C in 1 M KOH electrolyte	[108]	No data
$x = 2.25$	Spray pyrolysis, Co^{II} and Mn^{II} acetate, 3:1 ratio, aq. solution, sprayed on glass coating of fluorine-doped Sn-oxide, 150 °C for 1 h	$\eta = 110$ mV when $J = 13$ mA·cm^{-2} $A = 68$ mV·dec^{-1}, $C_{dl} = 60$ µF·cm^{-2} at T = 20 °C in 1 M KOH electrolyte		
$x = 2.5$	Spray pyrolysis, Co^{II} and Mn^{II} acetate, 5:1 ratio, aq. solution, sprayed on glass coating of fluorine-doped Sn-oxide, 150 °C for 1 h	$\eta = 65$ mV when $J = 18$ mA·cm^{-2} $A = 68$ mV·dec^{-1}, $C_{dl} = 60$ µF·cm^{-2} at T = 20 °C in 1 M KOH electrolyte		
$x = 2.73$	Hydrothermal, Co^{II} and Mn^{II} chloride, 10:1 ratio, aq. urea sol., 200 °C for 12 h, 400 °C for 5 h on platinized-titanium mesh	$\eta = 602$ mV when $J = 10$ mA·cm^{-2} in 0.5 M H_2SO_4 electrolyte	[79]	$S_{BET} = 10.57$ m^2 g^{-1}
$x = 2.75$	Spray pyrolysis, Co^{II} and Mn^{II} acetate, 11:1 ratio, aq. solution, sprayed on glass coating of fluorine-doped Sn-oxide, 150 °C for 1 h	$\eta = -10$ mV when $J = 22$ mA·cm^{-2} $A = 67$ mV·dec^{-1}, $C_{dl} = 60$ µF·cm^{-2} at T = 20 °C in 1 M KOH electrolyte	[108]	No data

Hong et al. studied the $MnCo_2O_4$ catalyst supported on an IrO_2/carbon cloth carrier (hydrothermal reaction of Co^{II} and Mn^{III} nitrate with aqueous urea and NH_4F solution at 130 °C for 12 h with subsequent calcination at 350 °C for 4 h) in OER. They studied the influence of the variation of the fluoride content of the catalyst that was set during the hydrothermal synthesis [82]. The obtained parameters were found to be better than those obtained with spinels without fluorine content. The stability of the fluorine-containing catalyst was found to be at least 100 h. Additionally, Xong et al. performed a solar-driven electrolyzer test also, actually a solar-to-hydrogen (STH) test, when light intensity was set to 1 ± 0.1 Sun, and the efficiency was 15.1% for at least 3000 s [82].

Du et al. studied the catalytic activity of a $CoMn_2O_4$-rGO composite in OERs [98]. The catalyst was prepared solvothermally from Co^{II} chloride and Mn^{II} acetate with oleylamine and stearic acid in xylene solution at 120 °C for 3 h on a reduced graphene oxide carrier. They found that the nanoparticles prepared in this way were good catalysts in OERs with the following parameters: $\eta = 1490$ mV when $J = 10$ mA·cm^{-2} $A = 56$ mV·dec^{-1} when the disk rotation was 1600 rpm and 0.1 M KOH electrolyte was used (Table 8) [98].

Xie et al. studied the OER performance (in 0.5 M H_2SO_4 electrolyte) of different $Co_xMn_{3-x}O_4$ ($x = 1.2$; 1.5 and 2) spinels with a hollow-box structure [105] prepared from manganese(II) [hexacyanocobaltate(III)] made in citric acid solution with subsequent heat treatment at 350 °C for 2 h. The spinel with $x = 1.5$ was found to be the best as far as the η (both at 10 and 50 mA·cm^{-2}), A, R_{CT}, and C_{dl} values are concerned. The long-term stability at 10 mA·cm^{-2} (in 0.5 M H_2SO_4 electrolyte) was 20 h (for simple Co_3O_4, it was only 4 h) [105]. Chueh et al. used a three-electrode electrochemical cell (Ag/AgCl reference electrode, platinum wires as counter electrode, and the catalyst-coated platinized-titanium mesh as the working electrode) for testing the OER efficiency of the $Co_xMn_{3-x}O_4$ with $x = 2.73$ in 0.5 M H_2SO_4 electrolyte [79]. The spinel was prepared in a hydrothermal reaction of metal chlorides and aqueous urea solution at 200 °C for 12 h, with heating at 400 °C for 5 h on platinized-titanium mesh. Neither the performance nor stability achieved the values for Co_3O_4, as those found by Wei et al. for the $x = 0.75$ catalyst applied in phosphate buffer (pH = 7.0) at 1 mA·cm^{-2} ($\eta = 710$ mV) [104]. Jiang et al. prepared $Co_{1.5}Mn_{1.5}O_4$ on carbon and FeN_x/C nanocomposite-supported catalysts for ORR [90]. They found that the best yield for H_2O_2 was obtained with the carbon-containing composite catalyst (16.1%).

Sugawara et al. tested the influence of the morphology (tunnel, layer, and spinel-type) of the $Co_{1.5}Mn_{1.5}O_4$ catalysts [99] prepared in the solvothermal reaction of cobalt(II) chloride and n-Bu$_4$NMnO$_4$ in isopropanol at 120 °C for 3 h. They studied both the OER and ORR performance of the Co-Mn oxides on a rotating disk electrode at 1600 rpm. The OER studies showed the order of linear shape, tunnel (7.8 times lower), and spinel type (26 times lower) catalysts. They were stable for 500 cycles even in 1 M KOH electrolyte [99]. The same order was found for the ORR as well [99] (Table 9).

Table 9. Photocatalytic water splitting and ORR performances of different $Co_xMn_{3-x}O_4$ spinels ($0 < x < 3$).

$Co_xMn_{3-x}O_4$	Precursors and Preparation Conditions	Catalyst Results and Conditions	Refs.	Remarks
$x = 0.75$	Hydrothermal, $K_3[Co(CN)_6]$ and Mn^{II} acetate, in 1:3 ratio, aqueous solution with polyvinylpyrrolidone, 450 °C for 2 h	O_2 evolution: 8.0 µmol yield: 21.3%, TOF = 2.3·10^{-4} mol$_{O2}$·mol$_{metal}^{-1}$·s^{-1}, 420 nm $\leq \lambda$, in 0.1 M phosphate	[104]	$S_{BET} = 61.4$ m^2 g^{-1},
$x = 1$	Solvothermal, Co^{II} and Mn^{II} nitrate, in 1:3 ratio, EtOH sol. on silica KIT-6, reflux, 350 °C for 3 h	O_2 evolution: TOF < 2.5·10^{-4} mol$_{O2}$·mol$_{metal}^{-1}$·s^{-1}, 400 nm $\leq \lambda$, at pH 5.8 in the presence of $[Ru(bpy)_3]^{2+}$ and $Na_2S_2O_8$	[93]	$S_{BET} = 93.8$ m^2 g^{-1},
$x = 1$	Solvothermal, Co^{II} chloride and Mn^{II} acetate, oleylamine, stearic acid in xylene, 120 °C for 3 h, reduced Graphene Oxide carrier	$E_{1/2} = 890$ mV at $J = 11.44$ mA·cm^{-2} $A \approx 0.5$ mV·dec^{-1} Mass activity: 254.7 A·g$_{metal}^{-1}$ TOF = 0.4 disk rot. 1600 rpm, in 0.1 M KOH electrolyte	[98]	Crystallite size distribution: 1.7–3 nm, $S_{BET} = 412$ m^2 g^{-1}, $d_p = 3.7$ nm

Table 9. *Cont.*

$Co_xMn_{3-x}O_4$	Precursors and Preparation Conditions	Catalyst Results and Conditions	Refs.	Remarks
$x = 1$	Hydrothermal, Co^{II} and Mn^{II} acetate, 1:2 ratio, aq. NH_3, 140 °C for 20 h	O_2 evolution: 10.56 μmol in 20 min $k = 5.76 \cdot 10^{-3}$ μmol·s^{-1} $k_{norm.} = 1.20 \cdot 10^{-4}$ μmol·g·s^{-1}·m^{-2} 300 W Xe-lamp, 400 nm $\leq \lambda$ in the presence of $[Ru(bpy)_3]^{2+}$ and $Na_2S_2O_8$ in Na_2SiF_6 buffer (pH = 6)	[31]	$S_{BET} = 48.05$ m^2 g^{-1},
$x = 1$	Electrospinning, Co^{II} and Mn^{II} acetate, 1:2 ratio, mixed with polyvinylpyrrolidone in DMF, 25 kV, 1.2 mL·h^{-1}, 600 °C 3 h	Mass activity: 3.3 A·g^{-1} Spec. surface activity: 20 mA·cm^{-2} $V = 0.7$ (V vs. RHE) in 0.1 M KOH electrolyte	[110]	$S_{BET} = 16.0$ m^2 g^{-1}, $d_p = 19.4$ nm, $V_P = 0.078$ cm^3 g^{-1}
$x = 1.2$	Solvothermal, Mn^{II} oleate and Co^{II} stearate, 1.5:1 ratio, 1-octadecene, 120 °C (vac.) for 1 h, 305 °C (N_2 atm.) for 0.5 h, $SrTiO_3$ carrier	Evolution: O_2: ~475 μmol and H_2: ~225 μmol disk rotation: 1600 rpm, 300 nm $\leq \lambda$ in 0.1 M phosphate electrolyte (pH = 6.9) under 5 h	[9]	No data
$x = 1.5$	Solvothermal, Co^{II} chloride and n-Bu_4NMnO_4, 1:1 molar ratio, 2-propanol, 3 h reflux, 120 °C	$A = 40$ mV·dec^{-1}, TOF: $2.8 \cdot 10^{-2}$ s^{-1}; disk rot. 1600 rpm, 0.63 cm$^2_{BET}$·cm$^{-2}_{disk}$ in 0.1 M KOH electrolyte	[99]	$S_{BET} = 223$ m^2 g^{-1}, TUNNEL
$x = 1.5$	Solvothermal, Co^{II} chloride and n-Bu_4NMnO_4, 1:1 molar ratio, 2-propanol, 3 h reflux, 120 °C	$A = 38$ mV·dec^{-1}, TOF: $2.0 \cdot 10^{-2}$ s^{-1}; disk rot. 1600 rpm, 0.63 cm$^2_{BET}$·cm$^{-2}_{disk}$ in 0.1 M KOH electrolyte	[99]	$S_{BET} = 35$ m^2 g^{-1}, LINEAR
$x = 1.5$	Solvothermal, Co^{II} chloride and n-Bu_4NMnO_4, 1:1 molar ratio, 2-propanol, 3 h reflux, 120 °C	$A = 52$ mV·dec^{-1}, TOF: $2.8 \cdot 10^{-2}$ s^{-1}; disk rot. 1600 rpm, 0.63 cm$^2_{BET}$·cm$^{-2}_{disk}$ in 0.1 M KOH electrolyte	[99]	$S_{BET} = 283$ m^2 g^{-1}, SPINEL
$x = 1.5$	Co^{II} and Mn^{II} nitrate, 1:1 ratio, aq. $NaHCO_3$, 24 h, 600 °C for 2 h	H_2 evaluation: 174 μmol, quantum yield: 1.47% in 10 min TOF = 0.175 μmol$_{H2}$·mg$_{cat.}^{-1}$·min^{-1}, $E_a = 0.18$ eV, W-lamp (19 mW·cm^{-2}), in $Na_2SO_4/Na_2S_2O_3$ electrolyte, pH = 13	[87]	$S_{BET} = 24.0$ m^2 g^{-1}, $d_p = 39$ nm, $V_P = 0.026$ cm^3 g^{-1}
$x = 1.5$	Trituration, Co^{II} and Mn^{II} oxalate, 1:1 ratio, aqueous NH_3., 400 °C for 2 h, ultrasonic processing activated C	$E_{1/2} = 760$ mV at $J = 6$ mA·cm^{-2} $A = 54$ mV·dec^{-1}, Yield(H_2O_2) = 16.1% Electron transfer number: 3.68, 1600 rpm, in 0.1 M KOH electrolyte	[108]	No data
$x = 1.5$	Trituration, Co^{II} and Mn^{II} oxalate, 1:1 ratio, aq. NH_3, 400 °C for 2 h, ultrasonic processing, FeN_x/C carrier	$E_{1/2} = 820$ mV at $J = 6$ mA·cm^{-2} $A = 35$ mV·dec^{-1}, Yield(H_2O_2) = 4.8% Electron transfer number: 3.90 $k = 11.7 \cdot 10^{-2}$ cm·s^{-1} 1600 rpm, in 0.1 M KOH electrolyte	[108]	No data
$x = 2$	Hydrothermal method, Co^{II} and Mn^{II} acetate, 2:1 ratio, aq. NH_3, 150 °C for 3 h	TOF (Clark electrode) = $1.8 \cdot 10^{-3}$ mol$_{O2}$·mol$_{metal}^{-1}$·s^{-1} TOF (Reactor/GC system) = $8.7 \cdot 10^{-4}$ mol$_{O2}$·mol$_{metal}^{-1}$·s^{-1} 300 W Xe-lamp, 400 nm $\leq \lambda$ Na_2SiF_6–$NaHCO_3$ buffer (pH = 5.8) in the presence of $[Ru(bpy)_3]^{2+}$, $Na_2S_2O_8$ and Na_2SO_4	[76]	$S_{BET} = 135$ m^2 g^{-1},

Table 9. *Cont.*

$Co_xMn_{3-x}O_4$	Precursors and Preparation Conditions	Catalyst Results and Conditions	Refs.	Remarks
$x = 2$	Hydrothermal, Co^{II} and Mn^{II} acetate, 2:1 ratio, aq. NH_3, 140 °C for 20 h	O_2 evolution: 13.04 µmol in 20 min $k = 7.22 \cdot 10^{-3}$ µmol·s^{-1} $k_{norm.} = 8.35 \cdot 10^{-5}$ µmol·g·s^{-1}·m^{-2} 300 W Xe-lamp, 400 nm $\leq \lambda$ in the presence of $[Ru(bpy)_3]^{2+}$ and $Na_2S_2O_8$ in Na_2SiF_6 buffer (pH = 6)	[31]	$S_{BET} = 86.46$ m^2 g^{-1},
$x = 2$	Electrospinning, Co^{II} and Mn^{II} acetate, 1:2 ratio, mixed with polyvinylpyrrolidone in DMF, 25 kV, 1.2 mL·h^{-1}, 600 °C for 3 h	Mass activity: 4.1 A·g^{-1}, spec. surface activity: 21 mA·cm^{-2} $V = 0.7$ (V vs. RHE) in 0.1 M KOH electrolyte	[110]	$S_{BET} = 19.3$ m^2 g^{-1}, $d_p = 21.1$ nm, $V_p = 0.102$ cm^3 g^{-1}

Jung et al. studied the OER and ORR activity of $Co_xMn_{3-x}O_4$ ($x = 1$; 2) catalyst prepared by electrospinning of Co^{II} and Mn^{II} acetate and polyvinylpyrrolidone in DMF (25 kV, 1.2 mL·h^{-1}, 600 °C 3 h) [110] and found that these spinels reduced the discharge/charge voltage gaps (improved the round-trip efficiency) of the Zn-air batteries [110]. Zhang et al. studied $Co_xMn_{3-x}O_4$ ($x = 1$; 2) spinels on carbon nanotubes (CNTs, prepared according to [71]) in OER and ORR processes in 0.1 M KOH solution [97]. Nissinen et al. found that the formation of $MnCo_2O_4$ on a carbon support could be controlled [126], and Zhang showed that better performance was achieved at 100 mA·cm^{-2} with $\eta = 450$ mV than with $\eta = 560$ mV [97], whereas the onset potentials were around 1.53 V. Yoshinaga et al. prepared Co-containing $Co_xMn_{3-x}O_4$ ($x = 0$–1.2) catalysts (solvothermal reaction, Mn^{II} oleate, and Co^{II} stearate, 1-octadecene, 120 °C in vacuum for 1 h, 305 °C in N_2 atmosphere for 0.5 h, $SrTiO_3$ carrier) and tested them in OER and photocatalytic O_2 evolution reaction. The results of the OER are summarized in Table 8. The best results in both OER and photocatalytic O_2 evolution reactions were obtained with the $Co_{1.2}Mn_{1.8}O_4$ catalyst [9].

Besides the OER studies, Wei et al. conducted detailed photocatalytic and Ce^{IV}-driven water oxidation reaction studies of cobalt manganese oxide spinels (prepared from manganese(II) [hexacyanocobaltate(III)] hydrothermally with polyvinylpyrrolidone and then calcined at 450 °C for 2 h) at room temperature with $[Ru(bpy)_3]^{2+}$ and $Na_2S_2O_8$ additives in 0.1 M phosphate buffer (pH 7.0) [104]. An LED light source with a glass filter (420 nm $\leq \lambda$) was used. The evolved amount of O_2 after 13 min was 8.0 µmol and 21.3% yield was observed with the $Co_xMn_{3-x}O_4$ ($x = 0.75$), which is lower than the 12.9 µmol O_2 with 34.4% yield obtained for Co_3O_4 [104]. Rosen et al. applied mesomorph $CoMn_2O_4$ catalyst (solvothermal synthesis from Co^{II} and Mn^{II} nitrate in ethanol, on silica KIT-6 sith and calcination at 350 °C for 3 h) in photocatalytic water oxidation (PWO) with a xenon lamp, with $[Ru(bpy)_3]^{2+}$ and $Na_2S_2O_8$ additives. Mn doping did not improve the catalysis performance [93]. Kihal et al. used a three-electrode electrochemical cell (Ag/AgCl reference electrode, platinum gauze as counter electrode, and the catalyst as the working electrode) for testing the photocatalytic H_2 evolution reaction efficiency of $Co_{1.5}Mn_{1.5}O_4$ spinel [87]. The catalyst was prepared from Co^{II} and Mn^{II} nitrate with aqueous $NaHCO_3$ with subsequent calcination at 600 °C for 2 h, and the resultant spinel had a p-type transition. They applied a tungsten lamp ($\phi = 1.39 \cdot 10^{19}$ photons $\cdot$ s^{-1}) and, in the presence of 1.25 mg/L catalyst, in 10 min, 174 µmol H_2 was produced (quantum yield: 1.47%) in a $Na_2SO_4/Na_2S_2O_3$ electrolyte, at pH = 13 [87].·

Zhang et al. tested a hydrothermally prepared $MnCo_2O_4$ catalyst (hydrothermal method, Co^{II} and Mn^{II} acetate, aqueous NH_3, 150 °C for 3 h) in a Clark electrode and Reactor/GC system [76]. Liu et al. [31] studied systems ($x = 1$ and 2, hydrothermal, Co^{II} and Mn^{II} acetate, aqueous NH_3, 140 °C for 20 h), the same as Rosen et al. [93] and

Zhang et al. [76], but the two studied catalysts ($Co_xMn_{3-x}O_4$, $x = 1$ and 2) gave a similar amount of O_2 (10.56 and 13.04 μmol, respectively) [31]. Du et al. studied the catalytic activity of $CoMn_2O_4$-rGO composite in ORRs and OERs. They found that the spinel nanoparticles prepared in a solvothermal way from Co^{II} chloride and Mn^{II} acetate in the presence of oleyl amine and stearic acid in xylene at 120 °C for 3 h were good catalysts in OERs (Table 8) and ORRs as well (Table 9) [98].

3.5. Application in Advanced Oxidation Processes (AOPs)

Recently, the removal of organic pollutants from wastewater, like pharmaceutical compounds or VOCs, has become an intensively studied area. The advanced oxidation processes (AOPs) with hydrogen peroxide (H_2O_2) or its precursors represent promising environmentally friendly strategies, including Fenton/Fenton-like oxidations [32–34], ozonization [36], or peroxymonosulfate mediated oxidations [35].

Wang et al. studied $Co_xMn_{3-x}O_4$ ($x = 1$ and 2) catalyst prepared from cobalt(II) nitrate and $KMnO_4$ with aqueous maleic acid with subsequent heating at 500 °C for 4 h and tested the AOP decomposition in toluene with and without peroxymonosulfate (PMS) oxidant [100]. Without PMS, the efficiency of the $Co_xMn_{3-x}O_4$ spinel catalysts was 38, 49, and 2% at $x = 1$, 2, and 3, respectively. The combination of the spinel catalysts with PMS increased the conversion up to 97.3% ($x = 1$), and the efficiency was 95% in 180 min. The concentration of CO_2 in the formed gaseous mixture was 80–205, 72–180, and 50–60 ppm, or the total amount of CO_2 in 180 min was 20,791, 16,580, and 7564 ppm for the spinels with $x = 1$, 2 and 3, respectively. In the presence of PMS, the exhausted gaseous mixture of the TOC (total organic carbon) and IC (inorganic carbon) values were 0.927 and 10.96 mg/L, 4.264 and 3.09 mg/L, and 5.284 and 4.635 mg/L, for the catalysts with x = 1, 2, and 3, respectively. The PMS concentration influenced the conversion (83.2, 97.3, and 96.5% at c = 0.02, 0.1, and 0.2 g/L, respectively) with the use of spinels with $x = 1$. The catalytic activity depended on pH; the best choice was an acidic media between pH = 3 and 7 (the conversion was 97%), but at pH = 1 and 11, the conversions were 55 and 27%, respectively. The catalyst was structurally stable even after 27 h at pH = 3. The authors tested the performance of the catalysts with tap water, then the efficiency dropped to 80.22% [100]. The best results were found at 30 ppm toluene concentration. These and some other catalytic results are summarized in Table 10.

Table 10. The effect of $Co_xMn_{3-x}O_4$ ($0 < x < 3$) in advanced oxidation processes (AOPs).

$Co_xMn_{3-x}O_4$	Precursors, Preparation Method, and Conditions	Catalyst Results and Conditions	Refs.	Remarks
$x = 0.75$	Carbon template cations-adsorption-calcination method, Co^{II} and Mn^{II} acetate, 1:3 ratio, 500 °C for 2 h	Norfloxacin (20 ppm) deg.: 94% and TOC: 42.5% in 80 min Xe lamp, 400 nm $\leq \lambda$, in the presence of PMS (0.2 g/L), $T = 45$ °C pH = 6.5	[62]	$S_{BET} = 29$ m^2 g^{-1}, $d_p = 3.2$ nm,
$x = 1$	Co^{II} nitrate and $KMnO_4$, 1:1 molar ratio, aq. maleic acid, 500 °C for 4 h	Toluene (*g*) degrad. (500 mL/min), 49% in 80 min, no PMS Toluene (*g*) degrad. (500 mL/min), 97.3% in 80 min, 95% in 180 min, TOC = 0.927 mg/L, IC = 10.96 mg/L in the presence of PMS 0.1 g/L	[100]	$S_{BET} = 98.21$ m^2 g^{-1}, $d_p = 7.818$ nm, $V_p = 0.254$ cm^3 g^{-1}
$x = 1$	Co-axial electrospinning, DC = 13 kV and 10 μL·min^{-1} flow rate, Co^{II} and Mn^{II} acetate, 1:2 ratio, 900 °C (N_2) for 100 min, 250 °C in the air for 30 min, on carbon nanofibers carrier (made form PMMA)	Rhodamine *B* degradation (50 μM), 100% in 40 min, $k_{deg} = 0.093$ min^{-1} in the presence of PMS at 20 °C, pH = 3.28 Rhodamine *B* degradation (50 μM), 100% in 5 min, $k_{deg} = 0.3406$ min^{-1} in the presence of PMS at 50 °C, pH = 3.28	[127]	$S_{BET} = 352.2$ m^2 g^{-1}, $d_p = 1.78$ nm, $V_p = 0.184$ cm^3 g^{-1}

Table 10. *Cont.*

$Co_xMn_{3-x}O_4$	Precursors, Preparation Method, and Conditions	Catalyst Results and Conditions	Refs.	Remarks
$x = 1$	Co^{II} and Mn^{II} nitrate, 1:2 molar ratio, aqueous solution, 90 °C for 2 h, 250 °C for 2 h, on HNT (w = 40%) carrier	CBZ degradation (21.16 μM), 100% in 20 min, TOC/IC = 60%, $k_{deg,CBZ}$ = 0.168 min^{-1}; SMZ and TC degradation 90%<, OFX = 65% in 30 min, in the presence of PMS (0.1 mM), pH = 5.8.	[72]	S_{BET} = 66.0 m^2 g^{-1}, V_P = 0.38 cm^3 g^{-1}
$x = 2$	Co^{II} nitrate and $KMnO_4$, 1:2 molar ratio, aq. solution, 70 °C for 70 min, 350 °C for 2 h	Ciprofloxacin degrad. (c = 10 mg/L): 81% in 100 min in the presence of 68 mg/L H_2O_2, $k_{deg,H2O2}$ = 0.284 min^{-1}	[101]	S_{BET} = 111.4 m^2 g^{-1}, d_p = 6.327 nm, V_P = 0.212 cm^3 g^{-1}
$x = 2$	Hydrothermal, Co and Mn chloride, 2:1 ratio, aq. solution urea, 120 °C 6 h, 400 °C 2 h	MB degradation (20 ppm): 96.7% (20 min) and 100% (25 min) in the presence of Oxone (500 ppm), T = 30 °C, pH = 3.5.	[80]	No data
$x = 2$	Hydrothermal, Co and Mn acetate, 2:1 ratio, aqueous monoethyleneglycol, pH = 10, 180 °C 8 h, thin-film annihilation on glass 550 °C 2 h	Acid Black 1 dye (11.10 ppm) Conv.: ~100% TOC: 91% in 2.5 h in water	[77]	No data

Kang et al. [127] tested a $Co_{1.5}Mn_{1.5}O_4$ catalyst prepared by coaxial electrospinning from Co^{II} and Mn^{II} acetates and fixed on a hollow activated carbon nanofibers carrier. The subject of the test was PMS-activated Rhodamine B (RhB) degradation at various pH (2.7–9.2), temperature (10–50 °C), RhB concentration (50–200 μM), and flow rate (1–2 mL·min^{-1}) settings in batch reaction and in continuous-flow tests, as well. The effect of the conditions on RhB degradation is summarized in Figure 9 [127].

The catalyst ($x = 1$) and PMS degraded RhB (c = 50 μM) after 80 min only in 2% and 12%, respectively; however, the catalyst and PMS together degraded RhB within 40 min and 5 min at room temperature and 50 °C, respectively, in 100% [127]. The best pH range was 2.7–7.2; the best PMS concentration was 3 mM. The RhB degradation after 5 cycles was 83.7% in 40 min; however, after regeneration with calcination at 150 °C for 1 h, 100% degradation could be reached in 40 min [127]. The continuous-flow test showed that the breakthrough of RhB occurs after 120 min at the flow rate of 2 mL·min^{-1} [127].

Dung et al. synthesized a Co_2MnO_4 catalyst with a hydrothermal method with urea as an ammonia precursor and studied its catalytic effect in the AOP degradation of organic dyes in detail [80]. The effect of catalyst dosage, oxidant and dye concentrations, temperature, and pH with different kinds of dyes (RhB, Metil orange, OG, and Methylene blue (MB)) were tested with and without peroxodisulfate (PS), peroxymonosulfate (PMS), and H_2O_2 [80] (Figure 10). It was found that the best order of adding the components of the AOP reaction was the MB–catalyst–PMS. The catalyst and PMS degraded MB only by 5.2% and 10.9%, respectively, whereas they together reached 96.7% degradation in 20 min [80]. With a six-fold increase in the catalyst dosage, the degradation of MB raised to 99.2% (from 37.8%) in 10 min, whereas an increase in the amount of Oxone (by 3 times) resulted in a change of degradation from 56.2 to 91.8% in 20 min [90]. In contrast, the increased dosage of the MB concentration had a negative effect on the degradation rate (69.2% degradation in 20 min). A shorter reaction time resulted from increasing the temperature since at 55 °C, 91.5% degradation was reached in 4 min. The best pH range was 4.5 to 7.0; in this range, at least 96% degradation was obtained in 14 min, whereas below or above this pH range, the degradation dropped. The degradation efficiency for different dyes decreased in the following order: MO > OG > RhB > MB (Figure 10) [80]. It was found that the concentration of Cl$^-$, even in 20 mM, inhibited the degradation and resulted in only 27.1% efficiency. For other anions, the following values were found in 20 min for MB: 96.7, 81.8, and 58.6% for NO_3^-, CO_3^{2-}, and HCO_3^-, respectively. If the PMS was changed to PS or H_2O_2, the degradation in 20 min dropped to 48.5 and 39.4%, respectively [80]. The photocatalytic decomposition of acid Black 1 textile dye catalyzed by Co_2MnO_4 prepared by a hydrothermal method from acetate salts in ethylene glycol was tested by Habibi et al. and found that after 2.5 min almost a complete decomposition and 91% TOC could be reached [77].

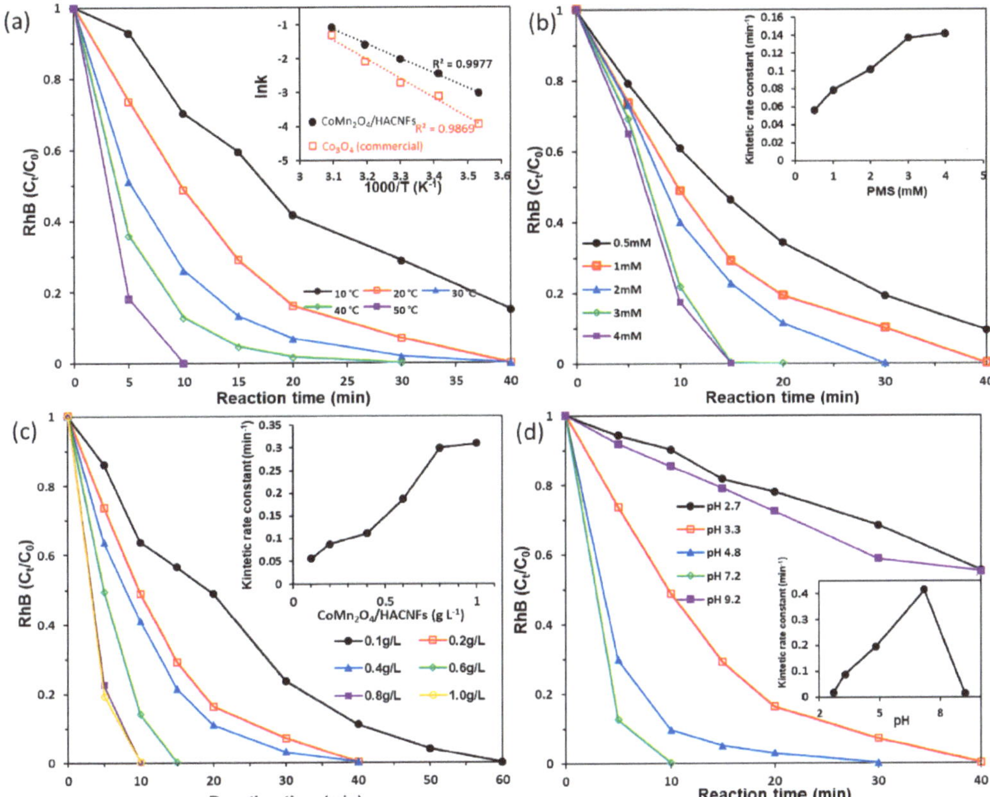

Figure 9. The effect of conditions on PMS-activated Rhodamine B (RhB) degradation in the presence of $Co_{1.5}Mn_{2.5}O_4$ catalyst on a hollow activated carbon nanofibers carrier as reported by Kang et al. [B35]. (**a**) Effect of T, when c_{RhB} = 50 μM, $c_{catalyst}$ = 0.02 g/L, c_{PMS} = 1 mM, pH 3.28, (**b**) effect of presence of PMS, when (c_{RhB} = 50 μM, $c_{catalyst}$ = 0.02 g/L, T = 25 °C, pH 3.28), (**c**) effect of different $c_{catalyst}$s, when c_{RhB} = 50 μM, T = 25 °C, c_{PMS} = 1 mM, pH 3.28, and (**d**) effect of pH, when c_{RhB} = 50 μM, $c_{catalyst}$ = 0.02 g/L, T = 25 °C, c_{PMS} = 1 mM. Reproduced from ref. [127].

Yang et al. prepared $CoMn_2O_4$ from nitrate salts with co-precipitation and fixed on halloysite, a natural mineral nanotube (HNT), and tested in PMS-supported AOP of pharmaceuticals such as carbamazepine (CBZ), ofloxacin (OFX), tetracycline (TC), or sulfamethoxazole (SMZ) in water [72]. Without PMS, the degradation of the CBZ was 6.2 and 6.7% when the HNT amount was 40 and 60%, respectively, after 30 min. When PMS was present, every $CoMn_2O_4$-HNT combination reached 80% CBZ conversion after 20 min, and the highest $k_{deg,CMZ}$ (0.168 min^{-1}) was obtained with the 40% HNT-containing catalyst. The increase in PMS concentration from 0.05 mM to 0.1 M almost completed the degradation; the optimal pH was found to be between 5.8 and 7.5. The degradation of CBZ was 90% after 3 cycles (each cycle was 45 min), and a small amount of Co and Mn were leached from the catalyst with a slight change in structure followed by the PXRD study [72]. Other pharmaceuticals, such as SMZ and TC, were degraded by over 90%, whereas OFX degradation was 65% in 30 min [72].

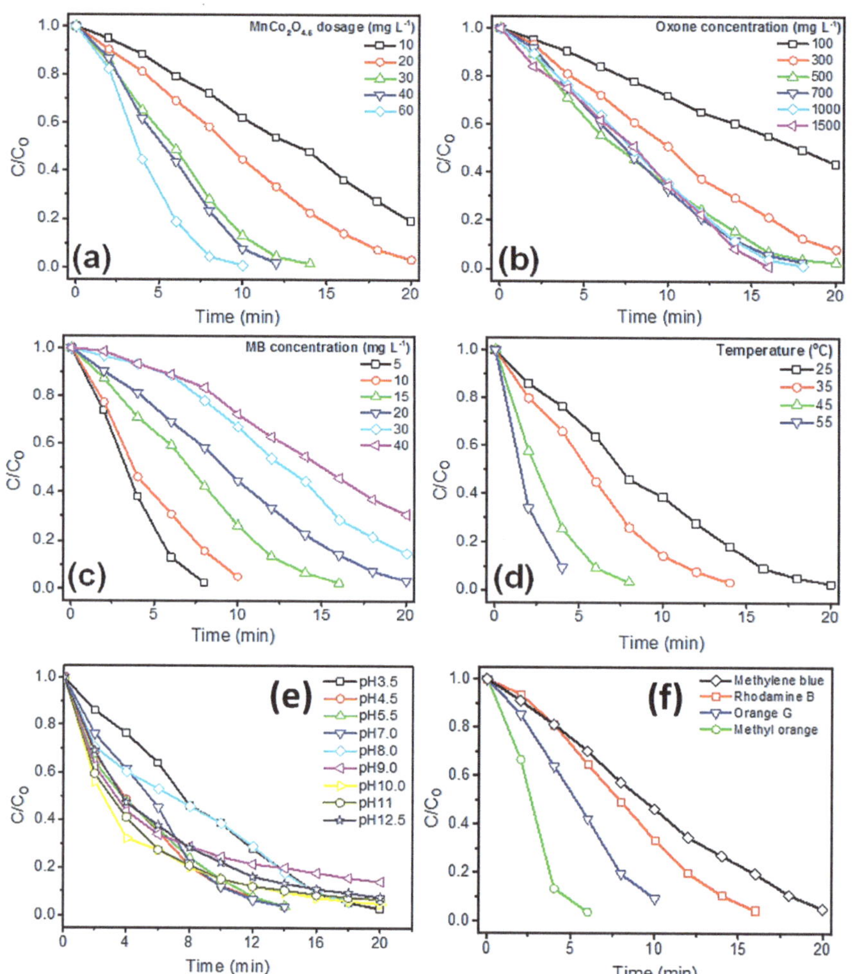

Figure 10. The effect of catalyst and reaction conditions on RhB, Methyl orange, OG, and Methylene blue degradation with and without PS, PMS, and H_2O_2 in the presence of Co_2MnO_4 catalyst (**a–e**) and different dyes (**f**). Reproduced from ref. [80].

Li et al. prepared different Co-containing Co-Mn spinel oxide ($Co_xMn_{3-x}O_4$ with $x = 0.75$, 1.5, and 2.25) photocatalysts from acetate salts on a carbon template for the AOP decomposition reaction of Norfloxacin (NOR) in the presence of PMS [62]. Exposition to visible light did not decompose NOR; however, in the presence of the catalysts, <5% degradation was achieved. The catalyst and PMS together resulted in 64% degradation in 80 min. The best result was achieved when PMS, a catalyst with $x = 0.75$, and light were applied together (93%) at pH 6.5; <0.2 g/L catalyst and 0.08 g/L PMS were used [62]. The TOC removal was 42.5% after 80 min [62].

3.6. $Co_xMn_{3-x}O_4$ (0 < x < 3) Catalysts in the Fischer–Tropsch Processes

One of the most essential topics of environmental sciences is to reduce the emission of greenhouse gases into the atmosphere to avoid global warming via decreasing fossil CO_2 emissions. The Fischer–Tropsch synthesis (FTS) [27,128,129] is one of the key routes to use renewable resources in fuel production because both H_2 and CO may be produced from biomass [130], which is usually used in households as heating material [131]. The Fischer–Tropsch method ensures the production of high-value basic chemicals, as well [26].

Paterson et al. studied different $Co_xMn_{3-x}O_4$ (x = 0–1.5) catalysts prepared from cobalt(II) nitrate and manganese(II) acetate and impregnated onto TiO_2 in FTS in a detailed manner and found that heavier alcohols and olefins were formed as the main products [26]. The selectivity of C_{5+} hydrocarbons decreased with the increase in Mn content in the catalyst (Figure 11) [26]. The results are summarized in Table 11. The active components of the catalyst (e.g., Co_2C) were formed via the decomposition of the starting spinel [26]. Li et al. observed similar Co_2C formation from Co_2MnO_4 prepared from Mn_3O_4 with manganese(II) nitrate impregnation, which resulted in acceptable conversions (Table 11) [25]. The pretreatment of the catalysts with H_2 before the FTS test resulted in Co and Mn oxides and Co_2C in larger amounts than without hydrogenation, and the selectivity of the FTS reaction shifted to the direction of linear paraffinic hydrocarbon formation [26]. Lindley et al. performed FTS tests with a small amount of Mn containing bimetallic catalysts ($Co_xMn_{3-x}O_4$ ($2.73 \leq x$)) and found that the mixed oxide slowly transformed into other materials, like CoO or elemental cobalt; however, the catalytic activity was good, since this results in 90% selectivity for C_{5+} products [28].

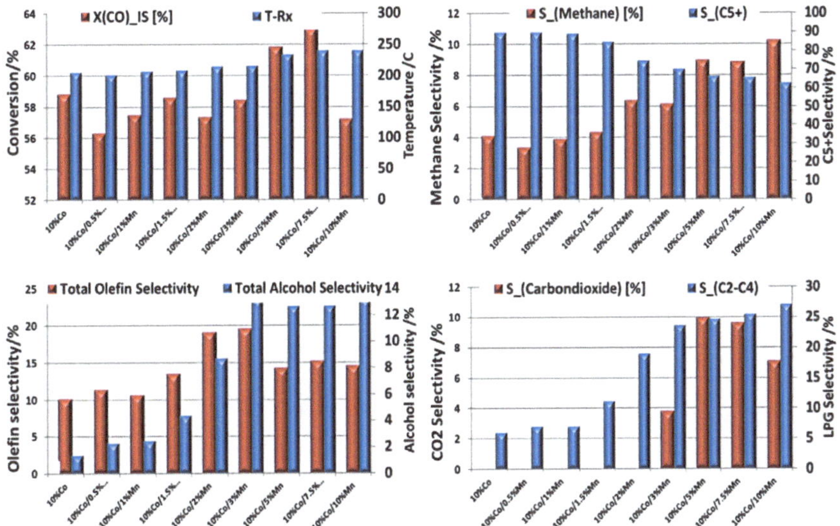

Figure 11. Catalyst performance data for different $Co_xMn_{3-x}O_4$ (x = 0–1.5) catalysts obtained by Paterson et al. under the following circumstances: H_2 to CO ratio 1.5, GHSV = 1500 h^{-1}, p = 42 bar. Reproduced from ref. [26].

Table 11. Catalytic performance of different $Co_xMn_{3-x}O_4$ ($0 < x < 3$) oxides in Fischer–Tropsch synthesis.

$Co_xMn_{3-x}O_4$	Precursors, Preparation Method, and Conditions	Catalyst results and conditions	Refs.	Remarks
$x = 1.5$	Co-precipitation with activated carbon, Co^{II} and Mn^{II} nitrate, 1:1 ratio, aq. 28–30% NH_3, 500 °C for 5 h,	CO conversion: 48.0%, selectivity.: CH_4: 7.0%, C_2: 4.3%, C_3: 16.1%, C_4: 7.6%, C_{5+}: 43.4%, CO_2: 20.4%, alcohols: 2.2% T =240 °C, CO:H_2 1:1, 6 bar, GHSV: 600 h^{-1}, t = 135 h	[29]	particle size: 9.3 nm
$x = 2$	Co^{II} and Mn^{II} nitrate, 2:1 ratio, aq. solution, 30 °C for 2 h, 330 °C for 3 h	1 bar, 250 °C, V: $N_2/H_2/CO$ = 3/64.6/3 12,000 mL·h^{-1}·g_{cat}^{-1},	[25]	S_{BET} = 22.3 m^2 g^{-1}, d_p = 16.9 nm, V_p = 0.11 cm^3 g^{-1}
$x = 2$	Impregnation, Co_3O_4 with Mn^{II} nitrate, 2:1 ratio, aq. solution, 35 °C for 2 h, 200 °C 3 h	1 bar, 250 °C, V: $N_2/H_2/CO$ = 3/64.6/3 12,000 mL·h^{-1}·g_{cat}^{-1},	[25]	S_{BET} = 49.8 m^2 g^{-1}, d_p = 9.5 nm, V_p = 0.13 cm^3 g^{-1}
$x = 2$	Co^{II} nitrate, Mn^{II} acetate, aqueous solution, impregnation on TiO_2	T_{max} = 219 °C, CO conversion: 35%, Selec.: CH_4: 11%, alcohol: 48% olefins: 4%, esters: 12% H_2 consumption: 2998 µmol·g^{-1} 1.8 H_2:CO, 30 barg, GHSV = 1500 h^{-1}	[26]	S_{BET} = 50.0 m^2 g^{-1}, d_p = 33 nm, V_p = 0.28 cm^3 g^{-1}
$X = 2.15$	Co^{II} nitrate, Mn^{II} acetate, aq. soln., impregnation on TiO_2	T_{max} = 218 °C, CO conversion: 37%, selec.: CH_4: 6%, alcohol: 51% olefins: 5%, esters: 10% H_2 consumption: 2944 µmol·g^{-1} 1.8 H_2:CO, 30 bar, GHSV = 1500 h^{-1}		
$x \approx 2.75$	Co-impregnation on TiO_2, aq. sol. of Co^{II} nitrate and Mn^{II} acetate, 10:1 ratio, 300 °C for 2 h	CO conversion: 62.3%, selec.: CH_4: 4.0%, C_2-C_4: 5.9%, C_{5+}: 67.0% 1.8 H_2:CO, 30 bar, GHSV = 1500 h^{-1}	[26]	S_{BET} = 50.0 m^2 g^{-1}, d_p = 33 nm, V_p = 0.28 cm^3 g^{-1}
$x > 2.75^a$	Co-impregnation on TiO_2, aqueous solution of Co^{II} nitrate and Mn^{II} acetate, 10:1 ratio, 300 °C for 2 h	CO conversion: 59.7%, selectivity: CH_4 8.8%, C_2-C_4: 25.2%, C_{5+}: 90.4% 1.8 H_2:CO, 30 bar, GHSV = 1500 h^{-1}		
$x = 2.78$	Co^{II} and Mn^{II} nitrate, 13:1 ratio, dried on CNTs at 60 °C, pyrolyzed at 360 °C for 2 h	Selectivity: CH_4: 23%, C_2-C_4: 31%, CH_{5+}: 46% ratios: $C_2^=/C_2$ = 0.37, $C_3^=/C_3$ = 4.9, $C_4^=/C_4$ = 2.5, $C_5^=/C_5$ = 1.5 $k_{m,CO,220°C}$ = 31·10^{-4} m^3·kg_{Co}^{-1}·s^{-1} E_a = 118 kJ·mol^{-1}, 1 bar, 220 °C, H_2 to CO 2:1 Selectivity: CH_4: 9%, C_2-C_4: 33%, CH_{5+}: 58% ratios: $C_2^=/C_2$ = 1.0, $C_3^=/C_3$ = 2.0, $C_4^=/C_4$ = 1.5, $C_5^=/C_5$ = 1.5 $k_{m,CO,220°C}$ = 0.9·10^{-4} m^3·kg_{Co}^{-1}·s^{-1} E_a = 163 kJ·mol^{-1}, 30 bar, 220 °C, H_2 to CO 2:1	[30]	S_{BET} = 205 m^2 g^{-1},

Iqbal et al. prepared $Co_{1.5}Mn_{1.5}O_4$ supported on activated carbon and studied its catalytic effect in FTS [27]. The stability of the catalytic performance was the same with and without support. However, the steady-state time was only 45 h for the carbon-supported one, compared to the 90 h of the bare catalyst, and the conversion values of CO were also different (48 and 36%, respectively) [27]. The selectivity was shifted to the direction of the C_{3+} products (67.1%) instead of 29.7% as an effect of the carbon support [27]. The best conversion and selectivity ratio were observed for the catalyst prepared at 500 °C; the selectivity for C_{5+} products was 89.6%, but the conversion was low (18.1%). The aging of the co-precipitated hydroxides negatively affected the conversion and selectivity [27].

Thiessen et al. studied low-Mn containing $Co_xMn_{3-x}O_4$ ($x \leq 2.78$) spinel oxides on a CNT carrier in FTS at 1 and 30 bar pressure; the effect of Pt or Ru co-catalysts was also tested

[B128]. The small amount of Mn doping positively affected the C5 selectivity. However, the trimetallic system was not found to be better than the simple Co-Mn-CNT composite [B128]. Similarly, the increase in pressure had no drastic effect on the reaction rate, but less CH_4 and more C_{5+} hydrocarbons were formed, and the amount of unsaturated hydrocarbons also decreased (Table 11) [30].

4. Conclusions and Outlook

Preparation methods and the catalytic activity of $Co_xMn_{3-x}O_4$ ($0 < x < 3$) spinels and multiphase materials with the formula $Co_xMn_{3-x}O_4$ for wide-scale industrial and environmental applications are summarized.

The variability of the preparation and annealing methods results in a critical influence on the composition, crystallinity, and surface layer chemistry. The range of compositions and preparation routes of cobalt manganese oxide spinels unambiguously shows that the catalytic properties of these spinels depend not only on the Co to Mn ratio but also on the distribution of the metal ions with their particular valence between the tetrahedral (A) and octahedral (B) spinel lattice sites. These metal valence and site distributions controlled by the synthesis conditions play a critical role in the catalytic activity of the $Co_xMn_{3-x}O_4$ ($0 < x < 3$) spinels. Thus, the catalytic features of $Co_xMn_{3-x}O_4$ ($0 < x < 3$) spinels can easily be controlled with the synthesis conditions, including the reaction routes, Co to Mn ratios, the valences of the metals in the precursor compounds, the counter-ions, reagent concentrations, temperature, time, and other reaction and annealing conditions.

The properties and catalytic activities of $Co_xMn_{3-x}O_4$ ($0 < x < 3$) catalysts prepared by various traditional solid-phase and solution-phase (precipitating reactions) methods for the removal of volatile organic compounds like toluene or inorganic gases such as CO and NO_x from the air, removal of textile dyes, and pharmaceuticals from water were reviewed. The Fischer–Tropsch reactions catalyzed with these cobalt manganese oxide spinels were also reviewed, together with oxidation reactions of organic compounds, e.g., to prepare 2,5-diformylfurane or vanillin, and other oxygen evolution and reduction reactions, as well.

In summary, in different catalytic reactions with various $Co_xMn_{3-x}O_4$ spinel catalysts, there are significant differences in applicability. For example, in toluene conversion and NH_3-SCR reactions, catalysts with $x < 1.5$ are preferred. In the case of advanced oxidation processes when $x = 1$, excellent degradation of organic pollutants can be achieved, whereas in the conversion of 5-hydroxymethylfurfural, the best results were obtained at $x = 2$. In the case of OERs or ORRs, the boundary line falls to $x = 1.5$, whereas a small amount of Mn or Co is enough to get a good conversion ratio with high selectivity for heavier hydrocarbons in the Fischer–Tropsch reactions.

The summarized results show the justification of research on new synthesis routes like the solid phase quasi-intramolecular redox reactions, which can produce cobalt manganese oxide spinel structures markedly different from those obtained by the well-known sol-gel or precipitation methods. One can prepare amorphous materials even below 150 °C, which shows better catalytic performance than the crystalline product in various electrocatalytic reactions due to their flexible structure and the presence of numerous active sites [132]. Additionally, the post-heat treatment of the amorphous products allows adjustment of the crystallite sizes and fine-tuning of the properties, e.g., phase purity, grain size, and the catalytic activity in a particular process.

Author Contributions: Conceptualization, L.K. writing—original draft preparation, K.A.B. and L.K.; writing—review and editing, L.K. and Z.H.; visualization, K.A.B.; supervision, Z.H. All authors have read and agreed to the published version of the manuscript.

Funding: This research received no external funding.

Data Availability Statement: Not applicable.

Conflicts of Interest: The authors declare no conflicts of interest.

References

1. Zhu, X.; Huang, A.; Martens, I.; Vostrov, N.; Sun, Y.; Richard, M.; Schülli, T.U.; Wang, L. High-Voltage Spinel Cathode Materials: Navigating the Structural Evolution for Lithium-Ion Batteries. *Adv. Mater.* **2024**, *36*, 2403482. [CrossRef]
2. Nayak, P.K.; Erickson, E.M.; Schipper, F.; Penki, T.R.; Munichandraiah, N.; Adelhelm, P.; Sclar, H.; Amalraj, F.; Markovsky, B.; Aurbach, D. Review on Challenges and Recent Advances in the Electrochemical Performance of High Capacity Li- and Mn-Rich Cathode Materials for Li-Ion Batteries. *Adv. Energy Mater.* **2018**, *8*, 1702397. [CrossRef]
3. Alcántara, R.; Pérez-Vicente, C.; Lavela, P.; Tirado, J.L.; Medina, A.; Stoyanova, R. Review and New Perspectives on Non-Layered Manganese Compounds as Electrode Material for Sodium-Ion Batteries. *Materials* **2023**, *16*, 6970. [CrossRef] [PubMed]
4. Kour, S.; Kour, P.; Sharma, A.L. Quantum Capacitance of MCo2O4 (M = Co, Zn, Mn, and Cr) Electrode Materials for Supercapacitor Applications: Insights from First Principles Calculations. *J. Energy Storage* **2025**, *107*, 114886. [CrossRef]
5. Shen, D.; Zhang, R.; Liu, S.; Yu, H.; Xia, D.; Niu, J.; Yang, Y.; Ren, Z.; Dong, W. Synthesis of Urchin-like MnCo2O4 Nanospheres as Electrode Material for Supercapacitors. *J. Energy Storage* **2025**, *105*, 114775. [CrossRef]
6. Ohji, T. *Advances in Materials Science for Environmental and Energy Technologies*; Ceramic transactions; Wiley: Hoboken, NJ, USA, 2012; ISBN 978-1-118-27342-5.
7. Wohlfahrt-Mehrens, M. Secondary Batteries–Lithium Rechargeable Systems–Lithium-Ion | Positive Electrode: Manganese Spinel Oxides. In *Encyclopedia of Electrochemical Power Sources*; Elsevier: Amsterdam, The Netherlands, 2009; pp. 318–327, ISBN 978-0-444-52745-5.
8. Huang, L.; Chen, K.; Zhang, W.; Zhu, W.; Liu, X.; Wang, J.; Wang, R.; Hu, N.; Suo, Y.; Wang, J. ssDNA-Tailorable Oxidase-Mimicking Activity of Spinel $MnCo_2O_4$ for Sensitive Biomolecular Detection in Food Sample. *Sens. Actuators B Chem.* **2018**, *269*, 79–87. [CrossRef]
9. Yoshinaga, T.; Saruyama, M.; Xiong, A.; Ham, Y.; Kuang, Y.; Niishiro, R.; Akiyama, S.; Sakamoto, M.; Hisatomi, T.; Domen, K.; et al. Boosting Photocatalytic Overall Water Splitting by Co Doping into Mn_3O_4 Nanoparticles as Oxygen Evolution Cocatalysts. *Nanoscale* **2018**, *10*, 10420–10427. [CrossRef]
10. Sabato, A.G.; Chrysanthou, A.; Salvo, M.; Cempura, G.; Smeacetto, F. Interface Stability between Bare, Mn Co Spinel Coated AISI 441 Stainless Steel and a Diopside-Based Glass-Ceramic Sealant. *Int. J. Hydrogen Energy* **2018**, *43*, 1824–1834. [CrossRef]
11. Molin, S.; Sabato, A.G.; Bindi, M.; Leone, P.; Cempura, G.; Salvo, M.; Cabanas Polo, S.; Boccaccini, A.R.; Smeacetto, F. Microstructural and Electrical Characterization of Mn-Co Spinel Protective Coatings for Solid Oxide Cell Interconnects. *J. Eur. Ceram. Soc.* **2017**, *37*, 4781–4791. [CrossRef]
12. Liu, Z.; Zhou, J.; Zhou, L.; Li, B.; Wang, T.; Liu, H. A Review on Mercury Removal in Chemical Looping Combustion of Coal. *Sep. Purif. Technol.* **2024**, *337*, 126352. [CrossRef]
13. Fergus, J.W. Recent Developments in Cathode Materials for Lithium Ion Batteries. *J. Power Sources* **2010**, *195*, 939–954. [CrossRef]
14. Puranen, J.; Laakso, J.; Kylmälahti, M.; Vuoristo, P. Characterization of High-Velocity Solution Precursor Flame-Sprayed Manganese Cobalt Oxide Spinel Coatings for Metallic SOFC Interconnectors. *J. Therm. Spray. Tech.* **2013**, *22*, 622–630. [CrossRef]
15. Bera, S.; Verma, K.D.; Kar, K.K. Traditional Electrode Materials for Supercapacitor Applications. In *Handbook of Nanocomposite Supercapacitor Materials IV*; Kar, K.K., Ed.; Springer Series in Materials Science; Springer International Publishing: Cham, Switzerland, 2023; Volume 331, pp. 19–64, ISBN 978-3-031-23700-3.
16. Gonçalves, J.M.; De Faria, L.V.; Nascimento, A.B.; Germscheidt, R.L.; Patra, S.; Hernández-Saravia, L.P.; Bonacin, J.A.; Munoz, R.A.A.; Angnes, L. Sensing Performances of Spinel Ferrites MFe_2O_4 (M = Mg, Ni, Co, Mn, Cu and Zn) Based Electrochemical Sensors: A Review. *Anal. Chim. Acta* **2022**, *1233*, 340362. [CrossRef]
17. Zhao, Q.P.; Jin, J.T.; Li, C.L.; Wang, Z.M.; Zhang, J.Y.; Ma, Q.; Cui, J.F. The Preparation and Study of Graphene Supported $Co_xMn_{3-x}O_4$ Nanocomposites as Advanced Oxygen Reduction Reaction Electrocatalyst. *Adv. Mater. Res.* **2013**, *652–654*, 348–351.
18. Zhao, H.; Wang, H.; Qu, Z. Synergistic Effects in Mn-Co Mixed Oxide Supported on Cordierite Honeycomb for Catalytic Deep Oxidation of VOCs. *J. Environ. Sci.* **2022**, *112*, 231–243. [CrossRef] [PubMed]
19. Zhang, L.; Shi, L.; Huang, L.; Zhang, J.; Gao, R.; Zhang, D. Rational Design of High-Performance DeNO$_x$ Catalysts Based on $Mn_xCo_{3-x}O_4$ Nanocages Derived from Metal–Organic Frameworks. *ACS Catal.* **2014**, *4*, 1753–1763. [CrossRef]
20. Jha, A.; Mhamane, D.; Suryawanshi, A.; Joshi, S.M.; Shaikh, P.; Biradar, N.; Ogale, S.; Rode, C.V. Triple Nanocomposites of $CoMn_2O_4$, Co_3O_4 and Reduced Graphene Oxide for Oxidation of Aromatic Alcohols. *Catal. Sci. Technol.* **2014**, *4*, 1771–1778. [CrossRef]
21. Pirogova, G.N.; Popova, N.N.; Korosteleva, I.R.; Voronin, Y.V. Reduction of Nitrogen(II) Oxide on Spinels and Perovskites. *Russ. J. Appl. Chem.* **2000**, *73*, 83–85.

22. Shi, Y.; Yi, H.; Gao, F.; Zhao, S.; Xie, Z.; Tang, X. Evolution Mechanism of Transition Metal in NH3-SCR Reaction over Mn-Based Bimetallic Oxide Catalysts: Structure-Activity Relationships. *J. Hazard. Mater.* **2021**, *413*, 125361. [CrossRef]

23. Ding, L.; Yang, W.; Chen, L.; Cheng, H.; Qi, Z. Fabrication of Spinel $CoMn_2O_4$ Hollow Spheres for Highly Selective Aerobic Oxidation of 5-Hydroxymethylfurfural to 2,5-Diformylfuran. *Catal. Today* **2020**, *347*, 39–47. [CrossRef]

24. Zulfugarova, S.; AziMova, G.R.; Aleskerova, S.Z.; TagiYev, D. Cobalt-Containing Oxide Catalysts Obtained by The Sol-Gel Method with Auto-Combustion in The Reaction of Low-Temperature Oxidation of Carbon Monoxide. *J. Turk. Chem. Soc. Sect. A Chem.* **2023**, *10*, 577–588. [CrossRef]

25. Li, Z.; Lin, T.; Yu, F.; An, Y.; Dai, Y.; Li, S.; Zhong, L.; Wang, H.; Gao, P.; Sun, Y.; et al. Mechanism of the Mn Promoter via CoMn Spinel for Morphology Control: Formation of Co_2C Nanoprisms for Fischer–Tropsch to Olefins Reaction. *ACS Catal.* **2017**, *7*, 8023–8032. [CrossRef]

26. Paterson, J.; Partington, R.; Peacock, M.; Sullivan, K.; Wilson, J.; Xu, Z. Elucidating the Role of Bifunctional Cobalt-Manganese Catalyst Interactions for Higher Alcohol Synthesis. *Eur. J. Inorg. Chem.* **2020**, *2020*, 2312–2324. [CrossRef]

27. Guse, K.; Papp, H. XPS Characterization of the Reduction and Synthesis Behaviour of Co/Mn Oxide Catalysts for Fischer-Tropsch Synthesis. *Fresenius. J. Anal. Chem.* **1993**, *346*, 84–91. [CrossRef]

28. Lindley, M.; Stishenko, P.; Crawley, J.W.M.; Tinkamanyire, F.; Smith, M.; Paterson, J.; Peacock, M.; Xu, Z.; Hardacre, C.; Walton, A.S.; et al. Tuning the Size of TiO_2 -Supported Co Nanoparticle Fischer–Tropsch Catalysts Using Mn Additions. *ACS Catal.* **2024**, *14*, 10648–10657. [CrossRef] [PubMed]

29. Iqbal, S.; Davies, T.E.; Morgan, D.J.; Karim, K.; Hayward, J.S.; Bartley, J.K.; Taylor, S.H.; Hutchings, G.J. Fischer Tropsch Synthesis Using Cobalt Based Carbon Catalysts. *Catal. Today* **2016**, *275*, 35–39. [CrossRef]

30. Thiessen, J.; Rose, A.; Meyer, J.; Jess, A.; Curulla-Ferré, D. Effects of Manganese and Reduction Promoters on Carbon Nanotube Supported Cobalt Catalysts in Fischer–Tropsch Synthesis. *Microporous Mesoporous Mater.* **2012**, *164*, 199–206. [CrossRef]

31. Liu, H.; Patzke, G.R. Cobalt-Manganese Spinel Oxides as Visible-Light-Driven Water Oxidation Catalysts. In *Ceramic Engineering and Science Proceedings*; Mathur, S., Hernandez-Ramirez, F., Kirihara, S., Widjaja, S., Eds.; Wiley: Hoboken, NJ, USA, 2013; pp. 75–86, ISBN 978-1-118-80762-0.

32. Khan, I.; Ibrahim, A.; Satake, S.; Béres, K.A.; Mohai, M.; Czigány, Z.; Sinkó, K.; Homonnay, Z.; Kuzmann, E.; Krehula, S.; et al. 57Fe-Mössbauer, XAFS and XPS Studies of Photo-Fenton Active $xMO\bullet40Fe_2O_3\bullet(60-x)SiO_2$ (M: Ni, Cu, Zn) Nano-Composite Prepared by Sol-Gel Method. *Ceram. Int.* **2024**, *50*, 55177–55189. [CrossRef]

33. Ali, A.S.; Khan, I.; Zhang, B.; Razum, M.; Pavić, L.; Šantić, A.; Bingham, P.A.; Nomura, K.; Kubuki, S. Structural, Electrical and Photocatalytic Properties of Iron-Containing Soda-Lime Aluminosilicate Glass and Glass-Ceramics. *J. Non-Cryst. Solids* **2021**, *553*, 120510. [CrossRef]

34. Khan, I.; Saito, H.; Ali, A.S.; Zhang, B.; Kubuki, S. Structural Characterization and Visible Light Activated Photocatalytic Ability of Glass–Ceramics Prepared from Municipal Solid Waste. *J. Mater. Cycles Waste Manag.* **2021**, *23*, 2266–2277. [CrossRef]

35. Berruti, I.; Nahim-Granados, S.; Abeledo-Lameiro, M.J.; Oller, I.; Polo-López, M.I. Peroxymonosulfate/Solar Process for Urban Wastewater Purification at a Pilot Plant Scale: A Techno-Economic Assessment. *Sci. Total Environ.* **2023**, *881*, 163407. [CrossRef] [PubMed]

36. Lim, S.; Shi, J.L.; Von Gunten, U.; McCurry, D.L. Ozonation of Organic Compounds in Water and Wastewater: A Critical Review. *Water Res.* **2022**, *213*, 118053. [CrossRef] [PubMed]

37. Yu, J.; Li, Z.; Liu, T.; Zhao, S.; Guan, D.; Chen, D.; Shao, Z.; Ni, M. Morphology Control and Electronic Tailoring of CoxAy (A = P, S, Se) Electrocatalysts for Water Splitting. *J. Chem. Eng.* **2023**, *460*, 141674. [CrossRef]

38. Aoki, I. Tetragonal Distortion of the Oxide Spinels Containing Cobalt and Manganese. *J. Phys. Soc. Jpn.* **1962**, *17*, 53–61. [CrossRef]

39. Wickham, D.G.; Croft, W.J. Crystallographic and Magnetic Properties of Several Spinels Containing Trivalent Ja-1044 Manganese. *J. Phys. Chem. Solids* **1958**, *7*, 351–360. [CrossRef]

40. Boucher, B.; Buhl, R.; Di Bella, R.; Perrin, M. Etude Par Des Mesures de Diffraction de Neutrons et de Magnétisme Des Propriétés Cristallines et Magnétiques de Composés Cubiques Spinelles $Co_{3-x}Mn_xO_4$ ($0,6 \leq x \leq 1,2$). *J. Phys. Fr.* **1970**, *31*, 113–119. [CrossRef]

41. Yamamoto, N.; Higashi, S.; Kawano, S.; Achiwa, N. Preparation of $MnCo_2O_4$ by a Wet Method and Its Metal Ion Distribution. *J. Mater. Sci. Lett.* **1983**, *2*, 525–526. [CrossRef]

42. Jabry, E.; Rousset, A.; Lagrange, A. Preparation and Characterization of Manganese and Cobalt Based Semiconducting Ceramics. *Ph. Transit.* **1988**, *13*, 63–71. [CrossRef]

43. Bordeneuve, H.; Guillemet-Fritsch, S.; Rousset, A.; Schuurman, S.; Poulain, V. Structure and Electrical Properties of Single-Phase Cobalt Manganese Oxide Spinels $Mn_{3-x}Co_xO_4$ Sintered Classically and by Spark Plasma Sintering (SPS). *J. Solid State Chem.* **2009**, *182*, 396–401. [CrossRef]

44. Franguelli, F.P.; Kováts, É.; Czégény, Z.; Bereczki, L.; Petruševski, V.M.; Barta Holló, B.; Béres, K.A.; Farkas, A.; Szilágyi, I.M.; Kótai, L. Multi-Centered Solid-Phase Quasi-Intramolecular Redox Reactions of [(Chlorido)Pentaamminecobalt(III)] Permanganate—An Easy Route to Prepare Phase Pure $CoMn_2O_4$ Spinel. *Inorganics* **2022**, *10*, 18. [CrossRef]

45. Bereczki, L.; Petruševski, V.M.; Franguelli, F.P.; Béres, K.A.; Farkas, A.; Holló, B.B.; Czégény, Z.; Szilágyi, I.M.; Kótai, L. [Hexaamminecobalt(III)] Dichloride Permanganate—Structural Features and Heat-Induced Transformations into $(Co^{II},Mn^{II})(Co^{III},Mn^{III})_2O_4$ Spinels. *Inorganics* **2022**, *10*, 252. [CrossRef]

46. Das, D.; Biswas, R.; Ghosh, S. Systematic Analysis of Structural and Magnetic Properties of Spinel CoB_2O_4 (B = Cr, Mn and Fe) Compounds from Their Electronic Structures. *J. Phys. Condens. Matter* **2016**, *28*, 446001. [CrossRef] [PubMed]

47. Han, H.; Lee, J.S.; Ryu, J.H.; Kim, K.M.; Jones, J.L.; Lim, J.; Guillemet-Fritsch, S.; Lee, H.C.; Mhin, S. Effect of High Cobalt Concentration on Hopping Motion in Cobalt Manganese Spinel Oxide ($Co_xMn_{3-x}O_4$, $x \geq 2.3$). *J. Phys. Chem. C* **2016**, *120*, 13667–13674. [CrossRef]

48. Elbaz, Y.; Rosenfeld, A.; Anati, N.; Caspary Toroker, M. Electronic Structure Study of Various Transition Metal Oxide Spinels Reveals a Possible Design Strategy for Charge Transport Pathways. *J. Electrochem. Soc.* **2022**, *169*, 040542. [CrossRef]

49. Frenzel, G. The Manganese Ore Minerals. In *The Geology and Geochemistry of Manganese*; Schweizerbarth: Stuttgart, Germany, 1980; p. 25.

50. Aukrust, E.; Muan, A. Phase Relations in the System Cobalt Oxide–Manganese Oxide in Air. *J. Am. Ceram. Soc.* **1963**, *46*, 511. [CrossRef]

51. Bulavchenko, O.A.; Afonasenko, T.N.; Ivanchikova, A.V.; Murzin, V.Y.; Kremneva, A.M.; Saraev, A.A.; Kaichev, V.V.; Tsybulya, S.V. In Situ Study of Reduction of $Mn_xCo_{3-x}O_4$ Mixed Oxides: The Role of Manganese Content. *Inorg. Chem.* **2021**, *60*, 16518–16528. [CrossRef] [PubMed]

52. Golikov, Y.V.; Tubin, S.Y.; Barkhatov, V.P.; Balakirev, V.F. Phase Diagrams of the Co-Mn-O System in Air. *J. Phys. Chem. Solids* **1985**, *46*, 539–544. [CrossRef]

53. Chaly, V.P.; Pashkova, E.V.; Krasan, Y.P. State Diagram of the Co-Mn-O System (Диаграмма Состояния Системы Co-Mn-O). *Ukr. Khim. Zh.* **1981**, *47*, 881–883.

54. Golikov, Y.V. Co–Mn–O Phase Diagram. *Izv. Akad. Nauk SSSR, Neorg. Mater.* **1988**, *24*, 1145–1149.

55. Koech, A.K.; Mwandila, G.; Mulolani, F.; Mwaanga, P. Lithium-Ion Battery Fundamentals and Exploration of Cathode Materials: A Review. *S. Afr. J. Chem. Eng.* **2024**, *50*, 321–339. [CrossRef]

56. Yankin, A.M.; Balakirev, V.F. Phase Equilibria in the Co–Mn–Ti–O System over Wide Ranges of Temperatures and Oxygen Pressures. *Inorg. Mater.* **2002**, *38*, 309–319. [CrossRef]

57. Martin De Vidales, J.L.; Vila, E.; Rojas, R.M.; Garcia-Martinez, O. Thermal Behavior in Air and Reactivity in Acid Medium of Cobalt Manganese Spinels $Mn_xCo_{3-x}O_4$ (1 .Ltoreq. x. Ltoreq. 3) Synthesized at Low Temperature. *Chem. Mater.* **1995**, *7*, 1716–1721. [CrossRef]

58. Dai, J.; Zhu, Y.; Tahini, H.A.; Lin, Q.; Chen, Y.; Guan, D.; Zhou, C.; Hu, Z.; Lin, H.-J.; Chan, T.-S.; et al. Single-Phase Perovskite Oxide with Super-Exchange Induced Atomic-Scale Synergistic Active Centers Enables Ultrafast Hydrogen Evolution. *Nat. Commun.* **2020**, *11*, 5657. [CrossRef] [PubMed]

59. Bhargava, A.; Chen, C.Y.; Dhaka, K.; Yao, Y.; Nelson, A.; Finkelstein, K.D.; Pollock, C.J.; Caspary Toroker, M.; Robinson, R.D. Mn Cations Control Electronic Transport in Spinel $Co_xMn_{3-x}O_4$ Nanoparticles. *Chem. Mater.* **2019**, *31*, 4228–4233. [CrossRef]

60. Wang, Z.; Liu, J.; Yang, Y.; Yu, Y.; Yan, X.; Zhang, Z. Regenerable $Co_xMn_{3-x}O_4$ Spinel Sorbents for Elemental Mercury Removal from Syngas: Experimental and DFT Studies. *Fuel* **2020**, *266*, 117105. [CrossRef]

61. Palvit, R.; Sreenivas, K.; Kumar, R. Dielectric Properties of Spinel $Co_{3-x}Mn_xO_4$ (x = 0.1, 0.4, 0.7, and 1.0) Ceramic Compositions. *IJPAP* **2014**, *52*, 625–631.

62. Li, J.; Li, X.; Wang, X.; Zeng, L.; Chen, X.; Mu, J.; Chen, G. Multiple Regulations of Mn-Based Oxides in Boosting Peroxymonosulfate Activation for Norfloxacin Removal. *Appl. Catal. A Gen.* **2019**, *584*, 117170. [CrossRef]

63. Lee, C.W.; Cazorla, C.; Zhou, S.; Zhang, D.; Xu, H.; Zhong, W.; Zhang, M.; Chu, D.; Han, Z.; Amal, R. Facet Engineering of Cobalt Manganese Oxide for Highly Stable Acidic Oxygen Evolution Reaction. *Adv. Energy Mater.* **2024**, 2402786. [CrossRef]

64. Han, D.; Ma, X.; Yang, X.; Xiao, M.; Sun, H.; Ma, L.; Yu, X.; Ge, M. Metal Organic Framework-Templated Fabrication of Exposed Surface Defect-Enriched Co_3O_4 Catalysts for Efficient Toluene Oxidation. *J. Colloid Interface Sci.* **2021**, *603*, 695–705. [CrossRef] [PubMed]

65. Thublaor, T.; Srihathai, P.; Wiman, P.; Muengjai, A.; Chandra-Ambhorn, S. Oxidation Behaviour of Mn-Co Spinel Coating on AISI 430 Ferritic Stainless Steel with and without Cu in $Ar-CO_2-H_2O$ Atmosphere. *J. Met. Mater. Miner.* **2023**, *33*, 29–37. [CrossRef]

66. Narita, T.; Ishikawa, T.; Karasawa, S.; Nishida, K. Oxidation Properties of Co-Mn Alloys at Temperatures from 1273 to 1473 K. *Oxid. Met.* **1987**, *27*, 267–282. [CrossRef]

67. Mansouri, M.; Atashi, H.; Tabrizi, F.F.; Mirzaei, A.A.; Mansouri, G. Kinetics Studies of Nano-Structured Cobalt–Manganese Oxide Catalysts in Fischer–Tropsch Synthesis. *J. Ind. Eng. Chem.* **2013**, *19*, 1177–1183. [CrossRef]

68. Béres, K.A.; Dürvanger, Z.; Homonnay, Z.; Bereczki, L.; Barta Holló, B.; Farkas, A.; Petruševski, V.M.; Kótai, L. Insight into the Structure and Redox Chemistry of [Carbonatotetraamminecobalt(III)] Permanganate and Its Monohydrate as Co-Mn-Oxide Catalyst Precursors of the Fischer-Tropsch Synthesis. *Inorganics* **2024**, *12*, 94. [CrossRef]

69. Kótai, L.; Béres, K.A.; Farkas, A.; Holló, B.B.; Petruševski, V.M.; Homonnay, Z.; Trif, L.; Franguelli, F.P.; Bereczki, L. An Unprecedented Tridentate-Bridging Coordination Mode of Permanganate Ions: The Synthesis of an Anionic Coordination Polymer—[Co^{III}(NH$_3$)$_6$]$_n$[(K(K^1-Cl)$_2$(M$^{2,2',2''}$-(K^3-O,O',O''-MnO$_4$)$_2$)$_{N\infty}$]—Containing Potassium Central Ion and Chlorido and Permanganato Ligands. *Molecules* **2024**, *29*, 4443. [CrossRef]

70. Klissurski, D.G.; Uzunova, E.L. Cation-Deficient Nano-Dimensional Particle Size Cobalt–Manganese Spinel Mixed Oxides. *Appl. Surf. Sci.* **2003**, *214*, 370–374. [CrossRef]

71. Tessonnier, J.; Becker, M.; Xia, W.; Girgsdies, F.; Blume, R.; Yao, L.; Su, D.S.; Muhler, M.; Schlögl, R. Spinel-Type Cobalt–Manganese-Based Mixed Oxide as Sacrificial Catalyst for the High-Yield Production of Homogeneous Carbon Nanotubes. *ChemCatChem* **2010**, *2*, 1559–1561. [CrossRef]

72. Yang, X.; Wei, G.; Wu, P.; Liu, P.; Liang, X.; Chu, W. Novel Halloysite Nanotube-Based Ultrafine CoMn$_2$O$_4$ Catalyst for Efficient Degradation of Pharmaceuticals through Peroxymonosulfate Activation. *Appl. Surf. Sci.* **2022**, *588*, 152899. [CrossRef]

73. Bai, M.; Chai, Y.; Chen, A.; Shao, J.; Zhu, S.; Yuan, J.; Yang, Z.; Xiong, J.; Jin, D.; Zhao, K.; et al. Co-Mn-Fe Spinel-Carbon Composite Catalysts Enhanced Persulfate Activation for Degradation of Neonicotinoid Insecticides: (Non) Radical Path Identification, Degradation Pathway and Toxicity Analysis. *J. Hazard. Mater.* **2023**, *460*, 132473. [CrossRef]

74. Naveen, A.N.; Selladurai, S. Tailoring Structural, Optical and Magnetic Properties of Spinel Type Cobalt Oxide (Co3O4) by Manganese Doping. *Phys. B Condens. Matter* **2015**, *457*, 251–262. [CrossRef]

75. Rajeesh Kumar, N.K.; Vasylechko, L.; Sharma, S.; Yadav, C.S.; Selvan, R.K. Understanding the Relationship between the Local Crystal Structure and the Ferrimagnetic Ordering of Co$_x$Mn$_{3-x}$O$_4$ (x = 0–0.5) Solid Solutions. *J. Alloys Compd.* **2021**, *853*, 157256. [CrossRef]

76. Zhang, Y.; Rosen, J.; Hutchings, G.S.; Jiao, F. Enhancing Photocatalytic Oxygen Evolution Activity of Cobalt-Based Spinel Nanoparticles. *Catal. Today* **2014**, *225*, 171–176. [CrossRef]

77. Habibi, M.H.; Bagheri, P. Spinel Cobalt Manganese Oxide Nano-Composites Grown Hydrothermally on Nanosheets for Enhanced Photocatalytic Mineralization of Acid Black 1 Textile Dye: XRD, FTIR, FESEM, EDS and TOC Studies. *J. Iran. Chem. Soc.* **2017**, *14*, 1643–1649. [CrossRef]

78. Mitran, G.; Chen, S.; Seo, D.-K. Selective Oxidation of *n* -Buthanol to Butyraldehyde over MnCo$_2$O$_4$ Spinel Oxides. *RSC Adv.* **2020**, *10*, 25125–25135. [CrossRef]

79. Chueh, L.-Y.; Hsu, Y.-W.; Wang, Z.-W.; Lin, H.-C.; Hung, S.-Y.; Chen, Y.-L.; Chen, H.-Y.; Pan, Y.-T. (Frank) Transition Metal (Sn, Sb, and Mn) Modification Effects on Co$_3$O$_4$ Spinel Structures: Enhancing Activity and Stability for Platinum Group Metal-Free Acidic Oxygen Evolution Reaction Catalyst. *Electrochim. Acta* **2024**, *497*, 144575. [CrossRef]

80. Dung, N.T.; Thu, T.V.; Van Nguyen, T.; Thuy, B.M.; Hatsukano, M.; Higashimine, K.; Maenosono, S.; Zhong, Z. Catalytic Activation of Peroxymonosulfate with Manganese Cobaltite Nanoparticles for the Degradation of Organic Dyes. *RSC Adv.* **2020**, *10*, 3775–3788. [CrossRef]

81. Cai, S.; Liu, J.; Zha, K.; Li, H.; Shi, L.; Zhang, D. A General Strategy for the in Situ Decoration of Porous Mn–Co Bi-Metal Oxides on Metal Mesh/Foam for High Performance de-NO$_x$ Monolith Catalysts. *Nanoscale* **2017**, *9*, 5648–5657. [CrossRef]

82. Hong, X.; Gao, Y.; Ji, M.; Li, J.; Ding, L.; Yu, Z.; Chang, K. Highly Valent Cobalt-Manganese Spinel Nanowires Induced by Fluorine-Doping for Durable Acid Oxygen Evolution Reaction. *J. Alloys Compd.* **2024**, *1007*, 176500. [CrossRef]

83. Jha, A.; Patil, K.R.; Rode, C.V. Mixed Co–Mn Oxide-Catalysed Selective Aerobic Oxidation of Vanillyl Alcohol to Vanillin in Base-Free Conditions. *ChemPlusChem* **2013**, *78*, 1384–1392. [CrossRef]

84. Rotko, M. Tlenkowe Katalizatory Kobaltowo-Manganowe Do Procesu Całkowitego Utleniania Metanu. *Przem. Chem.* **2019**, *1*, 95–97. [CrossRef]

85. Gruen, V.; Kajiyama, S.; Rosenfeldt, S.; Ludwigs, S.; Schenk, A.S.; Kato, T. One-Dimensional Assemblies of Co$_3$O$_4$ Nanoparticles Formed from Cobalt Hydroxide Carbonate Prepared by Bio-Inspired Precipitation within Confined Space. *Cryst. Growth Des.* **2022**, *22*, 5883–5894. [CrossRef]

86. Kovanda, F.; Rojka, T.; Dobešová, J.; Machovič, V.; Bezdička, P.; Obalová, L.; Jirátová, K.; Grygar, T. Mixed Oxides Obtained from Co and Mn Containing Layered Double Hydroxides: Preparation, Characterization, and Catalytic Properties. *J. Solid State Chem.* **2006**, *179*, 812–823. [CrossRef]

87. Kihal, I.; Douafer, S.; Lahmar, H.; Rekhila, G.; Hcini, S.; Trari, M.; Benamira, M. CoMn$_2$O$_4$: A New Spinel Photocatalyst for Hydrogen Photo-Generation under Visible Light Irradiation. *J. Photochem. Photobiol. A Chem.* **2024**, *446*, 115170. [CrossRef]

88. Dong, C.; Qu, Z.; Qin, Y.; Fu, Q.; Sun, H.; Duan, X. Revealing the Highly Catalytic Performance of Spinel CoMn$_2$ O$_4$ for Toluene Oxidation: Involvement and Replenishment of Oxygen Species Using In Situ Designed-TP Techniques. *ACS Catal.* **2019**, *9*, 6698–6710. [CrossRef]

89. Wang, W.; Huang, Y.; Rao, Y.; Li, R.; Lee, S.; Wang, C.; Cao, J. The Role of Cationic Defects in Boosted Lattice Oxygen Activation during Toluene Total Oxidation over Nano-Structured CoMnO$_x$ Spinel. *Environ. Sci. Nano* **2023**, *10*, 812–823. [CrossRef]

90. Jiang, R.; Tran, D.T.; McClure, J.P. Non-Precious Mn$_{1.5}$Co$_{1.5}$O$_4$–FeN$_x$/C Nanocomposite as a Synergistic Catalyst for Oxygen Reduction in Alkaline Media. *RSC Adv.* **2016**, *6*, 69167–69176. [CrossRef]

91. Xie, J.-Y.; Wang, F.-L.; Zhai, X.-J.; Li, X.; Zhang, Y.-S.; Fan, R.-Y.; Lv, R.-Q.; Chai, Y.-M.; Dong, B. Manganese Doped Hollow Cobalt Oxide Catalysts for Highly Efficient Oxygen Evolution in Wide pH Range. *Chem. Eng. J.* **2024**, *482*, 148926. [CrossRef]

92. Chiang, C.-Y.; Zhou, W. Formation Mechanism of $Mn_xCo_{3-x}O_4$ Yolk–Shell Structures. *RSC Adv.* **2021**, *11*, 29108–29114. [CrossRef] [PubMed]

93. Rosen, J.; Hutchings, G.S.; Jiao, F. Synthesis, Structure, and Photocatalytic Properties of Ordered Mesoporous Metal-Doped Co_3O_4. *Chin. J. Catal.* **2014**, *310*, 2–9. [CrossRef]

94. Xu, Z.; Zhang, Y.; Li, X.; Qin, L.; Meng, Q.; Zhang, G.; Fan, Z.; Xue, Z.; Guo, X.; Liu, Q.; et al. Template-Free Synthesis of Stable Cobalt Manganese Spinel Hollow Nanostructured Catalysts for Highly Water-Resistant CO Oxidation. *iScience* **2019**, *21*, 19–30. [CrossRef] [PubMed]

95. Acad, M. Nagiyev Institute of Catalysis and Inorganic Chemistry National Academy of Sciences of Azerbaijan; Mammadova, A.A. Interaction of Cobalt and Manganese Nitrates with Graphite Under Hydrothermal Conditions. *Chem. Probl.* **2022**, *20*, 35–39. [CrossRef]

96. Khanna, H.S.; Eddy, N.A.; Dang, Y.; Bhosale, T.S.; Shubhashish, S.; Suib, S.L.; Nandi, P. Nanoporous Co/Mn-Mixed Metal Oxides Templated via Polysulfones for Amine Oxidation. *ACS Appl. Nano Mater.* **2020**, *3*, 11923–11932. [CrossRef]

97. Zhang, Y.; Xie, X.; Zheng, Z.; He, X.; Du, P.; Zhang, R.; Guo, L.; Huang, K. Manganese-Cobalt Spinel Nanoparticles Anchored on Carbon Nanotubes as Bi-Functional Catalysts for Oxygen Reduction and Oxygen Evolution Reactions. *Appl. Sci.* **2023**, *13*, 12702. [CrossRef]

98. Du, J.; Chen, C.; Cheng, F.; Chen, J. Rapid Synthesis and Efficient Electrocatalytic Oxygen Reduction/Evolution Reaction of $CoMn_2O_4$ Nanodots Supported on Graphene. *Inorg. Chem.* **2015**, *54*, 5467–5474. [CrossRef]

99. Sugawara, Y.; Kobayashi, H.; Honma, I.; Yamaguchi, T. Effect of Metal Coordination Fashion on Oxygen Electrocatalysis of Cobalt–Manganese Oxides. *ACS Omega* **2020**, *5*, 29388–29397. [CrossRef]

100. Wang, S.; Liu, S.; Chen, X.; Guo, Y.; Xu, X.; Yang, L.; Zhao, Y.; Chen, C.; Liang, H.; Hao, R. Mn-Co Bimetallic Spinel Catalyst towards Activation of Peroxymonosulfate for Deep Mineralization of Toluene: The Key Roles of $SO_4^{\bullet-}$ and $O_2^{\bullet-}$ in the Ring-Opening and Mineralization of Toluene. *Chem. Eng. J.* **2023**, *453*, 139901. [CrossRef]

101. Wang, S.; Zhang, X.; Chen, G.; Liu, B.; Li, H.; Hu, J.; Fu, J.; Liu, M. Hydroxyl Radical Induced from Hydrogen Peroxide by Cobalt Manganese Oxides for Ciprofloxacin Degradation. *Chin. Chem. Lett.* **2022**, *33*, 5208–5212. [CrossRef]

102. Wang, C.; Chen, H.; Deng, J.; Li, L.; Zeng, Z.; Ma, X.; Wei, S. Enhanced Ability of Toluene Oxidation by Controlling Inversion Degree of Spinel Composed of Only Co, Mn. *J. Colloid Interface Sci.* **2024**, *658*, 943–951. [CrossRef]

103. Wang, C.; Xu, X.; Ma, X.; Chen, R.; Liu, B.; Du, Y.; Zeng, Z.; Li, L. Insight into an Invisible Active Site Presented in the Low Temperature Pyrolysis of Toluene by CoMn Spinel Catalyst. *J. Catal.* **2020**, *391*, 111–122. [CrossRef]

104. Wei, J.; Feng, Y.; Liu, Y.; Ding, Y. $M_xCo_{3-x}O_4$ (M = Co, Mn, Fe) Porous Nanocages Derived from Metal–Organic Frameworks as Efficient Water Oxidation Catalysts. *J. Mater. Chem. A* **2015**, *3*, 22300–22310. [CrossRef]

105. Xie, J.; Lü, Q.; Qiao, W.; Bu, C.; Zhang, Y.; Zhai, X.; Lü, R.; Chai, Y.; Dong, B. Enhancing Cobalt—Oxygen Bond to Stabilize Defective Co2MnO4 in Acidic Oxygen Evolution. *Acta Phys.-Chim. Sin.* **2024**, *40*, 2305021. [CrossRef]

106. Tian, Z.-Y.; Tchoua Ngamou, P.H.; Vannier, V.; Kohse-Höinghaus, K.; Bahlawane, N. Catalytic Oxidation of VOCs over Mixed Co–Mn Oxides. *Appl. Catal. B Environ.* **2012**, *117–118*, 125–134. [CrossRef]

107. Kamecki, B.; Karczewski, J.; Miruszewski, T.; Jasiński, G.; Szymczewska, D.; Jasiński, P.; Molin, S. Low Temperature Deposition of Dense $MnCo_2O_4$ Protective Coatings for Steel Interconnects of Solid Oxide Cells. *J. Eur. Ceram. Soc.* **2018**, *38*, 4576–4579. [CrossRef]

108. Rios, E.; Chartier, P.; Gautier, J.-L. Oxygen Evolution Electrocatalysis in Alkaline Medium at Thin $Mn_xCo_{3-x}O_4$ ($0 \leq x \leq 1$) Spinel Films on Glass/SnO_2: F Prepared by Spray Pyrolysis. *Solid State Sci.* **1999**, *1*, 267–277. [CrossRef]

109. Khramenkova, A.V.; Yakovenko, A.A.; Yuzhakova, K.R.; Mishurov, V.I.; Abdulvakhidov, K.G.; Polozhentsev, O.E. Study of Properties of the Cobalt–Manganese Spinel-Based Coatings Obtained by the Non-Stationary Electrolysis. *Russ. J. Electrochem.* **2023**, *59*, 707–713. [CrossRef]

110. Jung, K.-N.; Hwang, S.M.; Park, M.-S.; Kim, K.J.; Kim, J.-G.; Dou, S.X.; Kim, J.H.; Lee, J.-W. One-Dimensional Manganese-Cobalt Oxide Nanofibres as Bi-Functional Cathode Catalysts for Rechargeable Metal-Air Batteries. *Sci. Rep.* **2015**, *5*, 7665. [CrossRef]

111. Bhagwan, J.; Sivasankaran, V.; Yadav, K.L.; Sharma, Y. Porous, One-Dimensional and High Aspect Ratio Nanofibric Network of Cobalt Manganese Oxide as a High Performance Material for Aqueous and Solid-State Supercapacitor (2 V). *J. Power Sources* **2016**, *327*, 29–37. [CrossRef]

112. Li, K.; Lei, Y.; Liao, J.; Huang, S.; Zhang, Y.; Zhu, W. Design of Three Dimensional Flower-like MXene/Manganese-Cobalt Spinel Nanocomposites for Efficient Catalytic Thermal Decomposition of Ammonium Perchlorate. *Ceram. Int.* **2021**, *47*, 33269–33279. [CrossRef]

113. Risch, M. Reporting Activities for the Oxygen Evolution Reaction. *Commun. Chem.* **2023**, *6*, 221. [CrossRef] [PubMed]

114. Buthelezi, A.S.; Tucker, C.L.; Heeres, H.J.; Shozi, M.L.; Van De Bovenkamp, H.H.; Ntola, P. Fischer-Tropsch Synthesis Using Promoted, Unsupported, Supported, Bimetallic and Spray-Dried Iron Catalysts: A Review. *Results Chem.* **2024**, *9*, 101623. [CrossRef]

115. Linde, V.R.; Margolis, L.Y. Oxidation of Propylene on Cobalt-Manganese Spinels with Lithium, Titanium, and Copper Oxides Added. *Bull. Acad. Sci. USSR Div. Chem. Sci.* **1962**, *11*, 1639–1643. [CrossRef]

116. Dong, C.; Yang, C.; Ren, Y.; Sun, H.; Wang, H.; Xiao, J.; Qu, Z. Local Electron Environment Regulation of Spinel $CoMn_2O_4$ Induced Effective Reactant Adsorption and Transformation of Lattice Oxygen for Toluene Oxidation. *Environ. Sci. Technol.* **2023**, *57*, 21888–21897. [CrossRef]

117. Chen, Y.; Deng, J.; Yang, B.; Yan, T.; Zhang, J.; Shi, L.; Zhang, D. Promoting Toluene Oxidation by Engineering Octahedral Units via Oriented Insertion of Cu Ions in the Tetrahedral Sites of MnCo Spinel Oxide Catalysts. *Chem. Commun.* **2020**, *56*, 6539–6542. [CrossRef] [PubMed]

118. Bulavchenko, O.A.; Afonasenko, T.N.; Sigaeva, S.S.; Ivanchikova, A.V.; Saraev, A.A.; Gerasimov, E.Y.; Kaichev, V.V.; Tsybulya, S.V. The Structure of Mixed Mn–Co Oxide Catalysts for CO Oxidation. *Top. Catal.* **2020**, *63*, 75–85. [CrossRef]

119. Li, G.; Li, N.; Sun, Y.; Qu, Y.; Jiang, Z.; Zhao, Z.; Zhang, Z.; Cheng, J.; Hao, Z. Efficient Defect Engineering in Co-Mn Binary Oxides for Low-Temperature Propane Oxidation. *Appl. Catal. B Environ.* **2021**, *282*, 119512. [CrossRef]

120. Wright, P.A.; Natarajan, S.; Thomas, J.M.; Gai-Boyes, P.L. Mixed-Metal Amorphous and Spinel Phase Oxidation Catalysts: Characterization by x-Ray Diffraction, X-Ray Absorption, Electron Microscopy, and Catalytic Studies of Systems Containing Copper, Cobalt, and Manganese. *Chem. Mater.* **1992**, *4*, 1053–1065. [CrossRef]

121. Linde, V.R.; Margolis, L.Y.; Roginskii, S.Z. Reaction of Nitrous Oxide with Cobalt-Manganese Spinels Containing Admixed Lithium, Titanium and Copper Oxides. *Bull. Acad. Sci. USSR Div. Chem. Sci.* **1963**, *12*, 17–24. [CrossRef]

122. Mishra, D.K.; Lee, H.J.; Truong, C.C.; Kim, J.; Suh, Y.-W.; Baek, J.; Kim, Y.J. $Ru/MnCo_2O_4$ as a Catalyst for Tunable Synthesis of 2,5-Bis(Hydroxymethyl)Furan or 2,5-Bis(Hydroxymethyl)Tetrahydrofuran from Hydrogenation of 5-Hydroxymethylfurfural. *Mol. Catal.* **2020**, *484*, 110722. [CrossRef]

123. Shao, M.; Chang, Q.; Dodelet, J.-P.; Chenitz, R. Recent Advances in Electrocatalysts for Oxygen Reduction Reaction. *Chem. Rev.* **2016**, *116*, 3594–3657. [CrossRef] [PubMed]

124. Long, X.; Yu, P.; Zhang, N.; Li, C.; Feng, X.; Ren, G.; Zheng, S.; Fu, J.; Cheng, F.; Liu, X. Direct Spectroscopy for Probing the Critical Role of Partial Covalency in Oxygen Reduction Reaction for Cobalt-Manganese Spinel Oxides. *Nanomaterials* **2019**, *9*, 577. [CrossRef] [PubMed]

125. Junge, H.; Rockstroh, N.; Fischer, S.; Brückner, A.; Ludwig, R.; Lochbrunner, S.; Kühn, O.; Beller, M. Light to Hydrogen: Photocatalytic Hydrogen Generation from Water with Molecularly-Defined Iron Complexes. *Inorganics* **2017**, *5*, 14. [CrossRef]

126. Nissinen, T.; Leskelä, M.; Gasik, M.; Lamminen, J. Decomposition of Mixed Mn and Co Nitrates Supported on Carbon. *Thermochim. Acta* **2005**, *427*, 155–161. [CrossRef]

127. Kang, S.; Hwang, J. $CoMn_2O_4$ Embedded Hollow Activated Carbon Nanofibers as a Novel Peroxymonosulfate Activator. *Chem. Eng. J.* **2021**, *406*, 127158. [CrossRef]

128. Mahmoudi, H.; Mahmoudi, M.; Doustdar, O.; Jahangiri, H.; Tsolakis, A.; Gu, S.; LechWyszynski, M. A Review of Fischer Tropsch Synthesis Process, Mechanism, Surface Chemistry and Catalyst Formulation. *Biofuels Eng.* **2017**, *2*, 11–31. [CrossRef]

129. Teimouri, Z.; Abatzoglou, N.; Dalai, A.K. Kinetics and Selectivity Study of Fischer–Tropsch Synthesis to C_{5+} Hydrocarbons: A Review. *Catalysts* **2021**, *11*, 330. [CrossRef]

130. Jahangiri, H.; Bennett, J.; Mahjoubi, P.; Wilson, K.; Gu, S. A Review of Advanced Catalyst Development for Fischer–Tropsch Synthesis of Hydrocarbons from Biomass Derived Syn-Gas. *Catal. Sci. Technol.* **2014**, *4*, 2210–2229. [CrossRef]

131. Fáczán, E.; Megyeri, G.; Vojtela, T.; Béres, A. Comparison of the Flue Gas Emission Values of a Traditional Designed Biomass Boiler with the Specified Emission Standards. *J. Cent. Eur. Green Innov.* **2022**, *10*, 79–86. [CrossRef]

132. Zhang, L.; Jang, H.; Liu, H.; Kim, M.G.; Yang, D.; Liu, S.; Liu, X.; Cho, J. Sodium-Decorated Amorphous/Crystalline RuO_2 with Rich Oxygen Vacancies: A Robust pH-Universal Oxygen Evolution Electrocatalyst. *Angew. Chem. Int. Ed.* **2021**, *60*, 18821–18829. [CrossRef] [PubMed]

Review

Recent Advances in Methanol Steam Reforming Catalysts for Hydrogen Production

Mengyuan Zhang [1,2], **Diru Liu** [1,2], **Yiying Wang** [1,3], **Lin Zhao** [1,3], **Guangyan Xu** [1,2,*], **Yunbo Yu** [1,2,*] **and Hong He** [1,2]

[1] State Key Joint Laboratory of Environment Simulation and Pollution Control, Research Center for Eco-Environmental Sciences, Chinese Academy of Sciences, Beijing 100085, China; myzhang_st@rcees.ac.cn (M.Z.); drliu_st@rcees.ac.cn (D.L.); yywang@bjfu.edu.cn (Y.W.); dgzhaolin@bjfu.edu.cn (L.Z.); honghe@rcees.ac.cn (H.H.)

[2] University of Chinese Academy of Sciences, Beijing 100049, China

[3] College of Environmental Science and Engineering, Beijing Forestry University, Beijing 100083, China

* Correspondence: gyxu@rcees.ac.cn (G.X.); ybyu@rcees.ac.cn (Y.Y.)

Abstract: The pursuit of carbon neutrality has accelerated advancements in sustainable hydrogen production and storage methods, increasing the importance of methanol steam reforming (MSR) technology. Catalysts are central to MSR technology and are primarily classified into copper-based and noble metal-based catalysts. This review begins with an examination of the active components of these catalysts, tracing the evolution of the understanding of active sites over the past four decades. It then explores the roles of various supports and promoters, along with mechanisms of catalyst deactivation. To address the diverse perspectives on the MSR reaction mechanism, the existing research is systematically organized and synthesized, providing a detailed account of the reaction mechanisms associated with both catalyst types. The discussion concludes with a forward-looking perspective on MSR catalyst development, emphasizing strategies such as anti-sintering methods for copper-based catalysts, approaches to reduce byproduct formation in palladium-based catalysts, comprehensive research methodologies for MSR mechanisms, and efforts to enhance atomic utilization efficiency.

Keywords: methanol steam reforming; hydrogen production; copper-based catalysts; palladium-based catalysts; reaction mechanism

Academic Editors: Georgios Bampos, Paraskevi Panagiotopoulou and Eleni A. Kyriakidou

Received: 29 November 2024
Revised: 28 December 2024
Accepted: 31 December 2024
Published: 3 January 2025

Citation: Zhang, M.; Liu, D.; Wang, Y.; Zhao, L.; Xu, G.; Yu, Y.; He, H. Recent Advances in Methanol Steam Reforming Catalysts for Hydrogen Production. *Catalysts* **2025**, *15*, 36. https://doi.org/10.3390/catal15010036

1. Introduction

The extensive use of non-renewable fossil fuels such as coal, oil, and natural gas has caused the world to face significant challenges, including energy crises and climate change. To overcome such challenges, hydrogen, as an ideal renewable and clean energy source, has become a powerful alternative, although challenges in storage and transport remain [1–3]. Therefore, hydrogen production technologies at end-use locations, to sustainably supply this clean energy source with high calorific value, have become a major trend. The emergence of liquid organic hydrogen carriers (LOHCs), which can produce hydrogen by activating certain chemical bonds in the presence of catalysts, offers unlimited possibilities for *in situ* H_2 generation [4]. Methanol is an ideal LOHC, because it has a high hydrogen storage density of 99 kg m^{-3} and is inexpensive and widely available, being derived from biomass or CO_2 hydrogenation [2]. Hydrogen production from methanol can be carried out through the thermocatalytic process of methanol steam reforming (MSR), with the overall reaction represented by Equation (1) [5]. This includes two side reactions: methanol decomposition (MD, Equation (2)) and water-gas shift (WGS, Equation (3)).

$$CH_3OH + H_2O \leftrightarrow CO_2 + 3H_2 \qquad \Delta H^\theta = +49.2 \text{ kJ/mol} \qquad (1)$$

$$CH_3OH \leftrightarrow CO + 2H_2 \qquad \Delta H^\theta = +91.0 \text{ kJ/mol} \qquad (2)$$

$$CO + H_2O \leftrightarrow CO_2 + H_2 \qquad \Delta H^\theta = -41.1 \text{ kJ/mol} \qquad (3)$$

In practical applications, polymer electrolyte membrane fuel cells (PEMFCs) running on hydrogen require a CO-free hydrogen source to avoid poisoning of the anode catalysts of the cells [6,7]. In addition, a continuous and stable hydrogen source is a necessary guarantee. Therefore, developing MSR catalysts that exhibit high methanol conversion and hydrogen yield, low CO selectivity, and long-term stability is a critical and challenging task. To date, different kinds of MSR catalysts have been developed, among which copper-based (represented by $Cu/ZnO/Al_2O_3$) and palladium-based (represented by Pd/ZnO) catalysts predominate. These two types of catalysts have been proven to exhibit reliability and high efficiency under demanding conditions, making them the catalysts of choice for industrial hydrogen production via MSR. Furthermore, packed-bed reactors are the most commonly used [8], and reaction conditions such as the water-to-methanol ratio, reaction temperature, and weight hourly space velocity (WHSV) have a direct impact on the kinetics of the MSR reaction [9,10]. Generally, catalytic activity increases with higher water-to-methanol ratios and rising reaction temperature, up to a certain limit. The WHSV of methanol is usually in the range of 1–10 h^{-1}. Therefore, when evaluating the performance of MSR catalytic materials, it is essential to consider the reaction conditions to ensure a comprehensive assessment.

The research progress on copper-based and noble metal-based catalysts applied in methanol steam reforming for hydrogen production is examined in this paper, with a focus on developments from the late 20th century to the most recent findings (Figure 1). The analysis includes a detailed exploration of the catalytic components of these two types of catalysts, specifically the active metals, supports, and promoters, with an emphasis on their structure–activity relationships. Following this, the study delves into the reaction mechanisms and deactivation processes involved. The performance of both catalyst types is assessed, highlighting key scientific challenges that persist in the field. An outlook on potential strategies for overcoming these challenges is also provided, aiming to contribute to a deeper understanding of MSR catalysts.

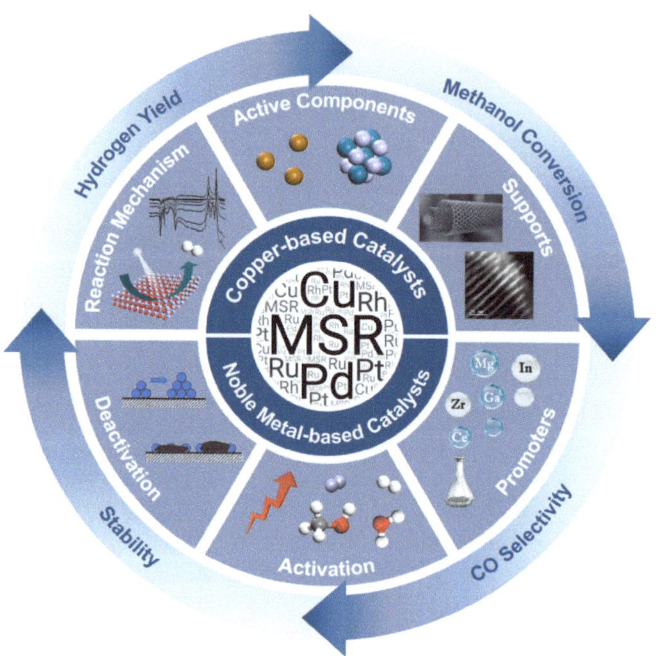

Figure 1. An overview of methanol steam reforming catalysts.

2. Copper-Based Catalysts

The development of copper-based catalysts for methanol steam reforming can be traced back to the late 20th century, when Takezawa et al. [11] discovered that metallic copper was active in this reaction. However, due to the complexity of the MSR process, the rich redox chemistry of copper, and the limitations of catalyst characterization techniques at that time, fundamental scientific questions such as the intrinsic active sites and structure–activity relationships of copper-based catalysts in MSR were not fully resolved [12]. Over the past decade, with the gradual emergence of related research findings (Table 1), significant progress has been made in the rational design of catalysts and the development of application technologies.

Table 1. Performance summary of representative copper-based catalysts for methanol steam reforming.

Catalysts	Temperature ($^\circ$C)	CH$_3$OH Conversion (%)	CO Selectivity (%)	H$_2$ Yield (mmol g^{-1} h^{-1})	Reaction Conditions	Stability	Ref.
Cu$_5$-Al	240	98	1.3 [b]	187.2	H$_2$ pretreatment; feed rate = 0.048 mL h^{-1}; H$_2$O/CH$_3$OH = 1.5	\	[13]
Cu/Cu(Al)O$_x$	240	99.5	1 [b]	398.88	H$_2$ pretreatment; feed rate = 2.4 mL h^{-1}; H$_2$O/CH$_3$OH = 2	240 $^\circ$C, 100 h, 14% drop in CH$_3$OH conversion	[12]
Cu/Al$_2$O$_3$	250	89.7	0.9 [a]	531.36	H$_2$ pretreatment; WHSV = 10.56 h^{-1}; H$_2$O/CH$_3$OH = 1	200 $^\circ$C, 100 h, 10% drop in H$_2$ production rate	[14]
CuZnAl	350	98	0	60.02	GHSV = 15,500 h^{-1}; H$_2$O/CH$_3$OH = 2	\	[15]
CuZnO/γ-Al$_2$O$_3$/Al	275	100	3.34 [a]	3580	GHSV = 4000 mL g^{-1} h^{-1}; H$_2$O/CH$_3$OH = 2	275 $^\circ$C, 100 h, 10% drop in CH$_3$OH conversion	[16]
Cu/ZnO/Al$_2$O$_3$	225	67	0.07 [a]	\	CH$_3$OH/H$_2$O/H$_2$ pretreatment; WHSV = 6 h^{-1}; H$_2$O/CH$_3$OH = 1.3	225 $^\circ$C, 40 h, 10% drop in CH$_3$OH conversion	[17]
Cu/ZnO/Al$_2$O$_3$	240	90	0	\	H$_2$ pretreatment; GHSV = 10,000 cm^3 g^{-1} h^{-1}; H$_2$O/CH$_3$OH = 1.5	240 $^\circ$C, 90 h, 30% drop in CH$_3$OH conversion	[18]
Cu/SiO$_2$	280	80	\	105	GHSV = 300 kg L^{-1} s^{-1}; H$_2$O/CH$_3$OH = 1.5	\	[19]
Cu-MCM-41	250	72.3	0.8 [b]	\	H$_2$ pretreatment; GHSV = 2838 h^{-1}; H$_2$O/CH$_3$OH = 3	250 $^\circ$C, 48 h, no drop	[20]

Table 1. *Cont.*

Catalysts	Temperature ($^\circ$C)	CH$_3$OH Conversion (%)	CO Selectivity (%)	H$_2$ Yield (mmol g^{-1} h^{-1})	Reaction Conditions	Stability	Ref.
CeCuZn/CNTs	300	94.2	2.6 [a]	H$_2$ yield = 98.2%	H$_2$ pretreatment; WHSV = 7.5 h^{-1}; H$_2$O/CH$_3$OH = 2	300 $^\circ$C, 48 h, 7% drop in CH$_3$OH conversion	[21]
Cu/Ce-Cu(BDC)	250	99	2 [a]	H$_2$ yield = 97%	WHSV = 9.2 h^{-1}; H$_2$O/CH$_3$OH = 2	250 $^\circ$C, 32 h, 7% drop in CH$_3$OH conversion	[22]
CuO/CeO$_2$	260	100	2.4 [a]	\	H$_2$ pretreatment; GHSV = 800 h^{-1}; H$_2$O/CH$_3$OH = 1.2	\	[23]
CuO/ZnO/CeO$_2$/Al$_2$O$_3$	200	100	0	\	H$_2$ pretreatment; GHSV = 10,000 cm^3 g^{-1} h^{-1}; H$_2$O/CH$_3$OH = 1.5	200 $^\circ$C, 24 h, no drop	[24]
ZrO$_2$/Cu	200	32	0	190	H$_2$ pretreatment; WHSV = 10 h^{-1}; H$_2$O/CH$_3$OH = 1.0	200 $^\circ$C, 200 h, no drop	[7]
Cu/ZnO/ZrO$_2$	250	88.6	0	12,600 mmol g$_{Cu}^{-1}$ h^{-1}	H$_2$ pretreatment; H$_2$O/CH$_3$OH = 1.0	\	[25]
Cu/Ce$_{1-x}$Zr$_x$O$_2$	240	23	0	316	H$_2$ pretreatment; WHSV = 27 h^{-1}; H$_2$O/CH$_3$OH = 1.5	240 $^\circ$C, 90 h, no drop	[26]
CuO/ZnO/CeO$_2$-ZrO$_2$	240	95	0.46 [a]	1836 mL g^{-1} h^{-1}	H$_2$ pretreatment; GHSV = 1200 h^{-1}; H$_2$O/CH$_3$OH = 1.2	230–260 $^\circ$C, 360 h, no drop	[27]
Cu/ZnO/CeO$_2$/ZrO$_2$/SBA-15	300	95.2	1.4 [b]	H$_2$ yield = 90%	H$_2$ pretreatment; WHSV = 43.68 h^{-1}; H$_2$O/CH$_3$OH = 2	300 $^\circ$C, 60 h, 12% drop in CH$_3$OH conversion	[28]
CuZnGaO$_x$	150	22.5	0	393.6 mL g^{-1} h^{-1}	H$_2$ pretreatment; feed rate = 6 mL h^{-1}; H$_2$O/CH$_3$OH = 2	\	[29]

Table 1. *Cont.*

Catalysts	Temperature (°C)	CH$_3$OH Conversion (%)	CO Selectivity (%)	H$_2$ Yield (mmol g^{-1} h^{-1})	Reaction Conditions	Stability	Ref.
CuGaZn	200	\	0.2 [a]	118.1	H$_2$ pretreatment; WHSV = 6 h^{-1}; H$_2$O/CH$_3$OH = 1.3	200 °C, 24 h, no drop	[30]
CuZnGaZr	250	42.9	0.3	10,620 mL g^{-1} h^{-1}	GHSV = 2200 h^{-1}; H$_2$O/CH$_3$OH = 1	275 °C, 44 h, 7% drop in CH$_3$OH conversion	[31]
Cu/MgAl$_2$O$_4$	300	96	2.8 [b]	\	H$_2$ pretreatment; WHSV = 8.5 h^{-1}; H$_2$O/CH$_3$OH = 1	200 °C, 30 h, 4% drop in CH$_3$OH conversion	[32]
Mg/Cu-Al spinel	255	96.5	3.8 [a]	H$_2$ yield = 96.54%	WHSV = 2.28 h^{-1}; H$_2$O/CH$_3$OH = 2.27	255 °C, 500 h, no drop	[33]
CuZnAlMg	200	68.5	0.88 [a]	172	H$_2$ pretreatment; WHSV = 3.84 h^{-1}; H$_2$O/CH$_3$OH = 1	350 °C, 8 h, 18% drop in CH$_3$OH conversion	[34]
CuCeMg/Al	250	100	0.29 [c]	H$_2$ yield = 29.1%	H$_2$ pretreatment; feed rate = 1 mL h^{-1}; H$_2$O/CH$_3$OH = 1	250 °C, 72 h, no drop	[35]
La(CoCuFeAlCe)$_{0.2}$O$_3$	600	98.9	8 [c]	436.8	LHSV = 20 h^{-1}; H$_2$O/CH$_3$OH = 4	600 °C, 50 h, no drop	[36]

a: $S_{CO} = \frac{y_{CO}}{y_{CO}+y_{CO_2}} \times 100\%$; b: $S_{CO} = \frac{y_{CO}}{y_{CO}+y_{CO_2}+y_{CH_4}} \times 100\%$; c: $S_{CO} = \frac{y_{CO}}{y_{CO}+y_{CO_2}+y_{H_2}} \times 100\%$.

2.1. Performance

2.1.1. Active Sites

Under reaction conditions, various copper species (Cu^0, $Cu^{\delta+}/Cu^+$) usually coexist on the copper-based catalysts, which adds complexity to identifying active sites. By using isotope-labeling experiments, *in situ* spectroscopy, and density functional theory (DFT) calculations [37], the role of Cu^0 species on Cu@mSiO$_2$ (CuO core/mesoporous silica shell) catalysts was elucidated (Figure 2a). It was shown that Cu^0 sites enabled the cleavage of the O-H bond and the C-H bond in methanol (CH_3OH), promoting the generation of the main intermediate methyl formate ($HCOOCH_3$) rather than the byproduct CO. Therefore, the activity and selectivity could be adjusted by controlling the ratio of Cu^0 and Cu^+. The process of transforming from methanol to methyl formate has generally been considered as a prototypical C_1 chemical reaction, such as in MSR. The 5 wt.%Cu5Zn10Al catalyst prepared by Mrad et al. [15] exhibited high activity, which was believed to be related to the presence of stable Cu^+ ions, regarded as the most active species in the MSR reaction. These conflicting views on active sites prompted more researchers to delve into this area. Ma et al. [38] constructed a catalyst with dual Cu^0 and Cu^+ sites on an SBA-15 support, achieving a hydrogen production rate of up to 1145 mol kg_{cat}^{-1} h^{-1}. Studies suggested that the significant performance enhancement was due to a Cu^+-dominated dual-site reaction pathway replacing the traditional Cu^0-dominated single-site pathway (Figure 2b). A synergistic effect was found between Cu^0 and Cu^+, with sufficient Cu^+ accelerating the reaction rate, while Cu^0 was crucial for H_2 desorption. Xu et al. [7] designed an inverse ZrO_2/Cu catalyst, achieving high activity, high stability, and CO-free production in MSR. On this inverse catalyst, experimental and theoretical studies indicated that its superior performance was closely related to the presence of ZrO(OH)-(Cu^+/Cu) composite sites. These sites promoted the formation of H_2 and CO_2 via the formate (HCOOH) intermediate, thus preventing the continuous dehydrogenation of methanol to form CO. Meng et al. [12] constructed Cu^0-Cu^+ dual sites on $Cu/Cu(Al)O_x$ catalysts, and, through *in situ* spectroscopic characterization and theoretical calculations, found that oxygen-containing intermediates, methoxy (CH_3O^*) and formate, adsorbed with moderate strength at these sites during the MSR reaction process. This promoted the transfer of electrons from the catalyst to surface species, significantly reducing the energy barrier for the cleavage of the C-H bond in the methoxy and formate intermediates (rate-determining steps). The optimized catalyst exhibited a methanol conversion rate of 99.5%, with a corresponding hydrogen production rate of 110.8 μmol s^{-1} g_{cat}^{-1}, and maintained stability at 240 °C for over 300 h. Ma et al. [26] designed $Cu/Ce_{1-x}Zr_xO_2$ solid solution catalysts and demonstrated that optimal catalytic performance was achieved when the ratio of Cu^+/Cu^0 was approximately 1.0. Through the integration of diffuse reflectance infrared Fourier transform spectroscopy (DRIFTS) experiments and DFT calculations, they elucidated that the CH_3OH molecule was activated upon adsorption at the Cu^+ site, while the H_2O molecule was activated at the Cu^0 site. Liu et al. [13] proposed that a dynamic $Cu^{0/+}$ site and an adjacent Cu^0 site were the active sites. Water is activated on the Cu^0 site, oxidizing it to Cu^+-OH species. These Cu^+-OH species subsequently promote the formation of reactive formate that produces H_2 and CO_2 upon dehydrogenation, while reducing Cu^+ back to Cu^0. The synergistic interaction between these two types of sites governs the overall reaction performance. As seen in the above studies, the perspective that the Cu^0-Cu^+ dual sites synergistically catalyze the MSR reaction seems to be more reliable.

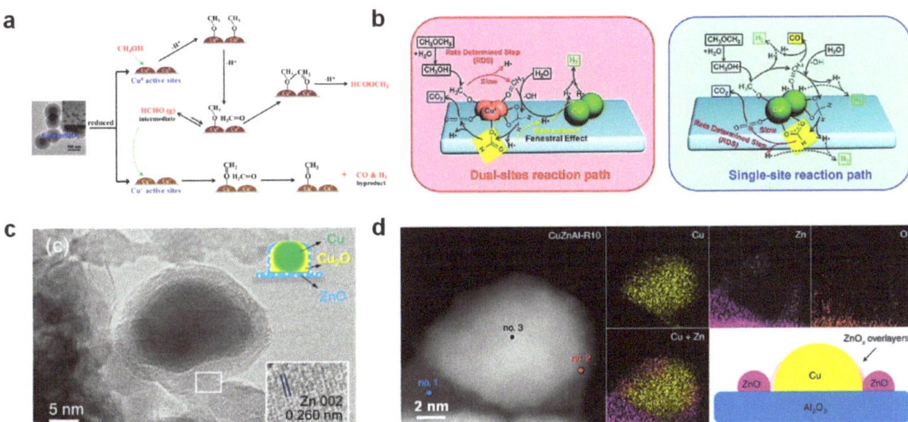

Figure 2. (**a**) Reaction pathways over different copper sites (Cu^0 or Cu^+) in methanol dehydrogenation (reproduced with permission from ref. [37], Copyright 2018 Wiley); (**b**) the hypothetical reaction paths on the dual sites of copper species (Cu^0 and Cu^+) and the single Cu^0 site [38]; (**c**) HRTEM images of the activated $CuZnO/\gamma\text{-}Al_2O_3/Al$ catalyst (reproduced with permission from ref. [39], Copyright 2021 Elsevier); (**d**) high-angle annular dark-field scanning transmission electron microscopy (HAADF-STEM) images and corresponding EDS elemental maps of the CuZnAl-R10 catalyst (reproduced with permission from ref. [17], Copyright 2022 Springer Nature).

In addition to the view that various Cu species are active sites for the MSR reaction, many studies have focused on interface sites. Through ambient-pressure X-ray photoelectron spectroscopy (AP-XPS) and Auger electron spectroscopy (AES), Rameshan et al. [6] demonstrated that the $Cu(Zn)^0/Zn(ox)$ interface was the active site on Cu/ZnO catalysts. The $Cu(Zn)^0$ region facilitated the selective dehydrogenation of methanol to formaldehyde (HCHO), while the redox-active $Cu(Zn)^0\text{-}Zn(ox)$ interface aided in the activation of water and the subsequent transfer of the resulting hydroxide to the formaldehyde, providing optimal conditions for high CO_2 activity and selectivity. Zhang et al. [39] prepared a series of $CuZnO/\gamma\text{-}Al_2O_3/Al$ catalysts for the MSR reaction with enhanced durability. Further studies involving high-resolution transmission election microscopy (HRTEM), X-ray photoelectron spectra (XPS), and DFT calculations revealed that ZnO maintained an optimal Cu^+/Cu^0 ratio and formed numerous Cu_2O/ZnO synergistic sites (Figure 2c). This significantly reduced the activation energy for the rate-determining step of methoxy dehydrogenation in the MSR reaction (1.17 eV), compared to Cu/ZnO (1.25 eV) and Cu (1.36 eV). Through an adsorbate-induced strong metal–support interaction [17], the migration of ZnO_x to the surface of Cu^0 nanoparticles occurred on commercial $Cu/ZnO/Al_2O_3$ catalysts, generating abundant $Cu\text{-}ZnO_x$ interface sites (Figure 2d), which doubled the catalytic activity for the MSR reaction. Mao et al. [14] elucidated the importance of the $Cu\text{-}Al_2O_3$ interface site for the MSR reaction using Cu/Al_2O_3 and inverse Al_2O_3/Cu catalysts. Combining *quasi-in situ* X-ray photoelectron spectroscopy, *in situ* CO DRIFTS, and *in situ* temperature-programmed DRIFTS methods, they demonstrated that the formate species adsorbed on the interfacial site (HCOO-CuAl) dissociated more rapidly to CO_2 and H_2 than those adsorbed on Al_2O_3 (HCOO-Al). The optimal sample with abundant $Cu\text{-}Al_2O_3$ interface sites thus showed a high hydrogen production rate of 147.6 μmol g^{-1} s^{-1} at 250 °C. The aforementioned studies highlight the significance of interface sites in the MSR reaction, and the formation of these sites is often associated with the carrier materials. For example, Jin et al. [40] identified the $Cu\text{-}O_V\text{-}Ce$ interface as the critical active site in their study of Cu/CeO_2 catalysts for the MSR reaction.

2.1.2. Supports

Although ZnO is generally used as a promoter in MSR catalysts at present, it was originally used as a support to load copper (Cu/ZnO), while it improved the MSR performance. Accordingly, many researchers investigated the promotion mechanism of ZnO [6]. To answer this question, different theoretical models have been developed for Cu/ZnO catalysts, including the spillover model, the morphology model, and the CuZn alloy model. The spillover model [41] suggested that ZnO served as a reservoir for hydrogen atoms, benefiting hydrogen spillover and back-spillover between Cu and ZnO, finally relating to the enhancement of catalytic activity. The morphology model [42] demonstrated that changes in the reaction atmosphere altered the interfacial free energy and affected oxygen vacancies at the Zn-O-Cu interface (Figure 3a). The partial reduction of ZnO enhanced its interactions with Cu, reduced the surface free energy, increased oxygen vacancies, and produced disk-like Cu particles with greater surface area and catalytic activity. The CuZn alloy model, supported by studies from Nakamura et al. [43], indicated that after reduction pretreatment, CuZn alloys were formed in the Cu/ZnO catalysts, increasing reaction activity. Additionally, ZnO could isolate Cu metal particles, preventing their agglomeration and sintering, thereby improving reaction stability. Later, when Al_2O_3 was introduced into the Cu/ZnO catalytic system as a support, the MSR performance was further enhanced [18,44–46]. Shokrani et al. [18] found through characterization by X-ray diffraction (XRD), Brunauer–Emmett–Teller (BET), and field emission scanning electron microscopy (FESEM) that adding Al_2O_3 to the traditional Cu/ZnO catalyst system reduced the crystallinity of the metal oxides, increased the specific surface area of the catalyst, and improved the dispersion of the metals. Performance test results also showed that the addition of Al_2O_3 increased methanol conversion and reduced CO production. Additionally, the proportions of Cu, Zn, and Al need to be carefully selected. Miyao et al. [47] adjusted the proportions of components in Cu/Zn/Al alloy catalysts, finding that a combination of 30 wt.% Cu, 20 wt.% Zn, and 50 wt.% Al achieved the highest hydrogen yield and selectivity.

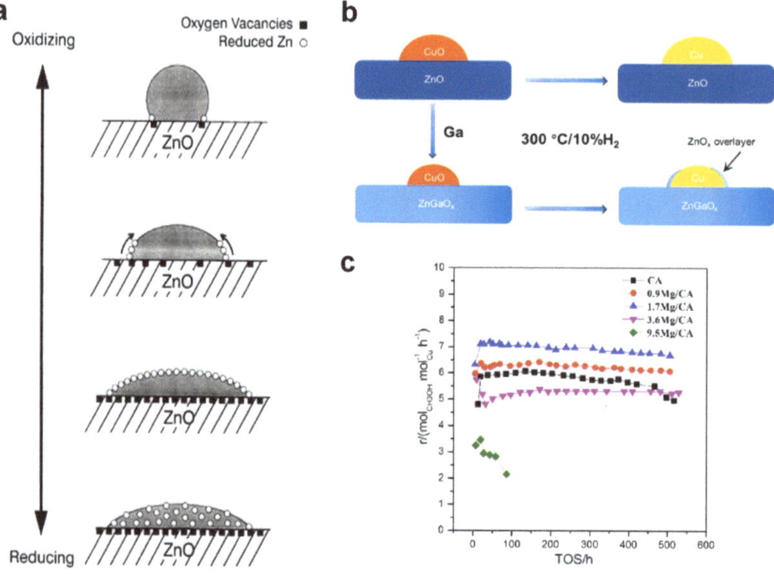

Figure 3. (**a**) The morphology model of the role of ZnO in copper-based catalysts (reproduced with permission from ref. [42], Copyright 2000 Elsevier); (**b**) the schematics of the Ga-doped Cu/ZnO catalyst

structures (reproduced with permission from ref. [30], Copyright 2024 American Chemical Society); (c) the long-time performance of Cu-Al spinel oxide (CA) and MgO modified CA catalysts (reaction conditions: 225 °C, WHSV = 2.184 h^{-1}) (reproduced with permission from ref. [33], Copyright 2020 Elsevier).

The selection of support materials for copper-based catalysts extends beyond traditional supports such as ZnO and Al_2O_3, encompassing SiO_2, TiO_2, SBA-15, MCM-41, and carbon nanotubes (CNTs) as well. Díaz-Pérez et al. [19] studied copper-based catalysts supported on SiO_2, Al_2O_3-SiO_2, TiO_2-rutile, and TiO_2-anatase, finding that their MSR activity varied in the order Cu/SiO_2 > Cu/TiO_2-rutile > Cu/Al_2O_3-SiO_2 > Cu/TiO_2-anatase. It should be noted that a loss of MSR activity was observed over Cu/SiO_2 and Cu/Al_2O_3-SiO_2 catalysts due to the growth of Cu particle size and coke deposition, respectively. In contrast, Cu particle size growth on Cu/TiO_2-rutile and Cu/TiO_2-anatase was not significant, attributed to the strong interaction between copper clusters and TiO_2. As a result, deactivation was hardly observed over these two samples. Mesoporous SBA-15 was also employed as a support to prepare copper-based catalysts [28]. Thanks to its high specific surface area, the distribution of metal particles was improved, making it a preferred support material. Among the tested samples, Cu/SBA-15 exhibited good catalytic performance, with 91% methanol conversion and 2.8% CO selectivity at 300 °C. Similarly to SBA-15, MCM-41 mesoporous molecular sieves have become a support choice for MSR reaction catalysts. For instance, Deshmane et al. [20] prepared Cu-MCM-41 catalysts for the MSR reaction. Performance results showed that the 15% Cu-MCM-41 sample achieved 89% methanol conversion and 0.8% CO selectivity at 300 °C with gas hourly space velocity (GHSV) of 2838 h^{-1}, maintaining good stability for 48 h. Carbon nanotubes have garnered significant attention as support materials due to several advantages [21,48–50]. Firstly, their mesoporous structure reduces mass transfer resistance, allowing reactant and product molecules to easily diffuse within the catalyst channels, thus enhancing reaction rates. Secondly, CNT-supported catalysts have much higher specific surface areas compared to conventional catalysts, exposing reactant molecules to both the inner and outer surfaces of the CNT-supported catalysts, thereby increasing catalytic conversion rates. Additionally, CNTs have uniform pore size distributions and high thermal stability, allowing for even heat dissipation during high-temperature reactions, preventing hot spots that could lead to catalyst sintering. Shahsavar et al. [21] prepared CeCuZn/CNT catalysts for the MSR reaction, which maintained long-term stability (>48 h) at 300 °C and a WHSV of 7.5 h^{-1}, attributed to the high dispersion of metal particles and the strong interaction between the metals and the CNT support.

2.1.3. Promoters

Besides the previously mentioned Zn, the elements of Ce, Zr, Ga, and Mg are most commonly used as MSR catalyst promoters with a view on enhancing catalytic performance such as hydrogen yield, activity, selectivity, and stability. It has been reported that the addition of CeO_2 can facilitate the formation of oxygen vacancies (V_0) and adsorbed oxygen molecules (O_0^x) on the catalyst surface, thereby promoting water dissociation through a hydration reaction ($V_0 + H_2O + O_0^x \rightarrow 2OH_0$) [22,51]. As we all know, the dissociation of H_2O is a crucial step in the MSR reaction [17]. In detail, when Ce^{4+} is converted to Ce^{3+}, oxygen vacancies are formed, accompanied by the production of adsorbed oxygen molecules [52,53]. This also increases the concentration of highly mobile oxygen species, which promotes carbon gasification and effectively enhances the anti-carbon stability of the MSR catalyst [54]. Men et al. [55] prepared the Cu/CeO_2/γ-Al_2O_3 catalyst and found that the copper–ceria interface was the active site for the MSR reaction, with its redox properties determining the catalytic activity. Oxygen vacancies at the interface served as the most active sites for water dissociation, whereas surface oxygen played a critical role in methanol decomposition. Liu et al. [56] enhanced the adsorption and dissociation of water in the MSR reaction by adding Ce and La to Cu-Al catalysts, which also strengthened the adsorption of the product CO_2. As a result, the WGS reaction was promoted, while the reverse-WGS reaction was suppressed, leading to a higher hydrogen production rate.

In the MSR reaction, the addition of ZrO_2 as a promoter significantly improved the stability of copper-based catalysts and reduced the concentration of CO byproducts. By using different characterization techniques, such as XRD, temperature-programmed oxidation (TPO), and temperature-programmed reduction (TPR), Agrell et al. [57] found that the introduction of ZrO_2 into the $Cu/ZnO/Al_2O_3$ catalyst effectively stabilized the copper by preventing crystal growth, thus enhancing catalyst stability. Similarly, Jeong et al. [58] noted that adding Zr to Cu/Zn catalysts increased copper dispersion, aiding in the formation of small copper particles on the catalyst surface. Wu et al. [59] suggested that in 10 wt.%ZrO_2/Cu catalysts, the presence of ZrO_2 increased the specific surface area of Cu, stabilized the grain size of Cu, and prevented the aggregation of Cu particles, thereby improving catalytic activity and stability. Additionally, the ZrO_2 surface had abundant hydroxyl groups that could stabilize Cu^+ species, which was a major reason for the superior performance of the ZrO_2/Cu catalyst compared to the ZnO/Cu catalyst. A subsequent research study performed by Liu et al. [31] also supported the role of Zr in enhancing stability. Specifically, the CuZnGaZr catalysts showed excellent stability of hydrogen production (44 h) for the MSR reaction. Furthermore, the role of ZrO_2 in reducing CO selectivity has garnered significant attention. Early studies by Lindström et al. [60] on the role of ZrO_2 in Cu/Zn/Zr catalysts showed that ZrO_2 doping reduced hydrogen yield but increased CO_2 selectivity, keeping CO byproduct concentration below 1%. Jeong and Liu et al. [31,58] found that ZrO_2 played a role in suppressing CO formation. They achieved low CO selectivity over $Cu/ZnO/ZrO_2/Al_2O_3$ (0.1%) and CuZnGaZr (0.3%) catalysts, respectively. Research by Sanches et al. [25] indicated that the reduced CO selectivity was due to the strong adsorption of CO by monoclinic zirconia over $Cu/ZnO/ZrO_2$ catalysts.

In the MSR reaction, the promotion effects of gallium and magnesium on copper-based catalysts have also been widely discussed. Tsang's team [29,61] pointed out that incorporating Ga into Cu-Zn oxides could promote the formation of a non-stoichiometric cubic spinel phase ($ZnGa_2O_4$). On the defect-rich surface of this $ZnGa_2O_4$ spinel, a large number of highly dispersed, extremely small copper clusters (5 Å) could be generated *in situ*, achieving a hydrogen yield of 393.6 mL g_{cat}^{-1} h^{-1} at 150 °C with almost no CO formation. Li et al. [30] found that the addition of Ga promoted the formation of a ZnO_x overlayer on the Cu/ZnO catalyst, enhancing strong metal–support interactions (SMSIs) and creating more abundant Cu-ZnO_x interfacial sites (Figure 3b), which increased the intrinsic activity of the MSR reaction by a factor of 4.6. In related research, the introduction of magnesium has often been associated with the formation of the spinel structure to improve MSR activity. Kamyar et al. [32] prepared four high-surface-area MAl_2O_4 (M = Ni, Co, Zn, Mg) spinel structures as supports for copper-loaded MSR catalysts. Among these, the $Cu/MgAl_2O_4$ catalyst exhibited a large specific surface area, high copper dispersion, and strong adsorption of reactants on its active sites, resulting in high methanol conversion (96%), high H_2 selectivity (86%), and low CO selectivity (2.8%). Hou et al. [33] discovered that although $CuAl_2O_4$ spinel catalysts could avoid the need for pre-reduction treatment before use, the catalyst activity and stability declined, as most of the Cu^{2+} in the lattice was released during the MSR reaction. To prevent this phenomenon, $CuAl_2O_4$ spinel catalysts were modified with MgO. The incorporation of Mg^{2+} into the spinel structure altered the environment around Cu^{2+}, reducing its release rate and promoting the formation of small copper particles. The slower release rate of Cu^{2+} and the high specific surface area of copper nanoparticles were the main reasons for the improved activity and stability of MgO-doped $CuAl_2O_4$ spinel catalysts (Figure 3c). In addition, Liu et al. [62] pointed out that this sustained-release catalytic system was fundamentally different from traditional metal catalytic systems. In conclusion, future research should focus on systematically investigating the interactions between various promoters to achieve an optimal balance among these components, thereby tailoring catalyst properties for specific industrial applications.

2.1.4. Preparation and Activation

Improvements in catalyst synthesis methods are often pursued to achieve better catalytic performance. For the MSR reaction, traditional preparation methods for copper-based catalysts include coprecipitation, sol–gel, and impregnation methods. Yang et al. [23] synthesized CeO_2 with different morphologies including nano-rod (R), nanoparticle (P), and sponginess (S) using hydrothermal, precipitation, and sol–gel methods, respectively, to serve as supports for copper-based catalysts (Figure 4a). The results showed that CuO/CeO_2-R had stronger interactions between copper oxide and the cerium dioxide support, as well as the highest content of surface oxygen vacancies, exhibiting better MSR catalytic activity than CuO/CeO_2-P and CuO/CeO_2-S. This indicated that adjusting the morphology of the support could enhance catalytic performance. In recent years, some new preparation methods and improved techniques of traditional methods have been used for synthesizing copper-based MSR catalysts. Wen et al. [63] developed a hydrolysis precipitation method to prepare Cu-ZnO@Al_2O_3 catalysts for the MSR reaction and compared them with those prepared by the traditional coprecipitation method. The former catalysts showed better catalytic performance due to their more developed pore organization, surface enrichment of Cu and Zn, and higher reducibility. Haghighi's team [24,64] synthesized CeO_2- and ZrO_2-doped $CuO/ZnO/Al_2O_3$ catalysts using the sonochemical coprecipitation method. Compared to traditional coprecipitation samples, the ultrasonic-assisted coprecipitation samples displayed clearer morphologies, more uniform active component distribution, smaller CuO crystallite size, and higher specific surface areas, resulting in higher methanol conversion rates (100%) at lower temperatures (200 °C) and lower CO byproduct concentrations. Additionally, $CuO/ZnO/Al_2O_3$ catalysts prepared by oxalate gel-coprecipitation [65] and urea hydrolysis homogeneous precipitation [66] methods have also shown advancements.

The dynamic structural evolution of heterogeneous catalysts during different stages of their life cycle (activation, reaction, and deactivation) is a common phenomenon [17]. Among them, the modulation of the activation process to improve the catalytic performance has great potential. Traditionally, copper-based catalysts require pretreatment to convert inactive components into active states through hydrogen reduction at high temperatures. Recently, Zhu's team [17] optimized the activation stage of commercial $Cu/ZnO/Al_2O_3$ catalysts and analyzed the structure–activity relationship. CuZnAl-R10, which was reduced in a H_2/N_2 mixture for 50 min and then in a $H_2/H_2O/CH_3OH/N_2$ mixture for 10 min, had a methanol conversion rate of 65.5% and a CO selectivity of 0.07% after 8 h at a reaction temperature of 225 °C (Figure 4b). However, CuZnAl-H, which was reduced in a H_2/N_2 mixture for 1 h, exhibited a methanol conversion rate of 55.1% and CO selectivity of 0.11% under the same reaction conditions. This indicated that the induced activation method improved the activity, selectivity, and stability of commercial $Cu/ZnO/Al_2O_3$ catalysts. A series of *in situ* characterization studies and DFT calculations revealed that a surface reconstruction process occurred during activation. This adsorbate-induced strong metal–support interaction accelerated the migration of ZnO_x species to the surface of Cu^0 nanoparticles, creating abundant Cu-ZnO_x interfacial sites (Figure 2d). These interfacial sites demonstrated faster CH_3O^* dehydrogenation and H_2O dissociation kinetics. Subsequently, they applied the same method by altering the activation gas for Cu/ZnO catalysts, thereby modulating the strong metal–support interaction between Cu and ZnO to enhance the catalytic activity for the MSR reaction [67]. During activation, more oxygen vacancies were generated on the ZnO support, promoting the dissociation of H_2O and hydroxylation of ZnO species, the occurrence of which facilitated the migration of ZnO to Cu, forming abundant interfacial sites and promoting the reaction. Zhang et al. [39] conducted an *in situ* self-activation process under reaction conditions ($CH_3OH/H_2O/N_2$) before evaluating the performance of $CuZnO/\gamma$-Al_2O_3/Al catalysts. This process resulted in the formation of a $Cu/Cu_2O/ZnO$ tri-layer core–shell structure (Figure 2c), where ZnO stabilized the optimal ratio of Cu^+/Cu^0 and created numerous Cu_2O/ZnO synergistic sites. These sites lowered the activation energy for the rate-determining step of CH_3O^* dehydrogenation in the MSR reaction (Figure 4c), thereby enhancing catalytic performance. This study of the process of improving catalysts, from the improvement of preparation methods to the activation pretreatment of catalysts, is an innovative quest for superior catalysts by researchers.

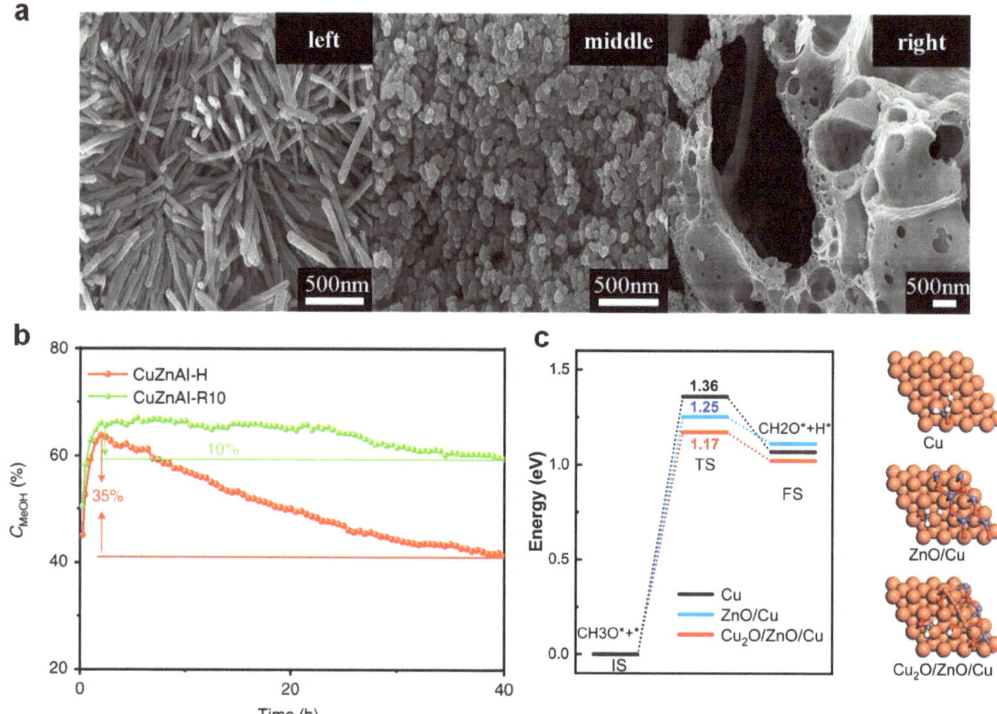

Figure 4. (**a**) SEM images of the CeO_2 materials with different morphologies, nano-rod (left), nanoparticle (middle), and sponginess (right) (reproduced with permission from ref. [23], Copyright 2019 Elsevier); (**b**) catalytic stabilities of CuZnAl-H and CuZnAl-R10 during the MSR reaction (reaction conditions: 100 mg of catalyst, H_2O/CH_3OH = 1.3, WHSV = 6 h^{-1}, 225 °C) (reproduced with permission from ref. [17], Copyright 2022 Springer Nature); (**c**) energy profile configurations for CH_3O^* dehydrogenation on the Cu, ZnO/Cu, and $Cu_2O/ZnO/Cu$, with a top view of three model transition states (*: adsorption sites available; IS: initial state; TS: transition state; FS: final state) (reproduced with permission from ref. [39], Copyright 2021 Elsevier).

2.2. Reaction Mechanism

After decades of development, two possible reaction pathways for methanol steam reforming on copper-based catalysts have been essentially identified as the following: the HCOO* route [14,17,68–72] and the HCOOCH₃* route [12,68–70,73,74] (Figure 5).

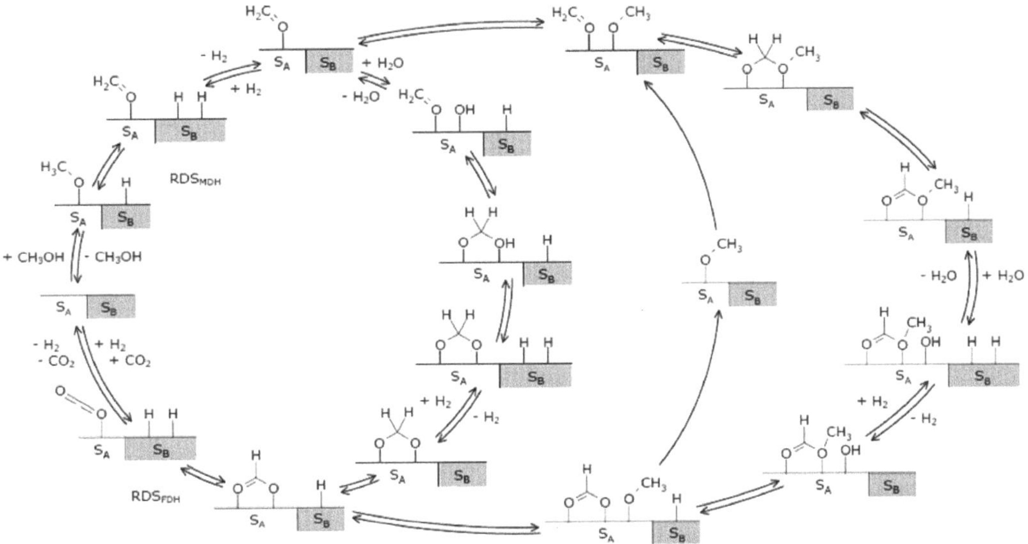

Figure 5. Schematic diagram of cycling process of methanol steam reforming reaction on Cu/ZnO/Al₂O₃ catalyst (reproduced with permission from ref. [68], Copyright 2007 Elsevier).

By using *in situ* temperature-programmed DRIFTS studies, Li et al. [17] elucidated the MSR reaction pathway over a commercial Cu/ZnO/Al₂O₃ catalyst, where surface methoxy species (*CH₃O) sequentially transform to formaldehyde (*CH₂O), methylene dioxygen (*CH₂OO), and formate (*CHOO), with formate ultimately decomposing to CO₂ and H₂. DFT calculations (Figure 6a) showed that the two steps with the highest energy barriers on Cu(111) were H₂O dissociation (1.06 eV) and the dehydrogenation of *CH₃O to *CH₂O (1.03 eV). Li et al.'s elucidation of the HCOO* route aligns with previous studies [71,72] and is supported by Mao et al. [14].

Meng et al. [12] conducted *operando* pulse experiments equipped with a mass spectrometer detector to elucidate the MSR reaction route on the Cu/Cu(Al)Oₓ catalyst. The MSR reaction on this catalyst follows the HCOOCH₃* route, consistent with previous studies [73,74], which involves the following steps: CH₃OH* firstly undergoes dehydrogenation to form CH₃O* and CH₂O* species; then, CH₂O* dimerizes or reacts with CH₃O* to generate HCOOCH₃*; subsequently, HCOOCH₃* hydrolyses to form HCOOH* and CH₃O*, and CH₃O* re-participates in the catalytic cycle; finally, the decomposition of HCOOH* occurs to produce CO₂ and H₂. Combined with DFT calculations (Figure 6b), it was demonstrated that the cleavage of the C-H bonds in the intermediates CH₃O* and HCOO* is the rate-determining step. Additionally, the study showed that water molecules promote the decomposition of HCOOCH₃* but do not directly participate in cleavage of the C-H bonds.

In fact, Frank et al. [68] had earlier summarized the HCOO* route and the HCOOCH₃* route (Figure 5) and pointed out that the ratio of methanol to water affects the selection of these pathways. When water is in excess, the HCOO* route is preferred, involving the one-step oxidation of HCHO* to HCOO* by hydroxyl groups or reactive oxygen species from H_2O dissociation, followed by the decomposition of HCOO* to produce CO_2 and H_2. Conversely, when methanol is in excess, the HCOOCH$_3$* route is preferred, involving the dehydrogenation of CH_3OH to $HCOOCH_3$, which is then hydrolyzed to HCOO*, followed by further decomposition to produce CO_2 and H_2.

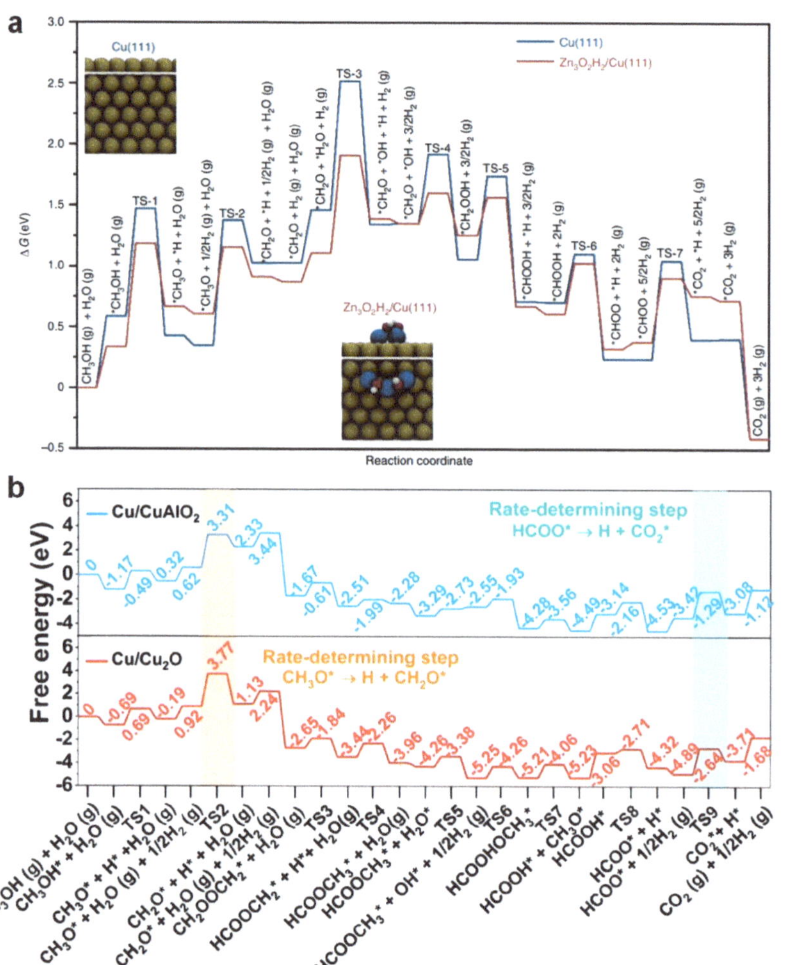

Figure 6. (a) The Gibbs free energy diagram for the MSR reaction on Cu(111) and $Zn_3O_2H_2$/Cu(111) (reproduced with permission from ref. [17], Copyright 2022 Springer Nature); (b) the full potential reaction pathway of the MSR reaction following the HCOOCH$_3$* mechanism over Cu/CuAlO$_2$ and Cu/Cu$_2$O (reproduced with permission from ref. [12], Copyright 2023 Springer Nature).

2.3. Deactivation

Improving the stability of copper-based catalysts and addressing issues including sintering and carbon deposition are crucial steps to enhance their practical application value in the MSR reaction. Sintering of metal nanoparticles proceeds through Ostwald ripening (OR), or through particle migration and coalescence (PMC) [75]. Once particles are sintered, it is almost impossible to restore the original catalytic activity of the catalysts through methods such as redispersion; so, it is necessary to fundamentally prevent sintering from occurring. From a thermodynamic perspective, the sintering of nanoparticles in catalysts is driven by the reduction in surface free energy, with the rate significantly increasing at higher temperatures [76]. Typically, the onset of sintering is estimated using the Hüttig temperature, where surface atoms become mobile, and the Tammann temperature, where bulk atom diffusion occurs [77]. Therefore, controlling the reaction to occur at lower temperatures is a direct strategy to inhibit the sintering of copper-based catalysts. However, for the endothermic MSR reaction, high activity is typically associated with elevated reaction temperatures up to a certain limit, making it challenging to prevent sintering.

To overcome the limitations of reaction temperature, other strategies to inhibit sintering include [76] controlling the uniformity of nanoparticles to eliminate chemical potential differences between particles, enhancing the chemical bonding between metal nanoparticles and the support, physically confining the metal nanoparticles, and constructing energy barriers to prevent the surface diffusion of metal nanoparticles. Li et al. [17] achieved surface reconstruction by adjusting the composition of the reducing agent, causing ZnO_x to migrate to the surface of Cu^0 nanoparticles in the $Cu/ZnO/Al_2O_3$ catalyst. This moderate encapsulation tripled the long-term stability of the copper-based catalyst (Figure 4b). Cheng et al. [34] improved the stability of $Cu/ZnO/Al_2O_3$ catalysts by adding Mg, which enhanced the Cu-ZnO synergistic effect, thus inhibiting the sintering of the Cu and ZnO phases. Sanches et al. [25] found that the presence of Zr in $Cu/ZnO/ZrO_2$ catalysts increased the microstrain in CuO and ZnO, reducing their grain size and limiting their growth. Siriruang et al. [78] prepared $Cu-Zn/ZrO_2$-doped Al_2O_3 catalysts using a urea impregnation method, which exhibited anti-sintering capability and maintained high hydrogen yield even after accelerated sintering treatment. Clearly, exploring more potential anti-sintering strategies based on fundamental scientific principles of sintering inhibition remains an area of ongoing effort for the future.

Carbon deposition can block catalyst pores and cover the active species on the MSR catalyst surface, leading to deactivation. Carbon deposits originate partly from hydrocarbons generated during the MSR reaction [79] or from carbon–oxygen species [54]; another source is elemental carbon produced from CO by the disproportionation reaction [54,79,80]. Unlike deactivation caused by sintering, catalysts deactivated by carbon deposition can be regenerated to restore their initial activity. A common regeneration strategy involves adding CeO_2 to copper-based catalysts, utilizing CeO_2's oxygen storage capacity to remove carbon deposits from the catalyst surface [27,35,54,81]. Specifically, under the reducing conditions of the MSR reaction, the partial reduction of CeO_2 generates mobile oxygen, which promotes the gasification of carbon deposits, thereby enhancing the stability of the catalysts. Industrial practices also highlight the critical role of catalyst regeneration techniques in maintaining performance and longevity. Methods such as advanced passivation and controlled-atmosphere treatments during handling and storage are widely employed to minimize sintering during downtime. To address carbon deposition, strategies including the optimization of reaction conditions (e.g., water-to-methanol ratio and operating temperature) and periodic catalyst regeneration through controlled oxidation cycles are commonly implemented [12]. Incorporating these industrial insights bridges the gap between fundamental research and practical application, providing effective solutions for ensuring the long-term stability and efficiency of catalysts.

In summary, copper-based catalysts remain the most prevalent catalysts used in methanol steam reforming. Regarding the identification of their active sites, many studies indicate a synergistic interaction between Cu^0 and Cu^+, although this conclusion often does not take into account the influence of the commonly used ZnO and Al_2O_3 supports. When these supports are present, Cu-Zn and Cu-Al interface sites are more likely to form. To elucidate the role of the widely used ZnO support in copper-based catalysts, three models are commonly referenced: the spillover model, the morphology model, and the CuZn alloy model. The discussion then shifted to $Cu/ZnO/Al_2O_3$ catalysts, emphasizing their advancements over Cu/ZnO catalysts. Additionally, the use of alternative supports, such as SiO_2, TiO_2, SBA-15, MCM-41, and CNTs, as well as the incorporation of promoters like Ce, Zr, Ga, and Mg, was also discussed. Furthermore, the impact of preparation methods and reaction pretreatment activation processes on catalytic performance was summarized. Two potential reaction pathways for copper-based catalysts in the MSR reaction were highlighted, namely the HCOO* route and the $HCOOCH_3$* route, with pathway selection influenced by factors like the water-to-alcohol ratio and the nature of the active metal in the catalyst. Finally, the challenges of sintering and coking in copper-based catalysts during the MSR reaction were analyzed, identifying contributing factors such as Ostwald ripening, particle migration and coalescence, and the origins of carbon species, while presenting practical solutions to address these issues.

3. Noble Metal-Based Catalysts

In addition to the copper-based catalysts, noble metal-based catalysts are another large class of MSR catalysts. The noble metals most often considered are Pd and Pt, followed by Ru and Rh. The performance of representative catalysts in this category is listed in Table 2.

Table 2. Performance summary of representative noble metal-based catalysts for methanol steam reforming.

Catalysts	Temperature ($^\circ$C)	CH_3OH Conversion (%)	CO Selectivity (%)	H_2 Yield (mmol g^{-1} h^{-1})	Reaction Conditions	Stability	Ref.
Pd/ZnO	400	94	0.5 [a]	1628	H_2 pretreatment; GHSV = 12,000 h^{-1}; H_2O/CH_3OH = 1.2	\	[82]
Pd/ZnAl$_2$O$_4$	250	35	3.0 [a]	41.04	H_2 pretreatment; $P_{methanol}$ = 6.4 mol%; H_2O/CH_3OH = 1.1	250 $^\circ$C, 100 h, no drop	[83]
ZnPd/MoC	160	40.3	0.9 [b]	68.9	CH_4/H_2 pretreatment; feed rate = 1.2 mL h^{-1}; H_2O/CH_3OH = 3	240 $^\circ$C, 170 h, initial deactivation only	[84]
Pd/Zn$_1$Zr$_1$O$_x$	330	46	0	\	H_2 pretreatment; GHSV = 17,000 h^{-1}; H_2O/CH_3OH = 1.3	330 $^\circ$C, 30 h, no drop	[85]
Pd/ZrO$_2$-TiO$_2$	300	98	37 [c]	\	H_2 pretreatment; GHSV = 30,000 h^{-1}; H_2O/CH_3OH = 0.16	\	[86]
Pd/In$_2$O$_3$/CeO$_2$	375	96	1.3 [c]	250	H_2 pretreatment; GHSV = 13,809.6 h^{-1}; H_2O/CH_3OH = 1.4	400 $^\circ$C, 30 h, no drop	[87]
Pd-Cu/ZnAl$_2$O$_4$	240	100	\	H_2 yield = 84%	H_2 pretreatment; GHSV = 2400 h^{-1}	\	[88]
CuPd/ZrO$_2$	220	63	5 [a]	86.3	H_2 pretreatment; GHSV = 295 mol g^{-1} h^{-1}; H_2O/CH_3OH = 1.4	240 $^\circ$C, 80 h, no drop	[89]
Pt/In$_2$O$_3$/CeO$_2$	325	98.7	2.6 [a]	333	feed rate = 1.2 mL h^{-1}; H_2O/CH_3OH = 1.4	325 $^\circ$C, 32 h, no drop	[90]
Pt/In$_2$O$_3$/CeO$_2$	350	99.9	2.5 [c]	H_2 yield = 64.7%	WHSV = 99,500 mL g^{-1} h^{-1}; H_2O/CH_3OH = 1.4	350 $^\circ$C, 100 h, 8% drop in CH_3OH conversion	[91]

Table 2. *Cont.*

Catalysts	Temperature (°C)	CH$_3$OH Conversion (%)	CO Selectivity (%)	H$_2$ Yield (mmol g^{-1} h^{-1})	Reaction Conditions	Stability	Ref.
Pt/MoC	200	100	3 [c]	\	CH$_4$/H$_2$ pretreatment; WHSV = 9000 cm^3 g^{-1} h^{-1}; H$_2$O/CH$_3$OH = 1	200 °C, 20 h, no drop	[92]
Zn-Pt/MoC	160	65.9	\	106.9	Carburizing treatment; feed rate = 1.2 mL h^{-1}; H$_2$O/CH$_3$OH = 3	120 °C, 25 h, 4% drop in CH$_3$OH conversion	[93]
Pt$_1$/ZnO	390	43	\	\	WHSV = 55,200 cm^3 g^{-1} h^{-1}; H$_2$O/CH$_3$OH = 1.5	\	[94]
Pt-K@S-1	250	15	<1.9% [c]	4308	H$_2$ pretreatment; WHSV = 45 h^{-1}; H$_2$O/CH$_3$OH = 3	400 °C, 50 h, no drop	[95]
Ru/TiO$_2$	300	98.9	5.4 [b]	\	H$_2$ pretreatment; WHSV = 1.8 h^{-1}; H$_2$O/CH$_3$OH = 1.2	\	[96]
RuCe	400	98	0.13 [c]	882 mmol cm^{-3} h^{-1}	feed rate = 3.47 mL h^{-1}; H$_2$O/CH$_3$OH = 2	400 °C, 115 h, no drop	[97]
Ru$_1$/CeO$_2$	350	25.6	2.2 [a]	139.6	feed rate = 3 mL h^{-1}; H$_2$O/CH$_3$OH = 3	350 °C, 72 h, no drop	[98]
Rh$_1$/CeO$_2$	350	21	36 [a]	100	feed rate = 3 mL h^{-1}; H$_2$O/CH$_3$OH = 3	\	[98]

a: $S_{CO} = \frac{y_{CO}}{y_{CO}+y_{CO_2}} \times 100\%$; b: $S_{CO} = \frac{y_{CO}}{y_{CO}+y_{CO_2}+y_{CH_4}} \times 100\%$; c: $S_{CO} = \frac{y_{CO}}{y_{CO}+y_{CO_2}+y_{H_2}+y_{CH_4}} \times 100\%$.

3.1. Palladium-Based Catalysts

The research on palladium-based catalysts for hydrogen production through methanol steam reforming began in the late 20th century. The Takezawa team [99] prepared a series of supported palladium-based catalysts using metal oxides as carriers, including Pd/SiO_2, Pd/Al_2O_3, Pd/La_2O_3, Pd/Nb_2O_5, Pd/Nd_2O_3, Pd/ZrO_2, and Pd/ZnO. Among the aforementioned catalysts, Pd/ZnO stood out for its high activity and selectivity in the MSR reaction. Therefore, the Takezawa team subsequently conducted a decade-long study on the Pd/ZnO catalyst [100–103]. They found that a PdZn alloy formed on the Pd/ZnO catalyst after high-temperature hydrogen reduction treatment, which greatly improved the MSR performance, especially in terms of selectivity.

3.1.1. Alloys

Notably, reduction temperature and palladium loading are important factors influencing the formation of the PdZn alloy. The PdZn alloy begins to form at 420 K, and as the reduction temperature increases, the proportion of the alloy increases, which corresponds to an enhancement in selectivity [100]. Additionally, the research indicated that an equivalent molar amount of ZnO to the palladium loading could be reduced to form the PdZn alloy; thus, the formation of the alloy was not limited by the palladium loading [101]. In the Takezawa team's series of studies, the palladium loading was mostly 10%. This prompted subsequent researchers to consider the possibility of forming alloys with lower palladium loadings. Studies by Halevi and Peterson et al. [104–106] indicated that the $PdZn_\beta$ phase (Pd/Zn molar ratio = 1) exhibited low CO selectivity, while the $PdZn_\alpha$ phase (Pd/Zn molar ratio > 1) favored CO formation during MSR. Meanwhile, the formation of different PdZn alloy phases was closely related to the palladium loadings. For example, the $PdZn_\alpha$ phase was predominant when the palladium loading was below 4.8% on the conventional Pd/ZnO catalyst [107]. This explains the use of high noble metal loadings in many studies on palladium-based catalysts for MSR. Based on these findings, Liu et al. [83] synthesized a PdZn alloy catalyst with a palladium loading of only 0.1% using $ZnAl_2O_4$ as the support. Due to the limited number of Zn atoms provided by $ZnAl_2O_4$ and the enhanced interaction between Pd and Zn by the polar facets exposed on $ZnAl_2O_4$, the $PdZn_\beta$ alloy predominated on the $Pd/ZnAl_2O_4$ catalyst. This enabled high CO_2 selectivity in MSR.

Subsequently, there have been more studies on PdZn alloy catalysts for MSR, focusing on the alloy formation process and the impact of alloy crystallite size on performance. Föttinger et al. [108] monitored the dynamic formation process of the active phase PdZn alloy for catalysis in real time by means of the *operando* quick-EXAFS (extended X-ray absorption fine structure) technique (Figure 7a). It was demonstrated that alloy formation started at the nanoparticle surface and then proceeded from the surface inward, leaving a metallic Pd core at the end. Moreover, alloying was reversible, and treatment with oxygen led to alloy decomposition and the formation of metallic Pd, which was due to the preferential oxidation of Zn. Similarly, Wang et al. [109,110] inferred that the reduction process followed the sequence $PdO/ZnO \rightarrow Pd/ZnO \rightarrow Pd/ZnO_{1-x} \rightarrow PdZn$ alloy/ZnO on the Pd/ZnO catalyst based on temperature-programmed reduction and X-ray diffraction results. Among them, the PdZn alloy and Pd/ZnO_{1-x} formed by partial oxidation during the reaction might be the real active species. Their study also suggested that the best MSR performance was achieved when the particle size of the PdZn alloy was 5–14 nm. Karim et al. [111] showed that high selectivity in the MSR reaction could still be achieved by reducing the size of the small particles of palladium to below 2 nm, despite the lack of complete PdZn alloying. In order to avoid the dissolution of ZnO or the alteration of ZnO morphology brought about by conventional aqueous impregnation, Dagle et al. [112]

synthesized Pd/ZnO catalysts using an organic preparation method. It was demonstrated that large-sized PdZn crystallites could significantly inhibit CO generation while exhibiting high activity for the MSR reaction. This conclusion was also supported by Lim et al. [113] and indicated that the decrease in defect sites due to the growth of PdZn alloy particles was the main reason for the suppression of CO generation. Although the size of the active metal components is the most prominent structural factor in the study of the structure–property relationship, their coordination environment, valence states, and geometric configuration are also changed along with the size [114]. Therefore, the effects of these structural change factors on catalytic performance need to be considered together.

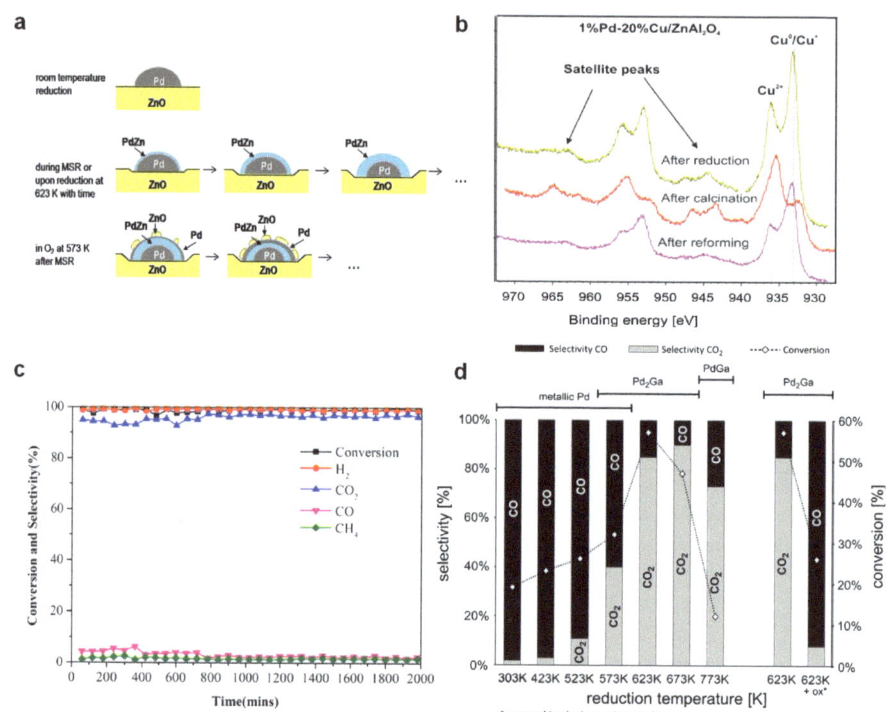

Figure 7. (**a**) An illustration of the suggested structural changes in Pd/ZnO in various environments (reproduced with permission from ref. [108], Copyright 2011 American Chemical Society); (**b**) Cu 2p XPS of the 1%Pd-20%Cu/ZnAl$_2$O$_4$ catalyst after calcination, after reduction, and after the reaction (reproduced with permission from ref. [88], Copyright 2014 Elsevier); (**c**) the stability test of the Pd/In$_2$O$_3$/CeO$_2$ (rod-shaped) catalyst for MSR (reaction conditions: 400 °C, S/C = 1.4, WHSV = 13,809.6 h^{-1}) (reproduced with permission from ref. [87], Copyright 2024 Elsevier); (**d**) the initial selectivity (bars) and total reaction rate (diamonds) during methanol conversion over Pd/Ga$_2$O$_3$ as a function of the reduction temperature (reaction conditions: 523 K, GHSV = 60,000 mL g^{-1} h^{-1}) (reproduced with permission from ref. [115], Copyright 2012 Elsevier).

The stability issues associated with copper-based catalysts and the high CO selectivity observed in palladium-based catalysts have led researchers to consider combining the advantages of both, resulting in the development of PdCu bimetallic catalysts. Azenha et al. [89,116] synthesized CuPd/ZrO$_2$ bimetallic catalysts via the wet impregnation method and found that the order of the impregnation of Cu and Pd affected the catalytic performance. When Pd was impregnated first, it facilitated a uniform distribution of Pd and Cu on the support surface, forming closely bonded PdCu nanoparticles or alloys. Additionally, the synergistic interaction between Pd and Cu adjusted the electronic structure of the system, thereby enhancing MSR activity and reducing CO selectivity. Ruano et al. [117] also demonstrated through *in situ* mass spectrometry that the formation of PdCu alloys was crucial for reducing CO selectivity. Mierczynski et al. [88] discovered that during the reduction process of the 1%Pd-20%Cu/ZnAl$_2$O$_4$ catalyst (300 °C, 5% H$_2$/Ar), hydrogen spillover from metallic Pd to Cu species promoted the reduction of copper oxide, leading to the formation of PdCu alloys. The formation of PdCu alloys increased the concentration of Cu^0 or Cu$^+$ active species (Figure 7b), thereby improving hydrogen yield and selectivity.

3.1.2. Supports and Promoters

Researchers have made many other attempts to develop palladium-based catalysts regarding supports (such as MoC, ZrO$_2$, TiO$_2$, and CeO$_2$) or promoters (such as In$_2$O$_3$ and Ga$_2$O$_3$). Tang et al. [84] introduced a small amount of Zn into Pd/MoC catalysts, which promoted the formation of the α-MoC$_{1-x}$ phase and improved the dispersion of Pd on the surface of MoC, thereby enhancing the low-temperature performance, including the methanol conversion and hydrogen production rate, although the selectivity still needed improvement. Pérez-Hernández et al. [86] prepared Pd/ZrO$_2$-TiO$_2$ catalysts using ZrO$_2$-TiO$_2$ synthesized by the sol–gel method as a support. Compared to palladium catalysts supported on either ZrO$_2$ or TiO$_2$ oxide, Pd/ZrO$_2$-TiO$_2$ showed higher activity, but the CO selectivity remained high, around 30%. Matsumura et al. [118] coprecipitated PdO/ZnO/Al$_2$O$_3$ on an amorphous ZrO$_2$ support, forming ultrafine PdO particles. The obtained PdZnAl/ZrO$_2$ with 3 wt.% Pd content had activity as high as that of 10 wt.% PdZnAl, and it remained stable even after the reaction at 550 °C. Wang et al. [85] prepared Pd/Zn$_1$Zr$_1$O$_x$ catalysts for the MSR reaction using nanoscale Zn$_1$Zr$_1$O$_x$ mixed oxides as supports. The presence of ZrO$_2$ facilitated the dispersion of ZnO clusters, which in turn favored PdZn alloying and alloy stability, resulting in excellent selectivity. Barrios et al. [119] prepared ZnO-CeO$_2$ nanocomposite-supported Pd catalysts and found that although the reaction activity and selectivity of Pd/ZnO-CeO$_2$ were lower than those of Pd/ZnO, they exhibited higher stability, possibly due to the reducibility of CeO$_2$ and its ability to generate oxygen vacancies. Zhang et al. [87] found that the morphology of the CeO$_2$ support significantly affected the MSR performance of Pd/In$_2$O$_3$ catalysts. A Pd/In$_2$O$_3$/CeO$_2$ (rod-shaped) catalyst promoted the reaction of the intermediate formaldehyde with hydroxyl to produce CO$_2$ and H$_2$, due to the presence of large palladium nanoparticles and a high density of oxygen vacancies created by the strong interaction between In and Ce, showing good reactivity and stability (Figure 7c).

Penner and Lorenz et al. [120,121] confirmed that Pd-Ga bimetallic particles formed on Pd/Ga$_2$O$_3$ at appropriate reduction temperatures effectively inhibited methanol dehydrogenation to CO, in comparison with Pd/ZnO catalysts. Haghofer et al. [115] explored the relationship between intermetallic compounds formed on Pd/Ga$_2$O$_3$ catalysts and MSR performance. Their research indicated that the intermetallic compound Pd$_2$Ga formed within the temperature range of 548–673 K favored the MSR reaction, while Ga-rich PdGa formed at a higher reduction temperature of 773 K resulted in poorer MSR performance (Figure 7d). Föttinger et al. [122] demonstrated that the formation of intermetallic compounds (Pd$_2$Ga or PdZn) with Ga or Zn was crucial for reducing the amount of the byproduct CO from methanol dehydrogenation on palladium-based catalysts. However, the surface degradation of these intermetallic compounds at low temperatures was one reason affecting their stability in catalyzing the MSR reaction. Rameshan et al. [123] found that on Pd-Ga$_2$O$_3$-In$_2$O$_3$ catalysts, as long as the appropriate intermetallic phases were present and exhibited optimized intermetallic-support phase boundary dimensions, the presence of various supported intermetallic InPd and GaPd$_2$ phases would not adversely affect the activity or selectivity in MSR. Overall, forming appropriate intermetallic compounds and optimizing the catalyst structure are crucial for improving MSR reaction performance and stability on palladium-based catalysts.

3.1.3. Reaction Mechanism

By comparison with the reaction mechanism of copper-based catalysts, Iwasa et al. [100,124] summarized three possible pathways of MSR on palladium-based catalysts (Figure 8a). On palladium alloys, pathways II and III, which primarily produced CO_2 and H_2, were preferred over pathway I. In contrast, for metallic palladium without alloy formation, a large amount of the byproduct CO, derived directly from the decomposition of HCHO, was produced. It appears that all the differences stem from HCHO, a key reaction intermediate of the MSR reaction. With the advancement of surface science research [125], researchers discovered that the adsorption configurations of formaldehyde on palladium alloys and metallic palladium differed (Figure 8b) [70,126]. This difference was likely the reason for the distinct reaction pathways observed on the two types of catalysts. As for the catalysts containing Pd alloys, the $\eta^1(O)$-HCHO configuration anchored on positively charged Pd sites shows preferential stability, where the carbonyl group is perpendicular to the surface and only oxygen interacts with the metal; this intermediate further reacts with the hydroxyl group to produce the formate species, which eventually decomposes into CO_2 and H_2. As for the catalysts containing metallic Pd, differently, formaldehyde adsorbs in an $\eta^2(C, O)$-HCHO configuration where the carbonyl group is parallel to the surface and both C and O atoms are bound to the metallic Pd. In this case, electrons from metallic Pd strongly donate to the π^*_{CO} antibonding orbital of formaldehyde, promoting the rapid decomposition of $\eta^2(C, O)$-HCHO to CO and H_2. Unfortunately, there is still a lack of microscopic evidence supporting the MSR mechanism over palladium-based catalysts through the formaldehyde adsorption configuration theory.

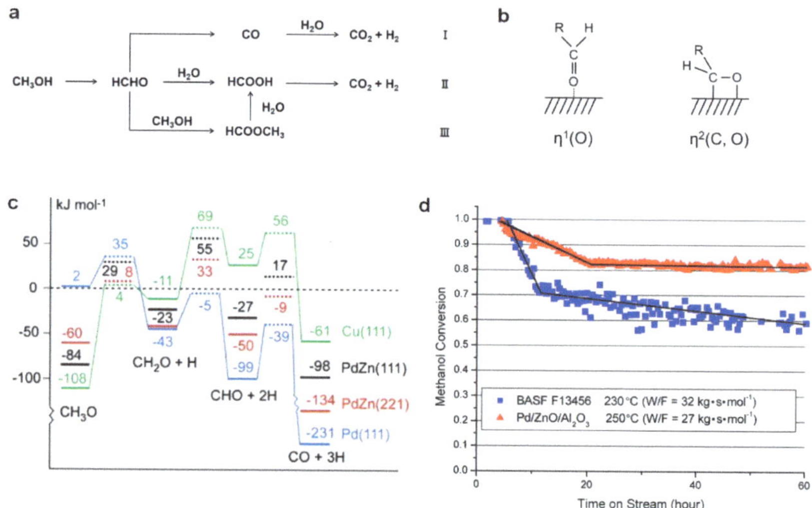

Figure 8. (**a**) Three reaction pathways for methanol steam reforming over palladium-based catalysts (reproduced with permission from ref. [100], Copyright 1995 Elsevier); (**b**) two configurations of aldehyde (reproduced with permission from ref. [70], Copyright 1997 Elsevier); (**c**) calculated energetics of formaldehyde dehydrogenation to formyl and then to CO on Pd, Cu, and PdZn surfaces (reproduced with permission from ref. [113], Copyright 2006 American Chemical Society); (**d**) activity over 60 h: comparison between a commercial copper-based catalyst and $Pd/Zn/Al_2O_3$ (reproduced with permission from ref. [127], Copyright 2008 Elsevier).

Although the preferred bonding configuration of the formaldehyde intermediate has been considered to be a reasonable mechanism for the high selectivity of the PdZn alloy, this hypothesis was contradicted by the DFT calculations of Chen et al. [113,128,129], based on the interaction between formaldehyde and the PdZn (111) surface. Their calculations indicated that the η^2(C, O) configuration was the most stable on the alloy surface, rather than the η^1(O) configuration proposed by Iwasa et al. [70,126]. These theoretical calculations also suggested that increasing the C-H bond dissociation barrier, thereby inhibiting formaldehyde dehydrogenation, was more likely due to the beneficial effect of forming the PdZn alloy. As shown in Figure 8c, the barrier for the dissociation of the C-H bond in formaldehyde on PdZn (111) is 40 kJ mol^{-1} higher than on Pd (111) [113]. Gu and Li [130] also confirmed through DFT calculations that on Pd(111), formaldehyde tended to undergo direct dehydrogenation. Later, the above results of computation were also verified by experimental means such as temperature-programmed desorption (TPD) and high-resolution electron energy loss spectroscopy (HREELS) by Jeroro et al. [131].

3.1.4. Deactivation

In the MSR reaction, the stability of palladium-based catalysts is widely recognized and confirmed to be superior to that of copper-based catalysts. Iwasa et al. [105] compared the stability of Pd/Zn/CeO$_2$ and Cu/ZnO catalysts. The results showed that at 623 K, Pd/Zn/CeO$_2$ did not deactivate within 180 min; in contrast, the hydrogen production of Cu/ZnO gradually decreased over time, dropping by 20% of the initial value after 180 min. Conant et al. [127] tested the 60 h stability of Pd/ZnO/Al$_2$O$_3$ and commercial Cu/ZnO/Al$_2$O$_3$ catalysts (Figure 8d). The results showed that the Cu/ZnO/Al$_2$O$_3$ catalyst rapidly deactivated within the first 12 h, with the methanol conversion decreasing by 40% after 60 h; meanwhile, the Pd/ZnO/Al$_2$O$_3$ catalyst deactivated by only about 17% in the initial 20 h and remained stable thereafter up to 60 h. Additionally, the Pd/ZnO/Al$_2$O$_3$ catalyst could recover its initial activity after redox treatment, whereas Cu/ZnO/Al$_2$O$_3$ could not, indicating that the deactivation of palladium-based catalysts may not be due to sintering.

Despite the relatively low extent of deactivation observed for palladium-based catalysts, investigation into the underlying causes of their deactivation remains an ongoing area of research. Suwa et al. [132] confirmed that the deactivation of Pd/ZnO catalysts was due to the active PdZn alloy sites being covered by zinc carbonate hydroxide (Zn$_4$CO$_3$(OH)$_6$·H$_2$O) formed from ZnO. Liu et al. [133] suggested two reasons for the deactivation of Pd/ZnO catalysts: the surface carbon deposition contamination and surface oxidation decomposition of the PdZn alloy (Pd-Zn $+ 2H_2O \leftrightarrow$ Pd $+$ Zn(OH$_2$)$+ H_2$). Therefore, the catalyst can be regenerated in an oxygen-containing atmosphere at relatively low temperatures to remove carbon deposits or in a hydrogen-containing atmosphere at higher temperatures to regenerate the PdZn alloy. From the above studies, it is clear that maintaining the stability of the PdZn alloy structure is crucial. Penner et al. [134] synthesized Pd/ZnO/SiO$_2$ catalysts, and the PdZn alloy particles on the surface could maintain good structural and thermal stability between 473 and 873 K. Even at temperatures above 873 K, the PdZn alloy only partially decomposed. This broad stability range is related to the strong interaction between Pd and ZnO and the high stability of the 1:1 PdZn alloy phase. These findings may explain the excellent stability of palladium-based catalysts, although their activity and selectivity still lag behind those of copper-based catalysts.

3.2. Other Catalysts

Following the success of palladium-based catalysts, researchers turned their attention to other noble metal-based catalysts, such as Pt, Ru, and Rh. Among them, Pt-based catalysts have proven to be promising and effective for the MSR reaction. Studies by the Men team [90,135,136] and Shanmugam et al. [91] have shown that the $Pt/In_2O_3/CeO_2$ catalyst exhibited low CO selectivity and good stability over 100 h of the MSR reaction. Over this catalyst, In_2O_3 improved the dispersion of Pt nanoparticles and enhanced the interaction between the metal and the support, while CeO_2 provided abundant active oxygen species with generating oxygen vacancies (Figure 9a). The combined effects of In_2O_3 and CeO_2 promoted the activation of water, favoring the main reaction and then reducing the formation of the byproduct CO. Additionally, many studies have used the α-MoC phase as support to increase the dispersion of loaded Pt atoms, achieving high performance in the MSR reaction [137]. Ma et al. [92] prepared Pt/α-MoC_{1-x}, which achieved 100% methanol conversion at the low temperature of 473 K. Cai et al. [93] modified Pt/MoC with Zn, promoting the formation of the α-MoC_{1-x} phase, thus enhancing Pt dispersion and improving the interaction between the metal and the support, thereby enhancing catalytic performance. The emergence of single-atom catalysts also offers the potential for further improving the MSR performance of Pt-based catalysts. Lin et al. [138] developed atomically dispersed Pt single-atom catalysts using α-MoC as the support. The well-dispersed Pt_1 atoms provided a high density of electron-deficient surface sites, promoting methanol adsorption and activation, while α-MoC offered high-activity sites for water dissociation, generating abundant surface hydroxyl groups. These resulted in excellent hydrogen production activity and stability. Gu et al. [94] reported that Pt single atoms were stabilized on the ZnO surface by lattice oxygen. This made the reaction intermediates bind more firmly to the active Pt_1 sites, altering the reaction energy and kinetics, and subsequently changing the reaction pathway. The turnover frequency (TOF) of the Pt_1 active sites was 1000 times higher than that of ZnO.

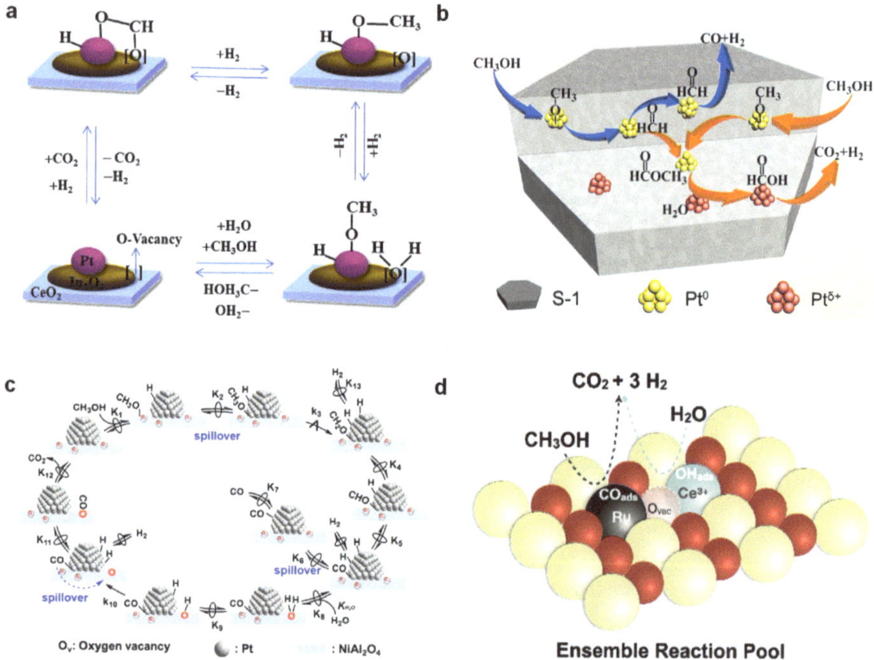

Figure 9. (**a**) Possible proposed mechanism for MSR over $Pt/In_2O_3/CeO_2$ catalyst (reproduced with permission from ref. [91], Copyright 2020 Elsevier); (**b**) proposed reaction scheme for MSR performed

on Pt^0 and $Pt^{\delta+}$ sites via methyl formate intermediate pathway (reproduced with permission from ref. [95], Copyright 2022 Elsevier); (**c**) scheme of reaction pathways of methoxyl dehydrogenation coupled with O_1H decomposition in methanol reforming over $Pt/NiAl_2O_4$ catalysts (reproduced with permission from ref. [139], Copyright 2023 American Chemical Society); (**d**) schematic diagram of MSR reaction on Ru_1/CeO_2 catalyst (reproduced with permission from ref. [98], Copyright 2021 American Chemical Society).

Many researchers have also made efforts to identify the active sites and reaction pathways of Pt-based catalysts in the MSR reaction. Shao et al. [95] analyzed the MSR reaction mechanism on Pt-K@S-1 catalysts using temperature-programmed surface reaction mass spectrometry (TPSR-MS) and DFT calculations (Figure 9b). The cleavage of the O-H bond in CH_3OH was activated over Pt^0 sites to produce $HCOOCH_3$, whereas $Pt^{\delta+}$ sites promoted the hydrolysis of $HCOOCH_3$ to $HCOOH$ and ultimately CO_2 and H_2. The synergy between Pt^0 and $Pt^{\delta+}$ on the S-1 support, along with the promotion by K, endowed the catalyst with excellent activity, selectivity, and stability for the MSR reaction. Wang et al. [139] elucidated the reaction mechanism on $Pt/NiAl_2O_4$ catalysts through spectroscopy, kinetics, and isotope studies. Their research suggested that the MSR reaction involved a tandem process of methanol dehydrogenation and water-gas shift; the interface sites between Pt and $NiAl_2O_4$ were active for methanol dehydrogenation, while the sites on $NiAl_2O_4$ were active for the water-gas shift reaction (Figure 9c). Jin et al. [140] analyzed the MSR reaction pathway on Pt (111) surfaces theoretically using DFT and kinetic Monte Carlo (kMC) calculations to reduce CO selectivity. Their studies indicated that the energy barrier difference between the steps $H_2O^* + * \rightarrow OH^* + H^*$ and $CH_3OH^* + * \rightarrow CH_2OH^* + H^*$ was critical for CO selectivity. When the energy barrier for the former was 0.30 eV lower than that for the latter, CO formation was significantly suppressed. These studies provide valuable insights for designing high-performance Pt-based catalysts for MSR.

Ru-based catalysts are widely used in homogeneous catalysis, such as low-temperature aqueous-phase methanol reforming, achieving efficient and stable hydrogen production [141–143]. By dispersing a liquid film over the large inner surface area of a porous solid, known as supported liquid phase (SLP) technology, these efficient Ru-based catalysts can be applied to the heterogeneous MSR reaction [144]. Schwarz et al. [144] deposited a basic and hygroscopic KOH coating on an Al_2O_3 support, effectively fixing the Ru-pincer complex on the support. In the temperature range of 130–170 °C, only trace amounts of CO were produced, demonstrating high hydrogen production activity without deactivation for 70 h. Tahay et al. [96] deposited metallic Ru on a monolithic TiO_2 support for the MSR reaction. Compared to other metals like Cu, Cu-Ni, and Pt, Ru/TiO_2 showed higher conversion and selectivity. This was attributed to the metal–support interaction, which enhanced the activity and dispersion of Ru metal particles on the TiO_2 support. Aouad et al. [97] prepared three catalysts, Ru/Ce, Ru/Al, and Ru/CeAl, using the impregnation method. Among them, the 5 wt.% Ru/Ce catalyst exhibited the best performance, with a CO selectivity of 0.81% at 400 °C, and remained active for 115 h without deactivation. This was due to the synergistic effect between Ru and Ce, promoting the formation of active sites with excellent redox properties. Chen et al. [98] studied the mechanism of the tandem MSR reaction on Ru_1/CeO_2 and Rh_1/CeO_2 single-atom catalysts (Figure 9d). The active centers, composed of metal single atoms and adjacent oxygen vacancies, exhibited a unique synergistic catalytic effect. Ru_1/CeO_2 showed a hydrogen production rate of up to 579 mL_{H2} g_{Ru}^{-1} s^{-1} with 99.5% CO_2 selectivity, followed closely by Rh_1/CeO_2. Additionally, Lytkina et al. [145–147] used detonation nanodiamond (DND) as a support to load Ru-Rh bimetals, achieving a high hydrogen production rate and low CO selectivity. Their studies provided a new insight into the design of RuRh bimetallic catalysts, although economic costs need careful consideration.

In summary, palladium-based catalysts are a focal point of research within noble metal catalysts for the MSR reaction. The analysis began with the well-established Pd/ZnO catalyst used in methanol steam reforming, examining the factors that influence the formation of the PdZn alloy and the effect of alloy size on catalytic performance. Other commonly used promoters and supports, such as MoC, ZrO_2, TiO_2, CeO_2, In_2O_3, and Ga_2O_3, were also discussed. Considering the advantages and disadvantages of both copper-based and palladium-based catalysts, the application of the PdCu alloy in MSR reactions was reviewed. However, current research primarily focuses on enhancing activity and selectivity, with comparatively less emphasis on catalyst stability and the underlying microscopic mechanisms. The possible reaction pathways on palladium-based catalysts appear to be similar to those on copper-based catalysts, although the former tend to produce higher amounts of CO. This difference between metallic palladium and PdZn alloys is attributed to two factors: the adsorption configuration of the key intermediate formaldehyde ($\eta^1(O)$-HCHO or $\eta^2(C, O)$-HCHO) and the energy barrier for C-H bond cleavage in formaldehyde. Despite the excellent stability exhibited by palladium-based catalysts, potential deactivation mechanisms were examined, including the coverage of active alloy sites, oxidation and decomposition of the alloy, and carbon deposition. Beyond palladium-based catalysts, other noble metal-based catalysts, such as those based on Pt, Ru, and Rh, were also discussed. The catalytic performance, active sites, and reaction mechanisms of Pt-based catalysts, as well as the applications of Ru-Rh bimetallic catalysts, were highlighted. Additionally, significant research interest has been drawn to the field of single-atom catalysts involving these noble metals.

4. Conclusions and Perspectives

The carbon neutrality initiative has driven the advancement of sustainable methods for hydrogen production and secure storage [1]. Methanol steam reforming technology for *in situ* hydrogen production perfectly aligns with the aforementioned requirements. In this technology, the catalysts serve as the core component. This review provides a detailed introduction to the two main categories of catalysts used in the MSR reaction—copper-based catalysts and noble metal-based catalysts. It not only covers the roles of each component but also delves into their active sites, structure–activity relationships, reaction mechanisms, and deactivation mechanisms.

In summary, copper-based catalysts, represented by the $Cu/ZnO/Al_2O_3$ catalyst, exhibit good low-temperature activity, high hydrogen yield, low CO selectivity as a byproduct, and low cost. However, they are prone to deactivation due to sintering and carbon deposition, and their stability needs further improvement. Noble metal-based catalysts, represented by Pd/ZnO catalysts, have high CO selectivity and high cost, but are less prone to deactivation and have excellent thermal stability as a standout advantage. In real-world applications, the selection between copper-based and noble metal-based catalysts is often influenced by the trade-off between cost and performance. Copper-based catalysts are typically the preferred choice for large-scale hydrogen production, where cost considerations are paramount. In contrast, noble metal-based catalysts are favored in niche applications that prioritize stability and durability under harsh reaction conditions. Hybrid systems that combine the cost effectiveness of copper with the high-temperature stability of noble metals may offer a balanced solution.

On the other hand, despite decades of research, the MSR reaction mechanism and the active sites of catalysts remain inconclusive, warranting further investigation. Firstly, the specific reaction intermediates, such as formate (HCOO*), methoxy (CH_3O^*), and hydroxyl groups (OH*), and their contributions to the reaction pathway are still debated. While formate species are often observed in experimental studies, their direct role in the formation of CO_2 versus CO remains unclear. Secondly, the interplay between copper nanoparticles and supports (e.g., ZnO, CeO_2, ZrO_2) significantly impacts catalytic activity and stability. However, the exact nature of active sites, particularly at the metal–support interface, and how support modifications influence the reaction pathway, requires further exploration. Finally, the factors governing selectivity toward CO or CO_2, including the contributions of methanol decomposition versus steam reforming pathways, remain insufficiently understood. This lack of understanding hinders the design of catalysts that can effectively suppress CO formation. To address these challenges, a combination of experimental and computational approaches should be applied systematically.

Given the current research status, future studies on MSR for hydrogen production could focus on the following areas:

(1) Improving the stability of copper-based catalysts based on structure–activity relationships: To mitigate sintering induced by Ostwald ripening, it is crucial to address the chemical potential difference between particles. This can be achieved by ensuring uniform copper particle size during catalyst preparation, which can be facilitated by methods such as strong electrostatic adsorption (SEA). Furthermore, selectively anchoring copper nanoparticles to specific oxide facets can enhance chemical bonding strength and reduce the likelihood of coalescence. Utilizing specific physical structures—such as core–shell, core–sheath, lamellar, and mesoporous matrices—can effectively limit copper particle growth and aggregation. Additionally, employing a dual-oxide support composed of size-controlled nanoscale domains of two different oxides can create energy barriers that impede copper nanoparticles from sintering via surface diffusion.

(2) Rationally designing palladium-based alloy catalysts to reduce CO selectivity: The previously mentioned PdZn and PdCu alloy catalysts have demonstrated excellent performance in reducing CO selectivity in the MSR reaction. This phenomenon may be related to the optimization of the electronic structure of the core metal atom Pd by the second metal in alloys, as well as the modulation of the interactions between the core metal and key intermediates. The choice of the second metal is not limited to Zn and Cu; other transition metals, such as Fe, Co, Ni, and W, are also ideal candidates.

(3) Establish methods to quickly and accurately identify the reaction mechanism of MSR: Most existing studies on the MSR reaction lack systematic and comprehensive approaches to quickly and accurately identify the active sites on catalysts and the reaction pathways. This may involve advanced *in situ* characterization techniques, such as *in situ* X-ray absorption spectroscopy (XAS), Fourier transform infrared spectroscopy (FTIR), and scanning transmission electron microscopy (STEM), which enable real-time monitoring of catalytic processes. In addition, density functional theory calculations and kinetic models can offer theoretical support by providing insights into reaction pathways and active sites on the catalyst surface. By integrating these experimental and computational approaches, a clearer understanding of the MSR mechanism can be achieved, leading to the development of more efficient catalysts.

(4) Enhancing the atomic utilization efficiency of MSR catalysts: From an economic perspective, improving the atomic utilization efficiency of MSR catalysts is beneficial, whether dealing with noble metals like palladium or transition metals like copper, which often require high loadings for enhanced activity. Single-atom alloy (SAA) catalysts, which feature an atomically dispersed metal within a bi- or multi-metallic complex, have emerged as a promising system for heterogeneous catalysis. Based on this, constructing PdCu single-atom alloy catalysts not only enhances atomic utilization efficiency but also potentially combines the advantages of both metals to achieve improved MSR performance.

In addition, methanol occupies a central role in sustainable energy systems, not only as a hydrogen carrier and feedstock for steam reforming but also as a platform molecule that can be produced from CO_2 and methane. The catalytic hydrogenation of CO_2 to methanol has emerged as a promising technology for carbon capture and utilization (CCU), transforming greenhouse gasses into valuable fuels and chemicals. Similarly, methane, the primary component of natural gas, can be selectively converted into methanol via oxidative or non-oxidative routes, offering a cleaner pathway for utilizing abundant natural gas reserves. These production methods create a closed-loop system where CO_2, captured from industrial emissions or the atmosphere, can be transformed into methanol and subsequently converted into hydrogen through MSR. This integrated cycle not only supports the transition to a hydrogen economy but also aligns with global efforts to mitigate climate change. Recent advances in catalyst design for CO_2 and methane conversion, such as bimetallic catalysts and single-atom catalysts, parallel the catalyst development strategies discussed in this review, highlighting opportunities for cross-disciplinary innovation.

Author Contributions: M.Z.: Writing—original draft preparation, Investigation, Resources, Visualization. D.L., Y.W., and L.Z.: Investigation, Resources. G.X.: Writing—review and editing, Visualization, Supervision, Funding acquisition. Y.Y. and H.H.: Supervision, Funding acquisition. All authors have read and agreed to the published version of the manuscript.

Funding: This research was funded by the National Key R&D Program of China (2022YFC3704400), the National Natural Science Foundation of China (22422609 and 22072179), the Strategic Priority Research Program of the Chinese Academy of Sciences (XDA0390103), and the project of eco-environmental technology for carbon neutrality (RCEES-TDZ-2021-6).

Data Availability Statement: Not applicable.

Conflicts of Interest: The authors declare no competing interests.

Abbreviations

MSR	methanol steam reforming
MD	methanol decomposition
WGS	water-gas shift
LOHCs	liquid organic hydrogen carriers
PEMFCs	polymer electrolyte membrane fuel cells
CCU	carbon capture and utilization
WHSV	weight hourly space velocity
GHSV	gas hourly space velocity
TOF	turnover frequency
SMSIs	strong metal–support interactions
OR	Ostwald ripening
PMC	particle migration and coalescence
SEA	strong electrostatic adsorption
SAA	single-atom alloy

References

1. Zhang, S.; Liu, Y.; Zhang, M.; Ma, Y.; Hu, J.; Qu, Y. Sustainable production of hydrogen with high purity from methanol and water at low temperatures. *Nat. Commun.* **2022**, *13*, 5527. [CrossRef] [PubMed]
2. Luo, S.; Lin, H.; Wang, Q.; Ren, X.; Hernández-Pinilla, D.; Nagao, T.; Xie, Y.; Yang, G.; Li, S.; Song, H.; et al. Triggering water and methanol activation for solar-driven H_2 production: Interplay of dual active sites over plasmonic ZnCu alloy. *J. Am. Chem. Soc.* **2021**, *143*, 12145–12153. [CrossRef] [PubMed]
3. Luo, S.; Song, H.; Ichihara, F.; Oshikiri, M.; Lu, W.; Tang, D.-M.; Li, S.; Li, Y.; Li, Y.; Davin, P.; et al. Light-induced dynamic restructuring of Cu active sites on TiO_2 for low-temperature H_2 production from methanol and water. *J. Am. Chem. Soc.* **2023**, *145*, 20530–20538. [CrossRef] [PubMed]
4. Preuster, P.; Papp, C.; Wasserscheid, P. Liquid organic hydrogen carriers (LOHCs): Toward a hydrogen-free hydrogen economy. *Acc. Chem. Res.* **2017**, *50*, 74–85. [CrossRef] [PubMed]

5. Ranjekar, A.M.; Yadav, G.D. Steam reforming of methanol for hydrogen production: A critical analysis of catalysis, processes, and scope. *Ind. Eng. Chem. Res.* **2021**, *60*, 89–113. [CrossRef]
6. Rameshan, C.; Stadlmayr, W.; Penner, S.; Lorenz, H.; Memmel, N.; Hävecker, M.; Blume, R.; Teschner, D.; Rocha, T.; Zemlyanov, D.; et al. Hydrogen production by methanol steam reforming on copper boosted by zinc-assisted water activation. *Angew. Chem. Int. Ed.* **2012**, *51*, 3002–3006. [CrossRef] [PubMed]
7. Xu, X.; Lan, T.; Zhao, G.; Nie, Q.; Jiang, F.; Lu, Y. Interface-hydroxyl enabling methanol steam reforming toward CO-free hydrogen production over inverse ZrO$_2$/Cu catalyst. *Appl. Catal. B Environ.* **2023**, *334*, 122839. [CrossRef]
8. Kang, J.; Song, Y.; Kim, T.; Kim, S. Recent trends in the development of reactor systems for hydrogen production via methanol steam reforming. *Int. J. Hydrogen Energy* **2022**, *47*, 3587–3610. [CrossRef]
9. Zhuang, X.; Xu, X.; Li, L.; Deng, D. Numerical investigation of a multichannel reactor for syngas production by methanol steam reforming at various operating conditions. *Int. J. Hydrogen Energy* **2020**, *45*, 14790–14805. [CrossRef]
10. Yang, W.; Ma, X.; Tang, X.; Dou, P.; Yang, Y.; He, Y. Review on developments of catalytic system for methanol steam reforming from the perspective of energy-mass conversion. *Fuel* **2023**, *345*, 128234. [CrossRef]
11. Takezawa, N.; Kobayashi, H.; Hirose, A.; Shimokawabe, M.; Takahashi, K. Steam reforming of methanol on copper-silica catalysts—Effect of copper loading and calcination temperature on the reaction. *Appl. Catal.* **1982**, *4*, 127–134. [CrossRef]
12. Meng, H.; Yang, Y.; Shen, T.; Yin, Z.; Wang, L.; Liu, W.; Yin, P.; Ren, Z.; Zheng, L.; Zhang, J.; et al. Designing Cu^0−Cu$^+$ dual sites for improved C−H bond fracture towards methanol steam reforming. *Nat. Commun.* **2023**, *14*, 7980. [CrossRef] [PubMed]
13. Liu, D.; Zhang, M.; Zhao, L.; Guo, X.; Xu, G.; He, H. Mechanistic insights into methanol steam reforming on copper catalysts: Dynamics of active sites and reaction pathway. *J. Catal.* **2025**, *442*, 115922. [CrossRef]
14. Mao, Q.; Gao, Z.; Liu, X.; Guo, Y.; Wang, Y.; Ma, D. The Cu−Al$_2$O$_3$ interface: An unignorable active site for methanol steam reforming hydrogen production. *Catal. Sci. Technol.* **2024**, *14*, 3448–3458. [CrossRef]
15. Mrad, M.; Gennequin, C.; Aboukaïs, A.; Abi-Aad, E. Cu/Zn-based catalysts for H$_2$ production via steam reforming of methanol. *Catal. Today* **2011**, *176*, 88–92. [CrossRef]
16. Shu, Q.; Zhang, Q.; Zhu, X. Enhancing activation and stability of core-shell CuZn catalyst by ZnOx oxygen vacancies for methanol steam reforming. *Appl. Catal. A Gen.* **2024**, *678*, 119652. [CrossRef]
17. Li, D.; Xu, F.; Tang, X.; Dai, S.; Pu, T.; Liu, X.; Tian, P.; Xuan, F.; Xu, Z.; Wachs, I.E.; et al. Induced activation of the commercial Cu/ZnO/Al$_2$O$_3$ catalyst for the steam reforming of methanol. *Nat. Catal.* **2022**, *5*, 99–108. [CrossRef]
18. Shokrani, R.; Haghighi, M.; Jodeiri, N.; Ajamein, H.; Abdollahifar, M. Fuel cell grade hydrogen production via methanol steam reforming over CuO/ZnO/Al$_2$O$_3$ nanocatalyst with various oxide ratios synthesized via urea-nitrates combustion method. *Int. J. Hydrogen Energy* **2014**, *39*, 13141–13155. [CrossRef]
19. Díaz-Pérez, M.A.; Moya, J.; Serrano-Ruiz, J.C.; Faria, J. Interplay of support chemistry and reaction conditions on copper catalyzed methanol steam reforming. *Ind. Eng. Chem. Res.* **2018**, *57*, 15268–15279. [CrossRef]
20. Deshmane, V.G.; Abrokwah, R.Y.; Kuila, D. Synthesis of stable Cu-MCM-41 nanocatalysts for H$_2$ production with high selectivity via steam reforming of methanol. *Int. J. Hydrogen Energy* **2015**, *40*, 10439–10452. [CrossRef]
21. Shahsavar, H.; Taghizadeh, M.; Kiadehi, A.D. Effects of catalyst preparation route and promoters (Ce and Zr) on catalytic activity of CuZn/CNTs catalysts for hydrogen production from methanol steam reforming. *Int. J. Hydrogen Energy* **2021**, *46*, 8906–8921. [CrossRef]
22. Varmazyari, M.; Khani, Y.; Bahadoran, F.; Shariatinia, Z.; Soltanali, S. Hydrogen production employing Cu(BDC) metal–organic framework support in methanol steam reforming process within monolithic micro-reactors. *Int. J. Hydrogen Energy* **2021**, *46*, 565–580. [CrossRef]
23. Yang, S.; Zhou, F.; Liu, Y.; Zhang, L.; Chen, Y.; Wang, H.; Tian, Y.; Zhang, C.; Liu, D. Morphology effect of ceria on the performance of CuO/CeO$_2$ catalysts for hydrogen production by methanol steam reforming. *Int. J. Hydrogen Energy* **2019**, *44*, 7252–7261. [CrossRef]
24. Hosseini, T.; Haghighi, M.; Ajamein, H. Fuel cell-grade hydrogen production from methanol over sonochemical coprecipitated copper based nanocatalyst: Influence of irradiation power and time on catalytic properties and performance. *Energy Convers. Manag.* **2016**, *126*, 595–607. [CrossRef]
25. Sanches, S.G.; Flores, J.H.; da Silva, M.I.P. Cu/ZnO and Cu/ZnO/ZrO$_2$ catalysts used for methanol steam reforming. *Mol. Catal.* **2018**, *454*, 55–62. [CrossRef]
26. Ma, J.; Mao, L.; Du, H.; Zhong, J.; Jiang, L.; Liu, X.; Xu, J.; Xu, X.; Fang, X.; Wang, X. Tracking the critical roles of Cu$^+$ and Cu^0 sites and the optimal Cu$^+$/Cu^0 ratio for CH$_3$OH steam reforming (MTSR) to manufacture H$_2$. *Chem. Eng. J.* **2024**, *496*, 154195. [CrossRef]
27. Zhang, L.; Pan, L.; Ni, C.; Sun, T.; Zhao, S.; Wang, S.; Wang, A.; Hu, Y. CeO$_2$–ZrO$_2$-promoted CuO/ZnO catalyst for methanol steam reforming. *Int. J. Hydrogen Energy* **2013**, *38*, 4397–4406. [CrossRef]
28. Tajrishi, O.Z.; Taghizadeh, M.; Kiadehi, A.D. Methanol steam reforming in a microchannel reactor by Zn-, Ce- and Zr-modified mesoporous Cu/SBA-15 nanocatalyst. *Int. J. Hydrogen Energy* **2018**, *43*, 14103–14120. [CrossRef]

29. Yu, K.M.K.; Tong, W.; West, A.; Cheung, K.; Li, T.; Smith, G.; Guo, Y.; Tsang, S.C.E. Non-syngas direct steam reforming of methanol to hydrogen and carbon dioxide at low temperature. *Nat. Commun.* **2012**, *3*, 1230. [CrossRef]

30. Li, D.; Yan, H.; Jiang, Z.; Qiu, R.; Liu, Q.; Zhu, M. Gallium-promoted strong metal-support interaction over a supported Cu/ZnO catalyst for methanol steam reforming. *ACS Catal.* **2024**, *14*, 9511–9520. [CrossRef]

31. Liu, X.; Toyir, J.; Ramírez de la Piscina, P.; Homs, N. Hydrogen production from methanol steam reforming over Al_2O_3- and ZrO_2-modified $CuOZnOGa_2O_3$ catalysts. *Int. J. Hydrogen Energy* **2017**, *42*, 13704–13711. [CrossRef]

32. Kamyar, N.; Khani, Y.; Amini, M.M.; Bahadoran, F.; Safari, N. Copper-based catalysts over A520-MOF derived aluminum spinels for hydrogen production by methanol steam reforming: The role of spinal support on the performance. *Int. J. Hydrogen Energy* **2020**, *45*, 21341–21353. [CrossRef]

33. Hou, X.; Qing, S.; Liu, Y.; Li, L.; Gao, Z.; Qin, Y. Enhancing effect of MgO modification of Cu–Al spinel oxide catalyst for methanol steam reforming. *Int. J. Hydrogen Energy* **2020**, *45*, 477–489. [CrossRef]

34. Cheng, Z.; Zhou, W.; Lan, G.; Sun, X.; Wang, X.; Jiang, C.; Li, Y. High-performance $Cu/ZnO/Al_2O_3$ catalysts for methanol steam reforming with enhanced Cu-ZnO synergy effect via magnesium assisted strategy. *J. Energy Chem.* **2021**, *63*, 550–557. [CrossRef]

35. Phongboonchoo, Y.; Thouchprasitchai, N.; Pongstabodee, S. Hydrogen production with a low carbon monoxide content via methanol steam reforming over $Cu_xCe_yMg_z/Al_2O_3$ catalysts: Optimization and stability. *Int. J. Hydrogen Energy* **2017**, *42*, 12220–12235. [CrossRef]

36. Zhao, T.; Shen, Q.; Andersson, M.; Li, S.; Yuan, J. Surface of high-entropy perovskite catalyst $La(CoCuFeAlCe)_{0.2}O_3$ (100): Experimental study of methanol steam reforming for hydrogen production and DFT mechanism research. *Appl. Surf. Sci.* **2025**, *684*, 161917. [CrossRef]

37. Yang, H.; Chen, Y.; Cui, X.; Wang, G.; Cen, Y.; Deng, T.; Yan, W.; Gao, J.; Zhu, S.; Olsbye, U.; et al. A highly stable copper-based catalyst for clarifying the catalytic roles of Cu^0 and Cu^+ species in methanol dehydrogenation. *Angew. Chem. Int. Ed.* **2018**, *57*, 1836–1840. [CrossRef]

38. Ma, K.; Tian, Y.; Zhao, Z.J.; Cheng, Q.; Ding, T.; Zhang, J.; Zheng, L.; Jiang, Z.; Abe, T.; Tsubaki, N.; et al. Achieving efficient and robust catalytic reforming on dual-sites of Cu species. *Chem. Sci.* **2019**, *10*, 2578–2584. [CrossRef]

39. Zhang, G.; Zhao, J.; Yang, T.; Zhang, Q.; Zhang, L. In-situ self-assembled Cu_2O/ZnO core-shell catalysts synergistically enhance the durability of methanol steam reforming. *Appl. Catal. A Gen.* **2021**, *616*, 118072. [CrossRef]

40. Jin, S.; Li, D.; Wang, Z.; Wang, Y.; Sun, L.; Zhu, M. Dynamics of the Cu/CeO_2 catalyst during methanol steam reforming. *Catal. Sci. Technol.* **2022**, *12*, 7003–7009. [CrossRef]

41. Burch, R.; Golunski, S.E.; Spencer, M.S. The role of copper and zinc oxide in methanol synthesis catalysts. *J. Chem. Soc. Faraday Trans.* **1990**, *86*, 2683–2691. [CrossRef]

42. Grunwaldt, J.D.; Molenbroek, A.M.; Topsøe, N.Y.; Topsøe, H.; Clausen, B.S. In situ investigations of structural changes in Cu/ZnO catalysts. *J. Catal.* **2000**, *194*, 452–460. [CrossRef]

43. Nakamura, J.; Choi, Y.; Fujitani, T. On the issue of the active site and the role of ZnO in Cu/ZnO methanol synthesis catalysts. *Top. Catal.* **2003**, *22*, 277–285. [CrossRef]

44. Shishido, T.; Yamamoto, Y.; Morioka, H.; Takehira, K. Production of hydrogen from methanol over Cu/ZnO and $Cu/ZnO/Al_2O_3$ catalysts prepared by homogeneous precipitation: Steam reforming and oxidative steam reforming. *J. Mol. Catal. A Chem.* **2007**, *268*, 185–194. [CrossRef]

45. Kurr, P.; Kasatkin, I.; Girgsdies, F.; Trunschke, A.; Schlögl, R.; Ressler, T. Microstructural characterization of $Cu/ZnO/Al_2O_3$ catalysts for methanol steam reforming—A comparative study. *Appl. Catal. A Gen.* **2008**, *348*, 153–164. [CrossRef]

46. Cheng, Z.; Wang, M.; Jiang, C.; Zhang, L.; He, X.; Sun, X.; Lan, G.; Qiu, Y.; Li, Y. Tuning lattice strain of copper particles in $Cu/ZnO/Al_2O_3$ catalysts for methanol steam reforming. *Energy Fuels* **2024**, *38*, 15611–15621. [CrossRef]

47. Miyao, K.; Onodera, H.; Takezawa, N. Highly active copper catalysts for steam reforming of methanol. catalysts derived from Cu/Zn/Al alloys. *React. Kinet. Catal. Lett.* **1994**, *53*, 379–383. [CrossRef]

48. Zheng, J.; Chen, L.; Xie, X.; Tong, Q.; Ouyang, G. Polydopamine modified ordered mesoporous carbon for synergistic enhancement of enrichment efficiency and mass transfer towards phenols. *Anal. Chim. Acta* **2020**, *1095*, 109–117. [CrossRef] [PubMed]

49. Ahmed, A.M.; Mohamed, F.; Ashraf, A.M.; Shaban, M.; Aslam Parwaz Khan, A.; Asiri, A.M. Enhanced photoelectrochemical water splitting activity of carbon nanotubes@TiO_2 nanoribbons in different electrolytes. *Chemosphere* **2020**, *238*, 124554. [CrossRef]

50. Yang, Y.; Chiang, K.; Burke, N. Porous carbon-supported catalysts for energy and environmental applications: A short review. *Catal. Today* **2011**, *178*, 197–205. [CrossRef]

51. Khani, Y.; Tahay, P.; Bahadoran, F.; Safari, N.; Soltanali, S.; Alavi, A. Synergic effect of heat and light on the catalytic reforming of methanol over Cu/x-TiO_2(x=La, Zn, Sm, Ce) nanocatalysts. *Appl. Catal. A Gen.* **2020**, *594*, 117456. [CrossRef]

52. He, H.; Dai, H.X.; Au, C.T. Defective structure, oxygen mobility, oxygen storage capacity, and redox properties of RE-based (RE = Ce, Pr) solid solutions. *Catal. Today* **2004**, *90*, 245–254. [CrossRef]

53. Park, J.B.; Graciani, J.; Evans, J.; Stacchiola, D.; Senanayake, S.D.; Barrio, L.; Liu, P.; Sanz, J.F.; Hrbek, J.; Rodriguez, J.A. Gold, copper, and platinum nanoparticles dispersed on $CeO_x/TiO_2(110)$ surfaces: High water-gas shift activity and the nature of the mixed-metal oxide at the nanometer level. *J. Am. Chem. Soc.* **2010**, *132*, 356–363. [CrossRef] [PubMed]

54. Patel, S.; Pant, K.K. Activity and stability enhancement of copper–alumina catalysts using cerium and zinc promoters for the selective production of hydrogen via steam reforming of methanol. *J. Power Sources* **2006**, *159*, 139–143. [CrossRef]

55. Men, Y.; Gnaser, H.; Zapf, R.; Hessel, V.; Ziegler, C. Parallel screening of $Cu/CeO_2/\gamma\text{-}Al_2O_3$ catalysts for steam reforming of methanol in a 10-channel micro-structured reactor. *Catal. Commun.* **2004**, *5*, 671–675. [CrossRef]

56. Liu, T.; Han, X.; Li, T.; Li, S.; Yin, C.; Wang, Y. Methanol steam reforming using Ce and La modified low-Cu catalysts for on-board hydrogen production. *Mol. Catal.* **2025**, *570*, 114663. [CrossRef]

57. Agrell, J.; Birgersson, H.; Boutonnet, M.; Melián-Cabrera, I.; Navarro, R.M.; Fierro, J.L.G. Production of hydrogen from methanol over Cu/ZnO catalysts promoted by ZrO_2 and Al_2O_3. *J. Catal.* **2003**, *219*, 389–403. [CrossRef]

58. Jeong, H.; Kim, K.I.; Kim, T.H.; Ko, C.H.; Park, H.C.; Song, I.K. Hydrogen production by steam reforming of methanol in a micro-channel reactor coated with $Cu/ZnO/ZrO_2/Al_2O_3$ catalyst. *J. Power Sources* **2006**, *159*, 1296–1299. [CrossRef]

59. Wu, G.-S.; Mao, D.-S.; Lu, G.-Z.; Cao, Y.; Fan, K.-N. The role of the promoters in Cu Based Catalysts for methanol steam reforming. *Catal. Lett.* **2009**, *130*, 177–184. [CrossRef]

60. Lindström, B.; Pettersson, L.J. Steam reforming of methanol over copper-based monoliths: The effects of zirconia doping. *J. Power Sources* **2002**, *106*, 264–273. [CrossRef]

61. Tong, W.; West, A.; Cheung, K.; Yu, K.-M.; Tsang, S.C.E. Dramatic effects of gallium promotion on methanol steam reforming Cu–ZnO catalyst for hydrogen production: Formation of 5 Å Copper Clusters from $Cu\text{-}ZnGaO_x$. *ACS Catal.* **2013**, *3*, 1231–1244. [CrossRef]

62. Liu, Y.-J.; Kang, H.-F.; Hou, X.-N.; Qing, S.-J.; Zhang, L.; Gao, Z.-X.; Xiang, H.-W. Sustained release catalysis: Dynamic copper releasing from stoichiometric spinel $CuAl_2O_4$ during methanol steam reforming. *Appl. Catal. B Environ.* **2023**, *323*, 122043. [CrossRef]

63. Wen, H.; Liu, Y.K.; Kong, A.; Zhou, C.; Wang, Z.; Guo, K.; Liu, D. Hydrolysis precipitation method for the preparation of $Cu\text{-}ZnO@Al_2O_3$ catalyst in methanol steam reforming. *ChemistrySelect* **2024**, *9*, e202304824. [CrossRef]

64. Ahmadi, F.; Haghighi, M.; Ajamein, H. Sonochemically coprecipitation synthesis of $CuO/ZnO/ZrO_2/Al_2O_3$ nanocatalyst for fuel cell grade hydrogen production via steam methanol reforming. *J. Mol. Catal. A Chem.* **2016**, *421*, 196–208. [CrossRef]

65. Zhang, X.-R.; Wang, L.-C.; Yao, C.-Z.; Cao, Y.; Dai, W.-L.; He, H.-Y.; Fan, K.-N. A highly efficient $Cu/ZnO/Al_2O_3$ catalyst via gel-coprecipitation of oxalate precursors for low-temperature steam reforming of methanol. *Catal. Lett.* **2005**, *102*, 183–190. [CrossRef]

66. Shishido, T.; Yamamoto, Y.; Morioka, H.; Takaki, K.; Takehira, K. Active Cu/ZnO and $Cu/ZnO/Al_2O_3$ catalysts prepared by homogeneous precipitation method in steam reforming of methanol. *Appl. Catal. A Gen.* **2004**, *263*, 249–253. [CrossRef]

67. Jin, S.; Zhang, Z.; Li, D.; Wang, Y.; Lian, C.; Zhu, M. Alcohol-induced strong metal-support interactions in a supported copper/ZnO catalyst. *Angew. Chem. Int. Ed.* **2023**, *62*, e202301563. [CrossRef]

68. Frank, B.; Jentoft, F.C.; Soerijanto, H.; Kröhnert, J.; Schlögl, R.; Schomäcker, R. Steam reforming of methanol over copper-containing catalysts: Influence of support material on microkinetics. *J. Catal.* **2007**, *246*, 177–192. [CrossRef]

69. Takahashi, K.; Kobayashi, H.; Takezawa, N. On the difference in reaction pathways of steam reforming of methanol over copper-silica and platinum-silica catalysts. *Chem. Lett.* **1985**, *14*, 759–762. [CrossRef]

70. Takezawa, N.; Iwasa, N. Steam reforming and dehydrogenation of methanol: Difference in the catalytic functions of copper and group VIII metals. *Catal. Today* **1997**, *36*, 45–56. [CrossRef]

71. Lin, S.; Johnson, R.S.; Smith, G.K.; Xie, D.; Guo, H. Pathways for methanol steam reforming involving adsorbed formaldehyde and hydroxyl intermediates on Cu(111): Density functional theory studies. *Phys. Chem. Chem. Phys.* **2011**, *13*, 9622–9631. [CrossRef] [PubMed]

72. Lin, S.; Xie, D.; Guo, H. Methyl formate pathway in methanol steam reforming on copper: Density functional calculations. *ACS Catal.* **2011**, *1*, 1263–1271. [CrossRef]

73. Takahachi, K.; Takezawa, N.; Kobayashi, H. The mechanism of steam reforming of methanol over a copper-silica catalyst. *Appl. Catal.* **1982**, *2*, 363–366. [CrossRef]

74. Takahashi, K.; Takezawa, N.; Kobayashi, H. Mechanism of formation of methyl formate from formaldehyde over copper catalysts. *Chem. Lett.* **1983**, *12*, 1061–1064. [CrossRef]

75. Hu, S.; Li, W.-X. Sabatier principle of metal-support interaction for design of ultrastable metal nanocatalysts. *Science* **2021**, *374*, 1360–1365. [CrossRef] [PubMed]

76. Dai, Y.; Lu, P.; Cao, Z.; Campbell, C.T.; Xia, Y. The physical chemistry and materials science behind sinter-resistant catalysts. *Chem. Soc. Rev.* **2018**, *47*, 4314–4331. [CrossRef] [PubMed]

77. Prieto, G.; Tüysüz, H.; Duyckaerts, N.; Knossalla, J.; Wang, G.-H.; Schüth, F. Hollow Nano- and Microstructures as Catalysts. *Chem. Rev.* **2016**, *116*, 14056–14119. [CrossRef] [PubMed]

78. Siriruang, C.; Charojrochkul, S.; Toochinda, P. Hydrogen production from methanol-steam reforming at low temperature over Cu–Zn/ZrO$_2$-doped Al$_2$O$_3$. *Monatsh. Chem.-Chem. Mon.* **2016**, *147*, 1143–1151. [CrossRef]

79. Thattarathody, R.; Artoul, M.; Digilov, R.M.; Sheintuch, M. Pressure, Diffusion, and S/M ratio effects in methanol steam reforming kinetics. *Ind. Eng. Chem. Res.* **2018**, *57*, 3175–3186. [CrossRef]

80. Słowik, G.; Rotko, M.; Ryczkowski, J.; Greluk, M. Hydrogen production from methanol steam reforming over Fe-modified Cu/CeO$_2$ catalysts. *Molecules* **2024**, *29*, 3963. [CrossRef]

81. Tonelli, F.; Gorriz, O.; Tarditi, A.; Cornaglia, L.; Arrúa, L.; Cristina Abello, M. Activity and stability of a CuO/CeO$_2$ catalyst for methanol steam reforming. *Int. J. Hydrogen Energy* **2015**, *40*, 13379–13387. [CrossRef]

82. Wang, H.; Fang, Z.; Wang, Y.; Meng, K.; Sun, S. The study of strong metal-support interaction enhanced PdZn alloy nanocatalysts for methanol steam reforming. *J. Alloys Compd.* **2024**, *986*, 174006. [CrossRef]

83. Liu, L.; Lin, Y.; Hu, Y.; Lin, Z.; Lin, S.; Du, M.; Zhang, L.; Zhang, X.-h.; Lin, J.; Zhang, Z.; et al. ZnAl$_2$O$_4$ spinel-supported PdZn$_\beta$ catalyst with parts per million Pd for methanol steam reforming. *ACS Catal.* **2022**, *12*, 2714–2721. [CrossRef]

84. Tang, J.; Qi, Y.; Zhang, R.; Cai, F. Promoting effect of Zn on Pd/MoC catalyst for the hydrogen production from methanol steam reforming. *Catal. Lett.* **2024**, *154*, 4768–4779. [CrossRef]

85. Wang, C.; Ouyang, M.; Li, M.; Lee, S.; Flytzani-Stephanopoulos, M. Low-coordinated Pd catalysts supported on Zn$_1$Zr$_1$O$_x$ composite oxides for selective methanol steam reforming. *Appl. Catal. A Gen.* **2019**, *580*, 81–92. [CrossRef]

86. Pérez-Hernández, R.; Avendaño, A.D.; Rubio, E.; Rodríguez-Lugo, V. Hydrogen production by methanol steam reforming over Pd/ZrO$_2$-TiO$_2$ catalysts. *Top. Catal.* **2011**, *54*, 572–578. [CrossRef]

87. Zhang, J.; Men, Y.; Wang, Y.; Liao, L.; Liu, S.; Wang, J.; An, W. Morphology effect of Pd/In$_2$O$_3$/CeO$_2$ catalysts on methanol steam reforming for hydrogen production. *Int. J. Hydrogen Energy* **2024**, *51*, 1185–1199. [CrossRef]

88. Mierczynski, P.; Vasilev, K.; Mierczynska, A.; Maniukiewicz, W.; Maniecki, T.P. Highly selective Pd–Cu/ZnAl$_2$O$_4$ catalyst for hydrogen production. *Appl. Catal. A Gen.* **2014**, *479*, 26–34. [CrossRef]

89. Azenha, C.; Lagarteira, T.; Mateos-Pedrero, C.; Mendes, A. Production of hydrogen from methanol steam reforming using CuPd/ZrO$_2$ catalysts—Influence of the catalytic surface on methanol conversion and CO selectivity. *Int. J. Hydrogen Energy* **2021**, *46*, 17490–17499. [CrossRef]

90. Liu, X.; Men, Y.; Wang, J.; He, R.; Wang, Y. Remarkable support effect on the reactivity of Pt/In$_2$O$_3$/MO$_x$ catalysts for methanol steam reforming. *J. Power Sources* **2017**, *364*, 341–350. [CrossRef]

91. Shanmugam, V.; Neuberg, S.; Zapf, R.; Pennemann, H.; Kolb, G. Hydrogen production over highly active Pt based catalyst coatings by steam reforming of methanol: Effect of support and co-support. *Int. J. Hydrogen Energy* **2020**, *45*, 1658–1670. [CrossRef]

92. Ma, Y.; Guan, G.; Shi, C.; Zhu, A.; Hao, X.; Wang, Z.; Kusakabe, K.; Abudula, A. Low-temperature steam reforming of methanol to produce hydrogen over various metal-doped molybdenum carbide catalysts. *Int. J. Hydrogen Energy* **2014**, *39*, 258–266. [CrossRef]

93. Cai, F.; Ibrahim, J.J.; Fu, Y.; Kong, W.; Zhang, J.; Sun, Y. Low-temperature hydrogen production from methanol steam reforming on Zn-modified Pt/MoC catalysts. *Appl. Catal. B Environ.* **2020**, *264*, 118500. [CrossRef]

94. Gu, X.-K.; Qiao, B.; Huang, C.; Ding, W.; Sun, K.; Zhan, E.; Zhang, T.; Liu, J.; Li, W.-X. Supported single Pt$_1$/Au$_1$ atoms for methanol steam reforming. *ACS Catal.* **2014**, *4*, 3886–3890. [CrossRef]

95. Shao, Z.; Zhang, S.; Liu, X.; Luo, H.; Huang, C.; Zhou, H.; Wu, Z.; Li, J.; Wang, H.; Sun, Y. Maximizing the synergistic effect between Pt^0 and Pt$^{\delta+}$ in a confined Pt-based catalyst for durable hydrogen production. *Appl. Catal. B Environ.* **2022**, *316*, 121669. [CrossRef]

96. Tahay, P.; Khani, Y.; Jabari, M.; Bahadoran, F.; Safari, N. Highly porous monolith/TiO$_2$ supported Cu, Cu-Ni, Ru, and Pt catalysts in methanol steam reforming process for H$_2$ generation. *Appl. Catal. A Gen.* **2018**, *554*, 44–53. [CrossRef]

97. Aouad, S.; Gennequin, C.; Mrad, M.; Tidahy, H.L.; Estephane, J.; Aboukaïs, A.; Abi-Aad, E. Steam reforming of methanol over ruthenium impregnated ceria, alumina and ceria-alumina catalysts. *Int. J. Energy Res.* **2016**, *40*, 1287–1292. [CrossRef]

98. Chen, L.; Qi, Z.; Peng, X.; Chen, J.-L.; Pao, C.-W.; Zhang, X.; Dun, C.; Young, M.; Prendergast, D.; Urban, J.J.; et al. Insights into the mechanism of methanol steam reforming tandem reaction over CeO$_2$ supported single-site catalysts. *J. Am. Chem. Soc.* **2021**, *143*, 12074–12081. [CrossRef]

99. Iwasa, N.; Kudo, S.; Takahashi, H.; Masuda, S.; Takezawa, N. Highly selective supported Pd catalysts for steam reforming of methanol. *Catal. Lett.* **1993**, *19*, 211–216. [CrossRef]

100. Iwasa, N.; Masuda, S.; Ogawa, N.; Takezawa, N. Steam reforming of methanol over Pd/ZnO: Effect of the formation of PdZn alloys upon the reaction. *Appl. Catal. A Gen.* **1995**, *125*, 145–157. [CrossRef]

101. Iwasa, N.; Ogawa, N.; Masuda, S.; Takezawa, N. Selective PdZn alloy formation in the reduction of Pd/ZnO catalysts. *Bull. Chem. Soc. Jpn.* **1998**, *71*, 1451–1455. [CrossRef]

102. Iwasa, N.; Mayanagi, T.; Nomura, W.; Arai, M.; Takezawa, N. Effect of Zn addition to supported Pd catalysts in the steam reforming of methanol. *Appl. Catal. A Gen.* **2003**, *248*, 153–160. [CrossRef]

103. Iwasa, N.; Nomura, W.; Mayanagi, T.; Fujita, S.; Arai, M.; Takezawa, N. Hydrogen production by steam reforming of methanol. *J. Chem. Eng. Jpn.* **2004**, *37*, 286–293. [CrossRef]

104. Halevi, B.; Peterson, E.J.; DeLaRiva, A.; Jeroro, E.; Lebarbier, V.M.; Wang, Y.; Vohs, J.M.; Kiefer, B.; Kunkes, E.; Havecker, M.; et al. Aerosol-Derived Bimetallic Alloy Powders: Bridging the Gap. *J. Phys. Chem. C* **2010**, *114*, 17181–17190. [CrossRef]

105. Peterson, E.J.; Halevi, B.; Kiefer, B.; Spilde, M.N.; Datye, A.K.; Peterson, J.; Daemen, L.; Llobet, A.; Nakotte, H. Aerosol synthesis and Rietveld analysis of tetragonal (β1) PdZn. *J. Alloys Compd.* **2011**, *509*, 1463–1470. [CrossRef]

106. Halevi, B.; Peterson, E.J.; Roy, A.; DeLariva, A.; Jeroro, E.; Gao, F.; Wang, Y.; Vohs, J.M.; Kiefer, B.; Kunkes, E.; et al. Catalytic reactivity of face centered cubic PdZn$_\alpha$ for the steam reforming of methanol. *J. Catal.* **2012**, *291*, 44–54. [CrossRef]

107. Zhang, H.; Sun, J.; Dagle, V.L.; Halevi, B.; Datye, A.K.; Wang, Y. Influence of ZnO facets on Pd/ZnO catalysts for methanol steam reforming. *ACS Catal.* **2014**, *4*, 2379–2386. [CrossRef]

108. Föttinger, K.; van Bokhoven, J.A.; Nachtegaal, M.; Rupprechter, G. Dynamic structure of a working methanol steam reforming catalyst: In situ Quick-EXAFS on Pd/ZnO nanoparticles. *J. Phys. Chem. Lett.* **2011**, *2*, 428–433. [CrossRef]

109. Wang, Y.; Zhang, J.; Xu, H. Interaction between Pd and ZnO during reduction of Pd/ZnO catalyst for steam reforming of methanol to hydrogen. *Chin. J. Catal.* **2006**, *27*, 217–222. [CrossRef]

110. Wang, Y.; Zhang, J.; Xu, H.; Bai, X. Reduction of Pd/ZnO catalyst and its catalytic activity for steam reforming of methanol. *Chin. J. Catal.* **2007**, *28*, 234–238. [CrossRef]

111. Karim, A.; Conant, T.; Datye, A. The role of PdZn alloy formation and particle size on the selectivity for steam reforming of methanol. *J. Catal.* **2006**, *243*, 420–427. [CrossRef]

112. Dagle, R.A.; Chin, Y.-H.; Wang, Y. The effects of PdZn crystallite size on methanol steam reforming. *Top. Catal.* **2007**, *46*, 358–362. [CrossRef]

113. Lim, K.H.; Chen, Z.-X.; Neyman, K.M.; Rösch, N. Comparative theoretical study of formaldehyde decomposition on PdZn, Cu, and Pd surfaces. *J. Phys. Chem. B* **2006**, *110*, 14890–14897. [CrossRef]

114. Guo, Y.; Wang, M.; Zhu, Q.; Xiao, D.; Ma, D. Ensemble effect for single-atom, small cluster and nanoparticle catalysts. *Nat. Catal.* **2022**, *5*, 766–776. [CrossRef]

115. Haghofer, A.; Föttinger, K.; Girgsdies, F.; Teschner, D.; Knop-Gericke, A.; Schlögl, R.; Rupprechter, G. In situ study of the formation and stability of supported Pd$_2$Ga methanol steam reforming catalysts. *J. Catal.* **2012**, *286*, 13–21. [CrossRef]

116. Azenha, C.S.R.; Mateos-Pedrero, C.; Queirós, S.; Concepción, P.; Mendes, A. Innovative ZrO$_2$-supported CuPd catalysts for the selective production of hydrogen from methanol steam reforming. *Appl. Catal. B Environ.* **2017**, *203*, 400–407. [CrossRef]

117. Ruano, D.; Pabón, B.M.; Azenha, C.; Mateos-Pedrero, C.; Mendes, A.; Pérez-Dieste, V.; Concepción, P. Influence of the ZrO$_2$ crystalline phases on the nature of active sites in PdCu/ZrO$_2$ catalysts for the methanol steam reforming reaction—An in situ spectroscopic study. *Catalysts* **2020**, *10*, 1005. [CrossRef]

118. Matsumura, Y. Enhancement in activity of Pd-Zn catalyst for methanol steam reforming by coprecipitation on zirconia support. *Appl. Catal. A Gen.* **2013**, *468*, 350–358. [CrossRef]

119. Barrios, C.E.; Bosco, M.V.; Baltanás, M.A.; Bonivardi, A.L. Hydrogen production by methanol steam reforming: Catalytic performance of supported-Pd on zinc-cerium oxides' nanocomposites. *Appl. Catal. B Environ.* **2015**, *179*, 262–275. [CrossRef]

120. Penner, S.; Lorenz, H.; Jochum, W.; Stöger-Pollach, M.; Wang, D.; Rameshan, C.; Klötzer, B. Pd/Ga$_2$O$_3$ methanol steam reforming catalysts: Part I. morphology, composition and structural aspects. *Appl. Catal. A Gen.* **2009**, *358*, 193–202. [CrossRef]

121. Lorenz, H.; Penner, S.; Jochum, W.; Rameshan, C.; Klötzer, B. Pd/Ga$_2$O$_3$ methanol steam reforming catalysts: Part II. catalytic selectivity. *Appl. Catal. A Gen.* **2009**, *358*, 203–210. [CrossRef]

122. Föttinger, K.; Rupprechter, G. In situ spectroscopy of complex surface reactions on supported Pd-Zn, Pd-Ga, and Pd(Pt)-Cu nanoparticles. *Acc. Chem. Res.* **2014**, *47*, 3071–3079. [CrossRef]

123. Rameshan, C.; Lorenz, H.; Armbrüster, M.; Kasatkin, I.; Klötzer, B.; Götsch, T.; Ploner, K.; Penner, S. Impregnated and Co-precipitated Pd-Ga$_2$O$_3$, Pd-In$_2$O$_3$ and Pd-Ga$_2$O$_3$-In$_2$O$_3$ catalysts: Influence of the microstructure on the CO$_2$ selectivity in methanol steam reforming. *Catal. Lett.* **2018**, *148*, 3062–3071. [CrossRef]

124. Iwasa, N.; Mayanagi, T.; Ogawa, N.; Sakata, K.; Takezawa, N. New catalytic functions of Pd-Zn, Pd-Ga, Pd-In, Pt-Zn, Pt-Ga and Pt-In alloys in the conversions of methanol. *Catal. Lett.* **1998**, *54*, 119–123. [CrossRef]

125. Davis, J.L.; Barteau, M.A. Spectroscopic identification of alkoxide, aldehyde, and acyl intermediates in alcohol decomposition on Pd(111). *Surf. Sci.* **1990**, *235*, 235–248. [CrossRef]

126. Iwasa, N.; Takezawa, N. New supported Pd and Pt alloy catalysts for steam reforming and dehydrogenation of methanol. *Top. Catal.* **2003**, *22*, 215–224. [CrossRef]

127. Conant, T.; Karim, A.M.; Lebarbier, V.; Wang, Y.; Girgsdies, F.; Schlogl, R.; Datye, A. Stability of bimetallic Pd–Zn catalysts for the steam reforming of methanol. *J. Catal.* **2008**, *257*, 64–70. [CrossRef]

128. Chen, Z.-X.; Neyman, K.M.; Lim, K.H.; Rösch, N. CH$_3$O decomposition on PdZn(111), Pd(111), and Cu(111). A Theoretical Study. *Langmuir* **2004**, *20*, 8068–8077. [CrossRef]

129. Chen, Z.-X.; Lim, K.H.; Neyman, K.M.; Rösch, N. Effect of steps on the decomposition of CH$_3$O at PdZn alloy surfaces. *J. Phys. Chem. B* **2005**, *109*, 4568–4574. [CrossRef]

130. Gu, X.-K.; Li, W.-X. First-principles study on the origin of the different selectivities for methanol steam reforming on Cu(111) and Pd(111). *J. Phys. Chem. C* **2010**, *114*, 21539–21547. [CrossRef]

131. Jeroro, E.; Vohs, J.M. Zn modification of the reactivity of Pd(111) toward methanol and formaldehyde. *J. Am. Chem. Soc.* **2008**, *130*, 10199–10207. [CrossRef] [PubMed]

132. Suwa, Y.; Ito, S.-i.; Kameoka, S.; Tomishige, K.; Kunimori, K. Comparative study between Zn–Pd/C and Pd/ZnO catalysts for steam reforming of methanol. *Appl. Catal. A Gen.* **2004**, *267*, 9–16. [CrossRef]

133. Liu, S.; Takahashi, K.; Fuchigami, K.; Uematsu, K. Hydrogen production by oxidative methanol reforming on Pd/ZnO: Catalyst deactivation. *Appl. Catal. A Gen.* **2006**, *299*, 58–65. [CrossRef]

134. Penner, S.; Jenewein, B.; Gabasch, H.; Klötzer, B.; Wang, D.; Knop-gericke, A.; Schlögl, R.; Hayek, K. Growth and structural stability of well-ordered PdZn alloy nanoparticles. *J. Catal.* **2006**, *241*, 14–19. [CrossRef]

135. Liu, D.; Men, Y.; Wang, J.; Kolb, G.; Liu, X.; Wang, Y.; Sun, Q. Highly active and durable $Pt/In_2O_3/Al_2O_3$ catalysts in methanol steam reforming. *Int. J. Hydrogen Energy* **2016**, *41*, 21990–21999. [CrossRef]

136. Liao, L.; Men, Y.; Wang, Y.; Xu, S.; Wu, S.; Wang, J.; Yan, Z.; Miao, X. Unravelling the morphology effect of Pt/In_2O_3 catalysts for highly efficient hydrogen production by methanol steam reforming. *Fuel* **2024**, *372*, 132221. [CrossRef]

137. Deng, Y.; Ge, Y.; Xu, M.; Yu, Q.; Xiao, D.; Yao, S.; Ma, D. Molybdenum carbide: Controlling the geometric and electronic structure of noble metals for the activation of O-H and C-H bonds. *Acc. Chem. Res.* **2019**, *52*, 3372–3383. [CrossRef]

138. Lin, L.; Zhou, W.; Gao, R.; Yao, S.; Zhang, X.; Xu, W.; Zheng, S.; Jiang, Z.; Yu, Q.; Li, Y.-W.; et al. Low-temperature hydrogen production from water and methanol using Pt/α-MoC catalysts. *Nature* **2017**, *544*, 80–83. [CrossRef]

139. Wang, X.; Li, D.; Gao, Z.; Guo, Y.; Zhang, H.; Ma, D. The nature of interfacial catalysis over $Pt/NiAl_2O_4$ for hydrogen production from methanol reforming reaction. *J. Am. Chem. Soc.* **2023**, *145*, 905–918. [CrossRef] [PubMed]

140. Jin, J.-Y.; Wang, Y.-F.; Zhang, R.-X.; Gao, Z.-H.; Huang, W.; Liu, L.; Zuo, Z.-J. Theoretical insight into hydrogen production from methanol steam reforming on Pt(111). *Mol. Catal.* **2022**, *532*, 112745. [CrossRef]

141. Hu, P.; Diskin-Posner, Y.; Ben-David, Y.; Milstein, D. Reusable homogeneous catalytic system for hydrogen production from methanol and water. *ACS Catal.* **2014**, *4*, 2649–2652. [CrossRef]

142. Nielsen, M.; Alberico, E.; Baumann, W.; Drexler, H.-J.; Junge, H.; Gladiali, S.; Beller, M. Low-temperature aqueous-phase methanol dehydrogenation to hydrogen and carbon dioxide. *Nature* **2013**, *495*, 85–89. [CrossRef]

143. Rodriguez-Lugo, R.E.; Trincado, M.; Vogt, M.; Tewes, F.; Santiso-Quinones, G.; Grützmacher, H. A homogeneous transition metal complex for clean hydrogen production from methanol-water mixtures. *Nat. Chem.* **2013**, *5*, 342–347. [CrossRef]

144. Schwarz, C.H.; Agapova, A.; Junge, H.; Haumann, M. Immobilization of a selective Ru-pincer complex for low temperature methanol reforming–Material and process improvements. *Catal. Today* **2020**, *342*, 178–186. [CrossRef]

145. Lytkina, A.A.; Mironova, E.Y.; Orekhova, N.V.; Ermilova, M.M.; Yaroslavtsev, A.B. Ru-Containing catalysts for methanol and ethanol steam reforming in conventional and membrane reactors. *Inorg. Mater.* **2019**, *55*, 547–555. [CrossRef]

146. Lytkina, A.A.; Orekhova, N.V.; Ermilova, M.M.; Petriev, I.S.; Baryshev, M.G.; Yaroslavtsev, A.B. Ru-Rh based catalysts for hydrogen production via methanol steam reforming in conventional and membrane reactors. *Int. J. Hydrogen Energy* **2019**, *44*, 13310–13322. [CrossRef]

147. Lytkina, A.A.; Orekhova, N.V.; Ermilova, M.M.; Belenov, S.V.; Guterman, V.E.; Efimov, M.N.; Yaroslavtsev, A.B. Bimetallic carbon nanocatalysts for methanol steam reforming in conventional and membrane reactors. *Catal. Today* **2016**, *268*, 60–67. [CrossRef]

 *catalysts*

Article

Effect of Calcination-Induced Oxidation on the Photocatalytic H_2 Production Performance of Cubic Cu_2O/CuO Composite Photocatalysts

Chi-Jung Chang *, Chun-Wen Kang and Arul Pundi

Department of Chemical Engineering, Feng Chia University, 100, Wenhwa Road, Seatwen, Taichung 40724, Taiwan; wainss9407278@gmail.com (C.-W.K.); apundi@fcu.edu.tw (A.P.)
* Correspondence: changcj@fcu.edu.tw

Abstract: This study explores the H_2 production performance of CuO/Cu_2O with different morphology (nanocubes) synthesized by different methods using different sacrificial reagent (lactic acid), compared with the other three reported CuO/Cu_2O photocatalysts used for H_2 production. A cubic Cu_2O photocatalyst was prepared using a hydrothermal method. It was then calcined at a certain temperature to form a cubic Cu_2O/CuO composite photocatalyst. XRD, TEM, and XPS spectra confirmed the successful synthesis of cubic Cu_2O/CuO composite photocatalysts by calcination-induced oxidation at a certain temperature. As the calcination temperature increases, the crystal phase of the photocatalyst changes from Cu_2O to Cu_2O/CuO and then to CuO. The effects of calcination-induced oxidation on morphology, light absorption, the separation of photoexcited carriers, and the H_2 production activity of photocatalysts were studied. EPR spectra were monitored to analyze the oxygen vacancies in different samples. Mott–Schottky and Tauc plots were utilized to establish the band structure of the composite photocatalyst. Cu_2O/CuO is a type II photocatalyst with a heterogeneous structure that helps to improve electron–hole separation efficiency. The H_2 production efficiency of Cu_2O/CuO composite photocatalyst reaches 11,888 μmol h^{-1}g^{-1}, 1.6 times that of Cu_2O. The formation of the Cu_2O/CuO heterojunction leads to enhanced light absorption, charge separation, and hydrogen production activity.

Keywords: photocatalyst; hydrogen production; Cu_2O; CuO; nanocube

Citation: Chang, C.-J.; Kang, C.-W.; Pundi, A. Effect of Calcination-Induced Oxidation on the Photocatalytic H_2 Production Performance of Cubic Cu_2O/CuO Composite Photocatalysts. *Catalysts* **2024**, *14*, 926. https://doi.org/10.3390/catal14120926

Academic Editors: Georgios Bampos, Paraskevi Panagiotopoulou and Eleni A. Kyriakidou

Received: 8 November 2024
Revised: 10 December 2024
Accepted: 13 December 2024
Published: 16 December 2024

1. Introduction

Hydrogen is the only carbon-free fuel with the highest energy content, considered the ideal renewable energy source. When hydrogen is used as a fuel in engines, the final product is only water, causing no environmental pollution [1]. Photocatalytic H_2 production has attracted increasing attention [2]. The activity of a photocatalyst can be improved by tuning morphology, loading the cocatalyst, doping, and adjusting the band structures of composite photocatalysts [3,4]. Usually, the composite photocatalysts are prepared through the one-step [5,6] or two-step [7,8] synthetic routes. In addition, the composite photocatalysts can also be obtained by thermal phase transition [9–12]. The changes in the morphology and crystal phase of photocatalysts during calcination were reported.

Cu_2O is a low-cost, non-toxic p-type semiconductor [13]. Various morphologies of Cu_2O photocatalysts have been successfully synthesized, such as nanowires [14], nanospheres [15], flakes [16], layered structures [17], cubes, and polyhedrons. Due to its narrow-band structure, Cu_2O can absorb most visible light. The position of its conduction and valence bands straddles the redox potential of water, allowing it to efficiently split water to produce hydrogen under light irradiation [18]. Yong et al. [19] used a hydrothermal method, selecting ascorbic acid as both a reducing agent and an etching agent. By etching the {1 0 0} crystal plane of Cu_2O with ascorbic acid, they successfully synthesized concave cubic Cu_2O with a concave {1 0 0} surface. They also synthesized octapod

Cu$_2$O by adjusting the water-to-n-butanol ratio in the precursor, improving hydrogen production efficiency through morphology modification. However, the poor charge separation ability of Cu$_2$O limits its photocatalytic applications [20]. Therefore, combining Cu$_2$O with other semiconductor materials is a feasible way to improve the separation rate of photoexcited electron–hole pairs. [21–23] The electron—hole recombination is a major challenge for many photocatalysts. The recombination problem can be reduced by combining appropriate semiconductor materials to form composite photocatalysts, thereby enhancing their photocatalytic activity. Composite photocatalysts can form type-I, type-II, or Z-scheme structures.

CuO is a p-type semiconductor that features high conductivity and good stability. It can absorb light in the near-infrared region [24]. CuO-based composites have been used for photocatalysis-related applications. [25–28] CuO/Cu$_2$O composites have been applied in various applications. Na Wang et al. [29] synthesized Cu$_2$O/CuO flower-like structures using a hydrothermal method as sensing materials for NO$_2$ sensors. The Cu$_2$O/CuO sensor has a detection limit as low as 5 ppb, with excellent selectivity, repeatability, and response speed. Animesh et al. [30] deposited Cu$_2$O/CuO on nickel foam through electrodeposition and annealing, applying it to the electrocatalytic reduction of CO$_2$ to methanol. They found that at an annealing temperature of 300 °C, the electrocatalyst exhibited the highest current density and better Faradaic efficiency. Bayat et al. [31] synthesized CuO/Cu$_2$O heterojunctions using copper powder as a precursor via chemical thermal oxidation, applying them to degrade methyl orange and methylene blue. They found that a higher Cu$_2$O content in the CuO/Cu$_2$O heterojunction led to better degradation efficiency. According to our literature survey, most of the Cu$_2$O/CuO composites were used as photocathodes for the photoelectrochemical H$_2$ production application [32–37]. Only a few papers reported the photocatalytic H$_2$ production performance of the Cu$_2$O/CuO composites with different morphology synthesized by various methods. The sacrificial solution was also different. Luévano-Hipólito et al. [38] reported the photocatalytic H$_2$ production by Cu$_2$O/CuO nanoneedles using seawater as the sacrificial solution. The Cu$_2$O/CuO photocatalysts were synthesized by the thermal oxidation of the copper mesh. Dubale et al. [39] studied the photocatalytic H$_2$ production performance of a carbon-supported chrysanthemum-like Cu$_2$O/CuO nanocomposite using an aqueous methanol solution as the sacrificial solution. Cu$_2$O/CuO nanocomposites were prepared from Cu-based metal–organic frameworks (MOF) by heat treatment in the air and Ar environment. Yoo et al. [40] reported the photocatalytic H$_2$ production by ZnO/Cu$_2$O-CuO photocatalysts using an aqueous Na$_2$S solution as the sacrificial solution. The Cu$_2$O-CuO oxide-decorated ZnO heterostructures were prepared by the thermal oxidation of the ZnO/CuS nanostructures.

In this study, cubic Cu$_2$O nanomaterials were synthesized using the hydrothermal method, followed by calcination to synthesize cubic Cu$_2$O, Cu$_2$O/CuO, and CuO photocatalysts. Their photocatalytic H$_2$ production performance using an aqueous lactic acid solution as the sacrificial solution was investigated. Compared with the reported CuO/Cu$_2$O photocatalysts, this study explored the effect of morphology (nanocubes, synthesized by different methods) and sacrificial solution on the photocatalytic H$_2$ production activity of CuO/Cu$_2$O photocatalysts.

2. Results and Discussion

2.1. Formation of Nanocomposites

2.1.1. Phase Change

XRD was applied to analyze the crystal phase of the photocatalysts. Figure 1 presents the XRD spectra of different photocatalysts calcined at various temperatures. The peaks of uncalcined Cu$_n$O photocatalyst at 29.6°, 36.4°, 42.3°, 52.5°, 61.3°, 73.5°, and 77.3° are assigned to the (110), (111), (200), (211), (220), (311), and (222) planes of Cu$_2$O (JCPDS no. 78-2076), respectively (Figure 1 (i)). After being calcined at 400 °C and 500 °C, the Cu$_n$O-400 and Cu$_n$O-500 photocatalysts show peaks at 2θ positions of 32.5°, 35.5°, 38.7°, 48.7°, 53.4°, 58.2°, 61.5°, 66.2°, 68.1°, 72.3°, and 75.2°, corresponding to the (110), (11-1),

(111), (20-2), (020), (202), (11-3), (31-1), (220), (311), and (22-2) planes of CuO (JCPDS no. 48-1548), respectively (Figure 1 (v) and (vi)). However, the Cu_nO-275 (Figure 1 (ii)), Cu_nO-300 (Figure 1 (iii)), and Cu_nO-350 (Figure 1 (iv)) photocatalysts exhibited peaks corresponding to both CuO and Cu_2O. When the calcination temperature increased from 275 °C to 350 °C, the peak intensity of CuO increased, indicating that the content of CuO in the photocatalyst increased with temperature. At 400 °C, the signal of Cu_2O disappeared, indicating that the photocatalyst had transformed to CuO. As the calcination temperature increased, the Cu_2O photocatalyst gradually oxidized into CuO, with a coexistence of both phases observed between 275 °C and 350 °C. The Cu_2O primarily grows on the (111) plane, located at $2\theta = 36.6°$. The CuO also grows mainly on the (111) plane, observed at 38.9° (2θ). These results indicate that Cu_2O mainly grows on the (111) plane, and CuO follows this pattern. Similar results were reported in the literature. Erdogan et al. [41] prepared Cu_2O, CuO, and Cu_2O/CuO through the heat treatment method and investigated the photoelectrochemical sensing performance of the materials. In their study, Cu_2O dominantly grew in the (111) plane, and CuO was found to follow suit.

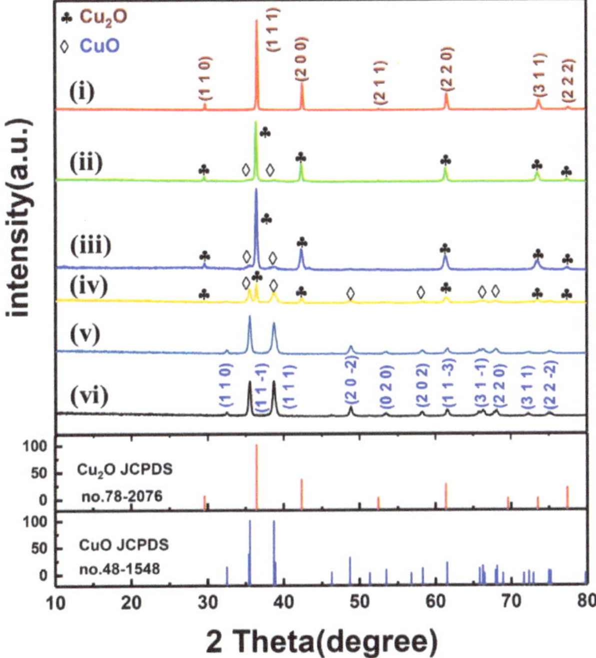

Figure 1. XRD spectra of (i) Cu_nO, (ii) Cu_nO-275, (iii) Cu_nO-300, (iv) Cu_nO-350, (v) Cu_nO-400, and (vi) Cu_nO-500 samples.

A series of photocatalysts were prepared using the calcination-induced oxidation method. As the calcination temperature increases, the crystal phase of the photocatalyst changes from Cu_2O to Cu_2O/CuO and then to CuO. XRD spectra of samples calcined at certain temperatures, including Cu_nO-275 and Cu_nO-350 samples, were analyzed to identify the phase change window. As the calcination temperatures range from 275 °C to 350 °C, both phases coexisted in the photocatalyst. Cu_2O/CuO composite calcined at 300 °C was used for the other tests based on the cost considerations of calcination. The Cu_nO, Cu_nO-300, and Cu_nO-500 photocatalysts are Cu_2O and Cu_2O/CuO composites and CuO photocatalysts, respectively. They were selected for the following properties and performance evaluation experiments.

The crystal sizes (D), dislocation density (δ), and strain (ε) of the annealed samples were evaluated using the XRD data. Table 1 presents the crystal sizes, dislocation density, and lattice strain of Cu_nO, Cu_nO-275, Cu_nO-300, Cu_nO-350, Cu_nO-400, and Cu_nO-500 samples. The crystal size was calculated using Scherrer's equation (1) [42].

$$D = K\lambda/(\beta \mathrm{Cos}\ \theta) \tag{1}$$

where "D" is the crystallite size, "K" is the Scherer constant, "λ" is the wavelength of the x-ray sources, "β" is the FWHM, and "θ" is the peak position.

Table 1. The crystal sizes, dislocation density, and lattice strain of Cu_nO, Cu_nO-275, Cu_nO-300, Cu_nO-350, Cu_nO-400, and Cu_nO-500 samples.

Sample	Peak Position ($2\theta°$)	FWHM ($\beta°$)	Size D (nm)	Dislocation Density ($\delta \times 10^{-3}$ nm^{-2})	Lattice Strain ($\varepsilon \times 10^{-3}$)
	36.5539	0.1726	48.472	0.425	2.280
	42.4475	0.2011	42.378	0.556	2.259
Cu_nO	61.5511	0.272	33.994	0.865	1.992
	73.7224	0.3364	29.516	1.147	1.957
	Average =		38.590	0.748	2.122
	36.4021	0.243	34.414	0.844	3.224
	42.293	0.272	31.315	1.019	3.068
Cu_nO-275 °C	61.3589	0.3375	27.369	1.334	2.482
	73.5085	0.4365	22.715	1.937	2.550
	Average =		28.953	1.284	2.831
	36.4483	0.2608	32.069	0.972	3.456
	42.2918	0.4364	19.518	2.624	4.922
Cu_nO-300 °C	61.3786	0.4873	18.957	2.782	3.582
	73.5256	0.7402	13.397	5.571	4.323
	Average =		20.985	2.987	4.071
	36.3986	0.284	29.445	1.153	3.769
	42.2767	0.2647	32.177	0.965	2.987
Cu_nO-350 °C	61.3876	0.5249	17.600	3.228	3.858
	73.5427	0.5461	18.160	3.032	3.188
	Average =		24.346	2.094	3.450
	35.5934	0.2813	29.660	1.136	3.823
	38.7622	0.4388	19.192	2.714	5.442
Cu_nO-400 °C	61.6702	0.4155	22.267	2.016	3.037
	72.4893	0.3033	32.477	0.948	1.805
	Average =		25.899	1.704	3.527
	35.5616	0.3157	26.426	1.431	4.295
	48.8574	0.3726	23.417	1.823	3.579
Cu_nO-500 °C	61.631	0.3446	26.843	1.387	2.520
	72.3625	0.4447	22.132	2.041	2.653
	Average =		24.704	1.671	3.262

The dislocation density and crystal strain were obtained using Equations (2) and (3), respectively [42].

$$\delta = 1/D^2 \tag{2}$$

$$\varepsilon = \beta/(4\tan\ \theta) \tag{3}$$

The Cu_nO-300 photocatalyst shows the highest dislocation values and strain among the six samples, as shown in Table 1.

2.1.2. Morphology Change

The surface morphology changes in the photocatalyst before and after calcination at different temperatures were observed by a field emission scanning electron microscope (FESEM). Figure 2a shows that the uncalcined Cu_2O photocatalyst has a cubic nanostructure with a smooth surface. Figure 2b shows that the Cu_nO-300 photocatalyst calcined at 300 °C also has a cubic nanostructure of varying sizes, with edge lengths ranging from 0.6 to 1.5 μm. The edges of the cubes remain straight, but the appearance changes. The surface of the photocatalyst becomes rough because of surface-loaded nanoparticles with diameters of about 20–100 nm. Figure 2c shows the FESEM image of the Cu_nO-500 photocatalyst. The Cu_nO-500 photocatalyst calcined at 500 °C still shows a cubic nanostructure, but it is composed of nanoparticles with diameters of 40–250 nm. The results indicate that high-temperature calcination causes changes in the surface morphology of the photocatalyst.

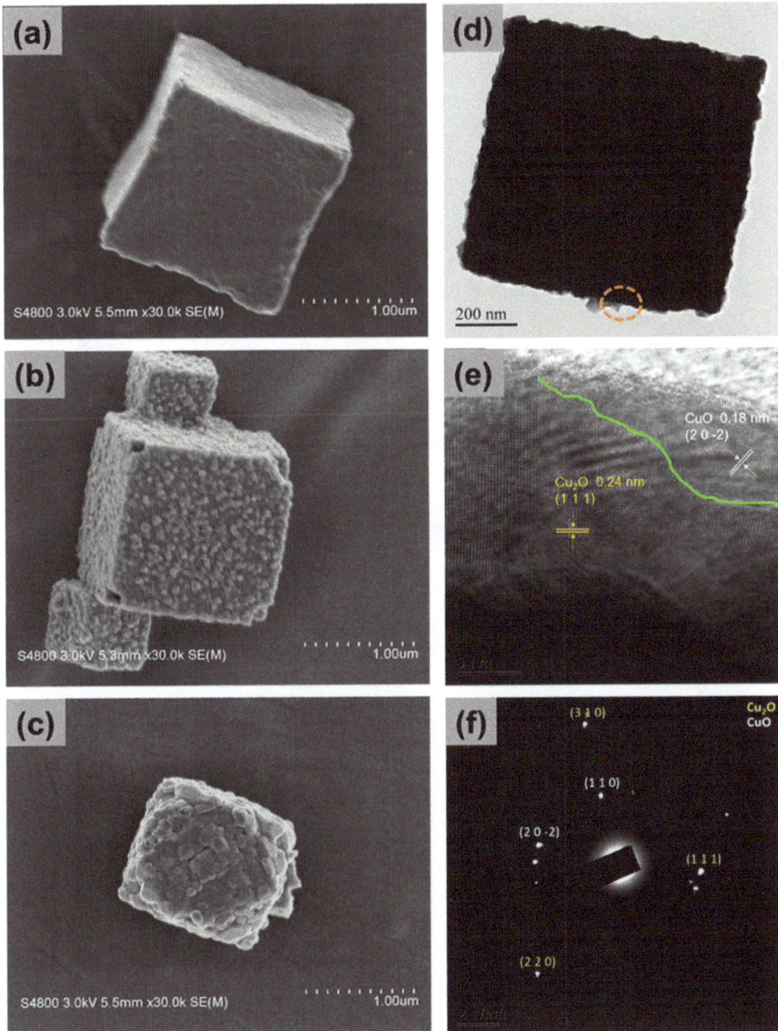

Figure 2. FESEM images of (**a**) Cu_nO, (**b**) Cu_nO-300, (**c**) Cu_nO-500; and the (**d**) TEM, (**e**) HR-TEM, and (**f**) SAED of Cu_nO-300 photocatalyst.

2.1.3. Crystal Phase of Cu_nO-300

A field emission transmission electron microscope (TEM) was applied to observe the lattice structure and morphology of the photocatalyst. Figure 2d shows the TEM image of the Cu_nO-300 sample calcined at 300 °C. The surface of the photocatalyst became rough, consistent with the FESEM result of the Cu_nO-300 photocatalyst shown in Figure 2b. The HRTEM image of Cu_nO-300 (Figure 2e) reveals regions with a lattice spacing of 0.24 nm, corresponding to the (111) crystal plane of Cu_2O (JCPDS no. 78-2076). A lattice spacing of 0.18 nm is assigned to the (202) crystal plane of CuO (JCPDS no. 48-1548). The selected area electron diffraction (SAED) pattern of Cu_nO-300 also exhibits diffraction spots related to the (111), (220), and (310) crystal planes of Cu_2O and to the (110) and (202) crystal planes of CuO (Figure 2f). These results confirm that the synthesized Cu_nO-300 is a photocatalyst with the coexisting Cu_2O and CuO phases.

2.1.4. EDX of Cu_nO-300 at Edges and Central Regions

The XRD and TEM analyses show that the Cu_nO-300 photocatalyst consists of Cu_2O and CuO phases. Figure 3 presents an analysis of the Cu and O contents at different positions (edges and central regions) of the Cu_nO-300 photocatalyst. The distribution of Cu_2O and CuO phases at various locations of the photocatalyst during calcination can be monitored by analyzing the Cu and O elemental composition. The elemental ratio at different positions (marked by red squares) shows that the Cu content increases towards the center of the cube. The Cu/O ratio at the surface of the cubic structure is close to 1:1 (Figure 3a). However, the Cu/O ratio approaches 2:1 in central regions (Figure 3b). The results indicate that the oxidation of Cu_2O starts from the outer layer during calcination. CuO forms gradually from the surface toward the center. Therefore, the Cu and O contents show gradient spatial distribution.

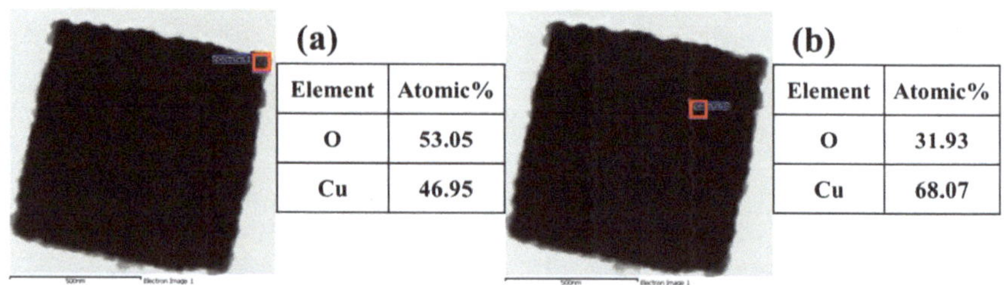

(a)

Element	Atomic%
O	53.05
Cu	46.95

(b)

Element	Atomic%
O	31.93
Cu	68.07

Figure 3. EDX of (**a**) edges and (**b**) center areas of the Cu_nO-300 photocatalyst.

2.2. Surface Chemistry

X-ray photoelectron spectroscopy (XPS) was used to analyze the surface elemental valence states and chemical composition of the photocatalyst. Figure 4 shows the XPS survey spectra of the (a) Cu_nO, (b) Cu_nO-300, and (c) Cu_nO-500 photocatalysts. Figure 4a displays the XPS Cu 2p spectrum of the Cu_nO photocatalyst. After deconvolution, the peaks at 932.2 and 952.5 eV are attributed to Cu^+ in Cu_2O at the Cu 2p3/2 and Cu 2p1/2 positions, respectively [43]. Figure 4b shows the XPS O 1s spectrum of the Cu_nO photocatalyst. After peak deconvolution, two peaks are observed at 529.6 and 531.7 eV, corresponding to the lattice oxygen (bonded Cu-O) and the adsorbed oxygen (oxygen species physically adsorbed on the photocatalyst surface), respectively. The results in Figure 4a,b indicate that the Cu_nO photocatalyst is Cu_2O. Figure 4c shows the XPS Cu 2p spectrum of the Cu_nO-300 photocatalyst after calcination at 300 °C. After the deconvolution of XPS data, the peaks at 932.2 and 952.5 eV correspond to Cu^+ in Cu_2O at the Cu 2p3/2 and Cu 2p1/2 positions, respectively. In addition, two smaller peaks are observed at 934.2 eV and 953.8 eV, which can be assigned to Cu^{2+} in CuO at the Cu 2p3/2 and Cu 2p1/2 positions, respectively [44]. Satellite peaks at 941.3, 943.7, and 962.2 eV are also observed, indicating the existence of Cu^{2+} in CuO [45]. Figure 4d illustrates the XPS O 1s spectrum of the Cu_nO-300 photocatalyst, with two peaks at 529.6 and 531.1 eV related to lattice oxygen and adsorbed oxygen, respectively. The results in Figure 4c,d indicate that the Cu_nO-300 photocatalyst contains both Cu_2O and CuO phases. Figure 4e presents the XPS Cu 2p spectrum of the Cu_nO-500 photocatalyst after calcination at 500 °C. It shows signals for Cu^{2+} in CuO at the Cu 2p3/2 and Cu 2p1/2 positions. The satellite peaks characteristic of Cu^{2+} in CuO are also observed. Figure 4f shows the XPS O 1s spectrum of the Cu_nO-500 photocatalyst, with two peaks related to lattice oxygen and adsorbed oxygen, respectively.

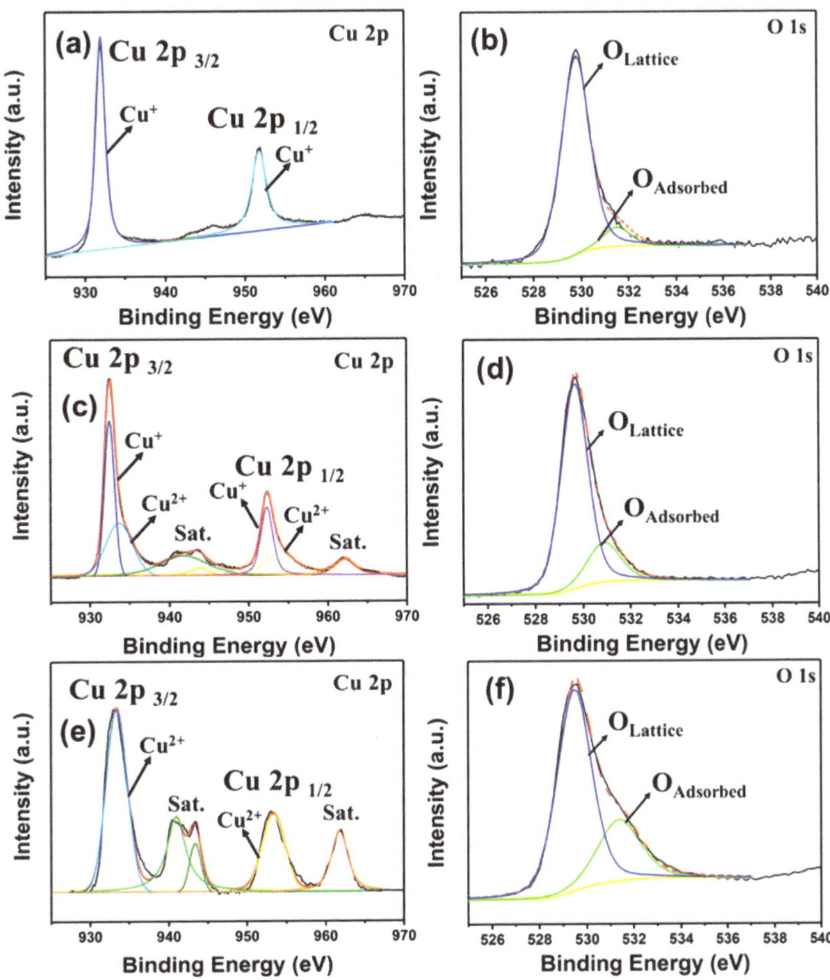

Figure 4. XPS (**a**) Cu2p and (**b**) O1s spectra of Cu_nO; XPS (**c**) Cu2p and (**d**) O1s spectra of Cu_nO-30; and XPS (**e**) Cu2p and (**f**) O1s spectra of Cu_nO-500.

2.3. Light Absorption and Band Gaps

Figure 5a shows the UV–visible diffuse reflectance spectra (DRS) of the uncalcined photocatalyst and samples calcined at different temperatures. The uncalcined Cu_nO photocatalyst appears as a red powder. The DRS reveal that the main absorption range of the Cu_nO photocatalyst lies between 300 and 650 nm, with the absorption edge at 650 nm. The calcined Cu_nO-300 and Cu_nO-500 photocatalysts possess enhanced UV and visible light absorption capabilities. The absorption edge of the Cu_nO-500 photocatalyst extends to 900 nm. Compared with the uncalcined Cu_nO, both Cu_nO-300 and Cu_nO-500 photocatalysts exhibit much higher absorptions at the 500–1200 nm range. Cu_nO-300 shows the highest light absorption among the three samples. Figure 5b presents the Tauc plots of Cu_nO, Cu_nO-300, and Cu_nO-500 photocatalysts. The Tauc plots of Cu_nO and Cu_nO-500 photocatalysts allow the estimation of the band gaps of Cu_2O and CuO, which are approximately 1.92 eV and 1.33 eV, respectively.

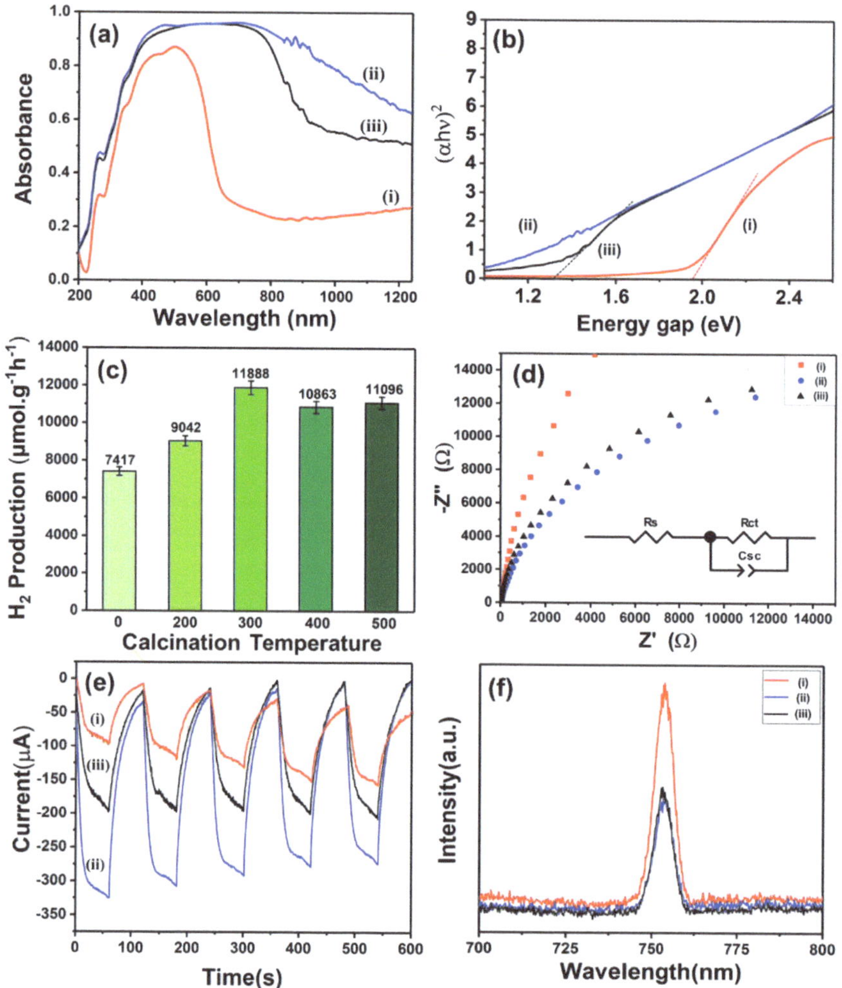

Figure 5. (**a**) DRS (**b**) Tauc plots of (i) Cu_nO, (ii) Cu_nO-300, (iii) Cu_nO-500; (**c**) photocatalytic hydrogen production activity of uncalcined Cu_nO, and Cu_nO-200, Cu_nO-300, Cu_nO-400, Cu_nO -500 photocatalysts calcined at different temperatures (200, 300, 400, and 500 °C). (**d**) EIS and (**e**) photocurrent–time curves under the light on/off illumination by the Xe lamp. (**f**) PL spectra of (i) Cu_nO, (ii) Cu_nO-300, and (iii) Cu_nO-500 samples.

2.4. Photocatalytic H_2 Production Performance

Figure 5c shows the hydrogen production performance of uncalcined Cu_nO and photocatalysts calcined at different temperatures (200, 300, 400, and 500 °C), labeled as Cu_nO, Cu_nO-200, Cu_nO-300, Cu_nO-400, and Cu_nO-500 using lactic acid as the sacrificial agent. The Cu_nO, Cu_nO-200, Cu_nO-300, Cu_nO-400, and Cu_nO-500 photocatalysts are Cu_2O, Cu_2O, the Cu_2O/CuO composite, CuO, and CuO photocatalysts, respectively. It is reported that if the Cu_2O/CuO heterostructures are used, a sacrificial agent should be added to prevent the decrease in the photocatalytic activity during the water-splitting reaction. Otherwise, the activity of Cu_2O/CuO photocatalysts will decrease because of the photoreduction of CuO to Cu_2O, competing with the photogenerated electrons [38]. The gas chromatogram for H_2 generation of the Cu_nO-300 photocatalyst is shown in Figure S1. The first peak observed during the gas chromatography at a retention time of 1.25 min with an abundance of 350,000 is assigned to H_2, revealing the successful H_2 production through photocatalysis under light irradiation. The effects of calcination temperatures on the photocatalytic H_2 production performance of the photocatalysts were studied. Cu_nO and Cu_nO-200 are Cu_2O. The Cu_nO-200 photocatalyst has higher hydrogen production activity (9042 μmol g^{-1}h^{-1}) than uncalcined Cu_nO (7417 μmol g^{-1}h^{-1}). The Cu_nO-300 photocatalyst (Cu_2O/CuO composite) achieves the highest hydrogen production activity (11,888 μmol g^{-1}h^{-1}). Further increasing the calcination temperature leads to the formation of CuO and a slight decrease in the activity. Cu_nO-400 and Cu_nO-500 become CuO. The Cu_nO-500 photocatalyst has slightly higher hydrogen production activity (10,863 μmol g^{-1}h^{-1}) than Cu_nO-400 (11,096 μmol g^{-1}h^{-1}). The results reveal that the hydrogen production activity follows the trend: Cu_2O/CuO composite > CuO > Cu_2O. In addition, comparing CuO (Cu_nO-400 and Cu_nO-500) and Cu_2O (Cu_nO and Cu_nO-200), photocatalysts calcined at higher temperatures (Cu_nO-200 and Cu_nO-500) exhibit higher activity than their analogs (Cu_nO and Cu_nO-400). In this study, the Cu_nO-300 photocatalyst with higher dislocation and microstrains (Table 1) exhibits better photocatalytic H_2 production performance.

Table 2 presents a comparison table of the photocatalyst morphology, photocatalyst crystal structure, sacrificial agent (reaction solution), light source, and photocatalytic H_2 production activity of Cu_2O-based and Cu_2O-CuO-based composite photocatalysts. Cu_2O-based composite photocatalysts, such as Cu_2O/rGO [46] and Zn-Doped Cu_2O/Pt [47], show good H_2 production activity (4530 and 3817 μmol·h^{-1}g^{-1}). Only a few papers reported the photocatalytic H_2 production performance of the Cu_2O/CuO composites with different morphology synthesized by various methods. The sacrificial solution was also different. Luévano-Hipólito et al. [38] reported that the photocatalytic H_2 production activity via Cu_2O/CuO nanoneedles under LED irradiation was 230 μmol·h^{-1}g^{-1} using seawater as the sacrificial solution. The photocatalysts were synthesized by the thermal oxidation of copper meshes. Dubale et al. [39] reported that the photocatalytic H_2 production performance of a carbon-supported chrysanthemum-like Cu_2O/CuO nanocomposite under Xe lamp exposure using an aqueous methanol solution as the sacrificial solution reached 26,700 μmol·h^{-1}g^{-1}. Yoo et al. [40] reported that the photocatalytic H_2 production activity through ZnO/Cu_2O-CuO photocatalysts using an aqueous Na_2S solution as the sacrificial solution was 1092.5 μmol·h^{-1}g^{-1}. These photocatalysts were prepared by different methods, including the thermal oxidation of the copper mesh [38], the heat treatment of a Cu-based metal–organic frameworks (MOF) [39], and the thermal oxidation of the ZnO/CuS nanostructures [40]. The crystal structures of the photocatalysts consist of cubic Cu_2O and monoclinic CuO. In this study, Cu_2O/CuO photocatalysts were synthesized using a hydrothermal method and the calcination process. The photocatalytic H_2 production activity by cubic Cu_2O/CuO photocatalysts under Xe lamp irradiation reached 11,888 μmol·h^{-1}g^{-1} using an aqueous lactic acid solution as the sacrificial solution.

Table 2. The photocatalyst morphology, photocatalyst crystal structure, sacrificial agent (reaction solution), light source, and photocatalytic H_2 production activity of Cu_2O-based and Cu_2O-CuO-based composite photocatalysts.

Samples	Morphology	Crystal Structure	Reaction Solution	Light Source	Activity (μmol·h^{-1}g^{-1})	Ref.
Cu_2O-CuO	Nanoneedle		Seawater	LED lamp	230	[38]
C@Cu_2O/CuO	Chrysanthemum-like	cubic Cu_2O, monoclinic CuO	Methanol/water	350 W Xe lamp (UV cutoff filter)	26,700	[39]
ZnO/Cu_2O-CuO	Pseudo-spherical ZnO with Cu_2O-CuO nanoparticle	cubic Cu_2O, monoclinic CuO	Na_2S/water	Standard solar irradiation	1092.5	[40]
Cu_2O/rGO	Nanoparticle	cubic phase Cu_2O	TEOA + water	300 W Xe lamp	4530	[46]
Zn-Doped Cu_2O/Pt	Hollow microcube	cubic phase Cu_2O	Glucose + water	300 W Xe lamp	3817	[47]
Cu_2O/CuO	Nanocube	cubic Cu_2O, monoclinic CuO	Lactic acid + water	300 W Xe lamp	11,888	This study

2.5. Electrochemical Impedance Spectroscopy (EIS)

The Nyquist plot of EIS is used to assess the charge transfer resistance of different photocatalysts. Figure 5d shows the EIS spectra for Cu_nO, Cu_nO-300, and Cu_nO-500 photocatalysts. Among these photocatalysts, the Cu_nO photocatalyst exhibits a larger Nyquist plot radius, indicating higher charge transfer resistance. Calcination helps reduce the Nyquist plot radius of the photocatalyst. The Cu_nO-300 photocatalyst has the smallest Nyquist plot radius, indicating the lowest charge transfer resistance. The Nyquist plot radius of the Cu_nO-500 photocatalyst is similar to that of the Cu_nO-300 photocatalyst. Calcination reduces the charge transfer resistance of the photocatalyst. The fitted equivalent circuit is demonstrated in the inset of Figure 5d. The R_{CT}, C_{SC}, and τ represent the series resistance, charge-transfer resistance at the electrolyte/photocatalyst interface, space-charge capacitance, and the lifetime of electrons, respectively. The lifetime of electrons at the depletion layer was obtained using Equation (4) [48]. The R_{CT}, CSC, and τ of Cu_nO, Cu_nO-300, and Cu_nO-500 samples are listed in Table 3.

$$\tau = R_{CT} \times C_{SC} \tag{4}$$

R_{ct} is in the decreasing order of $Cu_nO > Cu_nO$-500 $> Cu_nO$-300. τ decreases in the order of $Cu_nO > Cu_nO$-500 $> Cu_nO$-300. Cu_nO-300 has the lowest interface resistance. Cu_nO-300 exhibits the shortest lifetime, indicating that the electrons of the Cu_nO-300 photocatalyst spend less time in the depletion layer, resulting in the lowest recombination probability and the highest photocatalytic performance among the three samples. The recombination rate of photoexcited electron–hole pairs was reduced after forming the CuO/Cu_2O heterojunction.

Table 3. The charge-transfer resistance, space-charge capacitance, and electron lifetime of Cu_nO, Cu_nO-300, and Cu_nO-500 photocatalysts.

Photocatalysts	R_{ct} (Ω)	C_{sc} (F)	τ_n (s)
Cu_nO	1623	2.7×10^{-5}	4.4×10^{-2}
Cu_nO-300	1204	7.1×10^{-9}	8.6×10^{-6}
Cu_nO-500	1391	5.6×10^{-7}	5.6×10^{-7}

2.6. Photocurrent Response

The photocurrent response is used to evaluate the effectiveness of photogenerated electron–hole pair separation and the stability of repeated photocatalyst operations in different photocatalyst samples under instant illumination [49,50]. The photocatalysts are expected to display fast and high photocurrent response. [51] It was reported that enhanced light harvesting and close contact between the components may lead to enhanced photocurrent response [52–54]. Figure 5e shows the photocurrent response of (a) Cu_nO, (b) Cu_nO-300, and (c) Cu_nO-500 photocatalysts. The Cu_nO-300 photocatalyst has the highest negative photocurrent, while the uncalcined Cu_nO photocatalyst has the weakest negative photocurrent. The photocurrent intensity of the Cu_nO-300 photocatalyst calcined at 300 °C is about three times that of the uncalcined Cu_nO photocatalyst, indicating better photogenerated electron–hole pair separation and charge transfer efficiency. The negative photocurrent is observed because Cu_2O is a p-type semiconductor [55].

2.7. Photoluminescence (PL)

Figure 5f shows the fluorescence spectra of the uncalcined Cu_nO and the calcined photocatalysts Cu_nO-300 and Cu_nO-500. The fluorescence signals were observed at 753 nm for all samples when excited with a 500 nm wavelength light source at room temperature. The uncalcined Cu_nO exhibited the strongest fluorescence signal, indicating a higher electron–hole recombination rate. In contrast, the fluorescence signal intensity of the calcined Cu_nO-300 and Cu_nO-500 photocatalysts significantly decreased, suggesting that the CuO formed on the surface of Cu_2O after calcination effectively reduced the electron–

hole recombination rate. The fluorescence signal intensity of Cu_nO-300 was lower than that of Cu_nO-500. The formation of interfacial contact between CuO and Cu_2O resulted in improved charge separation and photocatalytic activity. Similar results were observed by Bayat et al. [31] in their study on enhancing the photocatalytic activity of Cu_2O/CuO by controlling Cu_2O content for the degradation of methyl orange and methylene blue.

2.8. Mott–Schottky Plot

Figure 6a,b shows the Mott–Schottky plots for Cu_2O and CuO, respectively. Using the Hg/HgO (1 M NaOH) electrode, the flat band potentials of Cu_2O and CuO obtained from the Mott–Schottky plots are 0.85 V and 1.08 V, respectively. The potential of Hg/HgO (1 M NaOH) is +0.14 V higher than that of the standard hydrogen electrode (SHE). Therefore, the flat band potentials of Cu_2O and CuO are 0.99 V and 1.22 V, respectively. The valence band (VB) potential of p-type semiconductors is 0.1–0.3 eV more positive than the flat band potential (E_{fb}) [56]. Hence, the valence bands of Cu_2O and CuO are calculated to be 1.09 V and 1.32 V, respectively. The band gaps of Cu_2O and CuO are 1.92 eV and 1.33 eV, respectively. Based on the relationship between the band gap, conduction band and valence band, the conduction bands of Cu_2O and CuO are −0.83 V and −0.01 V, respectively.

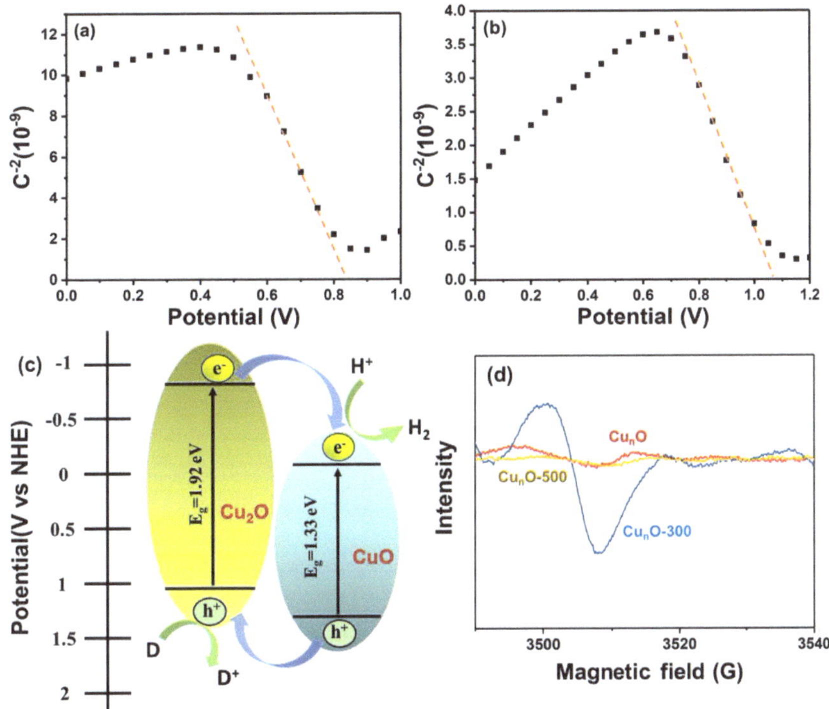

Figure 6. Mott–Schottky plots for (**a**) Cu_2O and (**b**) CuO; (**c**) band structure of Cu_2O and CuO; and (**d**) EPR spectra of Cu_nO, Cu_nO-300, and Cu_nO-500 samples.

2.9. Band Structure and Proposed Photocatalysis Mechanism

Using the flat band potentials (E_{fb}) from the Mott–Schottky plots and the band gaps (Eg) from the Tauc plots, the valence band (VB) and conduction band (CB) positions of Cu_2O and CuO are determined, and a band structure diagram is constructed. Figure 6c shows the band structure of Cu_2O and CuO and a schematic diagram of the hydrogen production mechanism. The VB of CuO is lower than that of Cu_2O, while the CB of Cu_2O is higher than that of CuO. The composite photocatalyst formed by combining Cu_2O

and CuO is a typical type II heterojunction. Upon illumination, electrons in Cu_2O are transferred from the CB to the CB of CuO. Meanwhile, holes move from the VB of CuO to the VB of Cu_2O. This complete separation of electrons and holes can effectively enhance the photocatalytic activity of the Cu_2O/CuO composite. The reaction mechanism can be inferred as follows: Cu_2O generates electron–hole pairs under light excitation (Equation (5)). CuO also generates electron–hole pairs under light irradiation (Equation (6)). The electrons in Cu_2O jump to the conduction band of CuO (Equation (7)). The CuO, which gains electrons, undergoes a reduction reaction with water to produce hydrogen (Equation (8)). The holes in CuO jump to the conduction band of Cu_2O. At this point, lactic acid acts as a hole scavenger and is oxidized to pyruvic acid (Equation (9)) [57].

$$Cu_2O + hv \rightarrow Cu_2O\left(e^- + h^+\right) \tag{5}$$

$$CuO + hv \rightarrow CuO\left(e^- + h^+\right) \tag{6}$$

$$Cu_2O\left(e^-\right) + CuO \rightarrow CuO\left(e^-\right) \tag{7}$$

$$2e^-\left(CuO\right) + H_2O \rightarrow H_2 + 2OH^- \tag{8}$$

$$2h^+\left(Cu_2O\right) + C_3H_6O_3 + 2OH^- \rightarrow C_3H_4O_3 + H_2O \tag{9}$$

2.10. Vacancy Analysis (Electron Paramagnetic Resonance, EPR)

EPR is a sensitive spectroscopic tool for investigating paramagnetic species. Figure 6d presents the EPR spectra of Cu_nO, Cu_nO-300, and Cu_nO-500. Cu_nO and Cu_nO-500 photocatalysts did not exhibit an EPR signal. However, the Cu_nO-300 photocatalyst had a significant EPR signal at g = 2.003 (Figure 6d), which can be attributed to oxygen vacancy [58,59]. Wei et al. found that there was a lattice mismatch at the interface of the heterojunction, leading to the formation of a lot of oxygen vacancies [60]. In summary, EPR spectra proved the formation of oxygen vacancy in the Cu_nO-300 (Cu_2O/CuO composite) sample.

XRD spectra (Figure 1) indicated that the Cu_nO, Cu_nO-300, and Cu_nO-500 photocatalysts are Cu_2O, the Cu_2O/CuO composite, and CuO photocatalysts, respectively. Cu_nO-300 shows the highest light absorption among the three samples (Figure 5a). The Cu_2O/CuO composite (Cu_nO-300 sample) exhibits higher photocatalytic activity than Cu_2O (Cu_nO) and fully oxidized CuO (Cu_nO-500 samples) (Figure 5c). Hence, the increased photocatalytic H_2 production activity of the Cu_nO-300 sample is due to the enhanced light absorption and the formation of CuO/Cu_2O heterojunctions that can improve the separation of photogenerated carriers (photocurrent response (Figure 5e) and PL spectra (Figure 5f)).

3. Materials and Methods

3.1. Chemicals

The copper sulfate pentahydrate (SHOWA), sodium citrate (J.T.Baker, Center Valley, PA, USA), sodium carbonate (Thermo Fisher Scientific, Ward Hill, MA, USA), polyvinylpyrrolidone (PVP, MW = 40,000, St. Louis, MO, USA) (Aldrich), glucose (Sigma, St. Louis, MO, USA), and lactic acid (TEDIA, Fairfield, CA, USA) were used as received.

3.2. Preparation of Cubic Cu_2O Photocatalyst

A total of 0.7 g of PVP was dissolved in 50 mL of deionized water. Then, 3 mL of 0.68 M $CuSO_4.5H_2O$ solution, 3 mL of 0.74 M sodium citrate solution, and 3 mL of 1.2 M sodium carbonate solution were sequentially added. The mixture was slowly added into 3 mL of 1.4 M glucose solution while stirring thoroughly. The solution was heated in an oil bath at 80 °C and stirred continuously for 2 h. After cooling, the product was collected by centrifugation. The product was washed several times using deionized water and ethanol. Cu_nO powder was obtained after drying.

3.3. Preparation of Cubic Cu₂O/CuO Photocatalyst

The Cu_2O powder was placed in a crucible and heated in a high-temperature furnace at a rate of 4 °C per hour, holding at the calcination temperatures for 3 h. Then, the obtained powder is the final product. The calcinated samples are called Cu_nO-T. T means that the samples were calcined at T °C.

3.4. Material Characterization

An X-ray diffraction diffractometer (XRD) (D8 SSS, Bruker, Karlsruhe, Germany) was used to identify the crystal structures of photocatalysts. Scanning electron microscopy (SEM, HITACHI S-4800, Hitachi, Tokyo, Japan) was utilized to determine the morphologies of samples. A JEOL/JEM-2100F (JEOL, Tokyo, Japan) transmission electron microscope (TEM) was applied to obtain high-resolution TEM (HRTEM) images and element mappings. X-ray photoelectron spectroscopy (XPS) and ultraviolet photoelectron spectroscopy (UPS) were carried out using a ULVAC PHI/Versa Probe 4 instrument (ULVAC, Chigasaki, Japan). The photoluminescence (PL) was measured by a fluorescence spectrophotometer (FluoroMax-Plus, HORIBA, Miyanohigashi, Japan). A UV/VIS/NIR diffuse reflectance spectrophotometer (V-700, JASCO, Tokyo, Japan) was employed for the diffuse reflectance spectra (DRS) measurements.

3.5. Photocatalytic H₂ Production

The photocatalyst (10 mg) was dispersed in a closed quartz reactor with 150 mL of aqueous solution containing 30 wt% lactic acid as the sacrificial agent. The quartz reactor was placed in a water bath and irradiated by a 300 W xenon lamp at the light intensity of 0.24 kw/m^2 (measured by a power meter, gigahertz-optik SN60306) under continuous stirring. After being irradiated for three hours, the volume of collected gas was determined. The concentration of H_2 was determined using a Shimadzu GC-2014 (Shimadzu, Kyoto, Japan) gas chromatography (GC) machine with a molecular-sieve column (5 Å) and a thermal conductivity detector, using argon as the carrier gas. The experimental setup is similar to that reported in our previous study [61].

3.6. Photoelectrochemical Tests

Electrochemical impedance spectroscopy (EIS) and the photocurrent response of the photocatalysts were obtained by a CHI 920D workstation and a three-electrode system. The photocatalyst-coated FTO (fluorine-doped tin oxide) glass was used as the working electrode. The platinum plate and Hg/HgO/1 M NaOH were used as the counter electrode and the reference electrode, respectively. The solution containing Na_2SO_4 (0.1 M) was prepared as the electrolyte for the measurement of EIS. The photocurrent response was investigated using a three-electrode system with sacrificial solution (aqueous solution containing 30 wt% lactic acid) as the electrolyte under the Xenon lamp irradiation. The photocurrent was measured under the irradiation of chopped light for five on/off cycles. The duration of light irradiation and dark is 60 s for each on/off cycle.

4. Conclusions

A cubic Cu_2O/CuO composite photocatalyst was prepared by the calcination-induced oxidation method. As the calcination temperature increases, the crystal phase of the photocatalyst changes from Cu_2O to Cu_2O/CuO and then to CuO. Meanwhile, the morphology changed from cubes with smooth surfaces to cubes with surface-loaded nanoparticles, and then cubes consisting of nanoparticles. As the calcination temperatures range from 275 °C to 350 °C, both phases coexist in the photocatalyst. XRD, XPS, and TEM results confirmed the successful synthesis of cubic Cu_2O/CuO composite photocatalyst Cu_nO-300. At 400 °C, the signal for Cu_2O disappeared, indicating that the photocatalyst had completely transformed into CuO. Photoluminescence spectra and photocurrent analysis showed the effective separation of photogenerated electrons and holes for the Cu_2O/CuO composite photocatalyst. DRS and EIS results showed that the Cu_2O/CuO composite photocatalyst exhibited better light absorption and lower charge transfer resistance than the Cu_2O photocatalyst. The best hydrogen production efficiency was observed with the Cu_2O/CuO photocatalyst calcined at 300 °C, achieving 11,888 $\mu mol\ g^{-1}h^{-1}$, which is 1.6 times higher than that of Cu_2O (7413 $\mu mol\ g^{-1}h^{-1}$). EPR spectra reveal that the Cu_2O/CuO catalyst (Cu_nO-300) contains rich oxygen vacancies (OVs), while pristine Cu_2O and CuO catalysts (Cu_nO and Cu_nO-500) do not. The Mott–Schottky and Tauc plots reveal that the composite photocatalyst. Cu_2O/CuO is a type II photocatalyst. The staggered band alignment (type-II heterojunction), oxygen vacancies, and improved light absorption lead to improved electron–hole separation efficiency and the activity of Cu_2O/CuO photocatalyst.

Supplementary Materials: The following supporting information can be downloaded at: https://www.mdpi.com/article/10.3390/catal14120926/s1. Figure S1: The gas chromatogram for H_2 generation of the Cu_nO-300 photocatalyst.

Author Contributions: C.-J.C.: conceptualization, methodology, writing—original draft, supervision, writing—review and editing; C.-W.K.: data curation, writing—original draft, formal analysis; A.P.: data curation, formal analysis. All authors have read and agreed to the published version of the manuscript.

Funding: National Science and Technology Council: MOST 111-2221-E-035-002-MY3.

Data Availability Statement: Dataset available on request from the authors.

Acknowledgments: The authors thank the National Science and Technology Council for the financial support under the MOST 111-2221-E-035-002-MY3 contract.

Conflicts of Interest: The authors declare no conflicts of interest.

References

1. Nikolaidis, P.; Poullikkas, A. A comparative overview of hydrogen production processes. *Renew. Sustain. Energy Rev.* **2017**, *67*, 597–611. [CrossRef]
2. Chang, C.J.; Chao, P.Y.; Chen, J.K.; Pundi, A.; Yu, Y.H.; Chiang, C.L.; Lin, Y.G. Metal complex/ZnS-modified Ni foam as magnetically stirrable photocatalysts: Roles of redox mediators and carrier dynamics monitored by operando synchrotron X-ray spectroscopy. *ACS Appl. Mater. Interfaces* **2022**, *14*, 41870–41882. [CrossRef] [PubMed]
3. Shah, A.A.; Bhatti, M.A.; Tahira, A.; Chandio, A.D.; Channa, I.A.; Sahito, A.G.; Chalangar, E.; Willander, M.; Nur, O.; Ibupoto, Z.H. Facile synthesis of copper doped ZnO nanorods for the efficient photo degradation of methylene blue and methyl orange. *Ceram. Int.* **2020**, *46*, 9997–10005. [CrossRef]
4. Chang, C.J.; Lin, Y.G.; Chen, J.; Huang, C.Y.; Hsieh, S.C.; Wu, S.Y. Lonic liquid/surfactant-hydrothermal synthesis of dendritic PbS@CuS core-shell photocatalysts with improved photocatalytic performance. *Appl. Surf. Sci.* **2021**, *546*, 149106. [CrossRef]
5. Reddy, N.L.; Shankar, M.V.; Sharma, S.C.; Nagaraju, G. One-pot synthesis of Cu–TiO$_2$/CuO nanocomposite: Application to photocatalysis for enhanced H_2 production, dye degradation & detoxification of Cr (VI). *Int. J. Hydrogen Energy* **2020**, *45*, 7813–7828.
6. Wang, Y.Q.; Yang, C.; Gan, L.H. Preparation of direct Z-scheme Bi$_2$WO$_6$/TiO$_2$ heterojunction by one-step solvothermal method and enhancement mechanism of photocatalytic H_2 production. *Int. J. Hydrogen Energy* **2023**, *48*, 19372–19384. [CrossRef]
7. Wu, A.; Tian, C.; Jiao, Y.; Yan, Q.; Yang, G.; Fu, H. Sequential two-step hydrothermal growth of MoS$_2$/CdS core-shell heterojunctions for efficient visible light-driven photocatalytic H_2 evolution. *Appl. Catal. B Environ.* **2017**, *203*, 955–963. [CrossRef]
8. Guo, Q.; Huang, Y.; Xu, H.; Luo, D.; Huang, F.; Gu, L.; Wei, Y.; Zhao, H.; Fan, L.; Wu, J. The effects of solvent on photocatalytic properties of Bi$_2$WO$_6$/TiO$_2$ heterojunction under visible light irradiation. *Solid State Sci.* **2018**, *78*, 95–106. [CrossRef]
9. Chao, P.Y.; Chang, C.J.; Lin, K.S.; Wang, C.F. Synergistic effects of morphology control and calcination on the activity of flower-like Bi$_2$WO$_6$-Bi$_2$O$_3$ photocatalysts prepared by an ionic liquid-assisted solvothermal method. *J. Alloys Compd.* **2021**, *883*, 160920. [CrossRef]
10. Wu, W.; Sun, Y.; Zhou, H. In-situ construction of β-Bi$_2$O$_3$/Ag$_2$O photocatalyst from deactivated AgBiO$_3$ for tetracycline degradation under visible light. *Chem. Eng. J.* **2022**, *432*, 134316. [CrossRef]
11. Xie, Y.; Wu, H.; Luo, J.; Zhang, S.; Wang, L.; Wang, X.; Ma, Y.; Ning, P. Phase transition guided V$_2$O$_5$/β-Bi$_2$O$_3$ Z-scheme heterojunctions for efficient photocatalytic Hg^0 oxidation. *Sep. Purif. Technol.* **2024**, *330*, 125318. [CrossRef]
12. Chang, C.J.; Chen, Y.C.; Tsai, Z.T. Effect of calcination induced phase transition on the photocatalytic hydrogen production activity of BiOI and Bi$_5$O$_7$I based photocatalysts. *Int. J. Hydrogen Energy* **2022**, *47*, 40777–40786. [CrossRef]

13. Ibrahim, A.M.; Abdel-wahab, M.S.; Elfayoumi, M.A.K.; Tawfik, W.Z. Highly efficient sputtered Ni-doped Cu_2O photoelectrodes for solar hydrogen generation from water-splitting. *Int. J. Hydrogen Energy* **2023**, *48*, 1863–1876. [CrossRef]

14. Sondors, R.; Gavars, D.; Sarakovskis, A.; Kons, A.; Buks, K.; Erts, D.; Andzane, J. Facile fabrication of Cu_2O nanowire networks with large Seebeck coefficient for application in flexible thermoelectrics. *Mater. Sci. Semicond. Process.* **2023**, *159*, 107391. [CrossRef]

15. Fang, Y.; Chen, L.; Zhang, Y.; Chen, Y.; Liu, X. Construction of Cu_2O single crystal nanospheres coating with brilliant structural color and excellent antibacterial properties. *Opt. Mater.* **2023**, *138*, 113724. [CrossRef]

16. Ma, G.; Syzgantseva, O.A.; Huang, Y.; Stoian, D.; Zhang, J.; Yang, S.; Luo, W.; Jiang, M.; Li, S.; Chen, C.; et al. A hydrophobic Cu/Cu_2O sheet catalyst for selective electroreduction of CO to ethanol. *Nat. Commun.* **2023**, *14*, 501. [CrossRef] [PubMed]

17. Okoye, P.C.; Azi, S.O.; Qahtan, T.F.; Owolabi, T.O.; Saleh, T.A. Synthesis, properties, and applications of doped and undoped CuO and Cu_2O nanomaterials. *Mater. Today Chem.* **2023**, *30*, 101513. [CrossRef]

18. Wang, X.; Dong, H.; Hu, Z.; Qi, Z.; Li, L. Fabrication of a $Cu_2O/Au/TiO_2$ composite film for efficient photocatalytic hydrogen production from aqueous solution of methanol and glucose. *Mater. Sci. Eng. B* **2017**, *219*, 10–19. [CrossRef]

19. Zhang, Y.H.; Jiu, B.B.; Gong, F.L.; Chen, J.L.; Zhang, H.L. Morphology-controllable Cu_2O supercrystals: Facile synthesis, facet etching mechanism and comparative photocatalytic H_2 production. *J. Alloys Compd.* **2017**, *729*, 563–570. [CrossRef]

20. Zhang, Y.H.; Liu, M.M.; Chen, J.L.; Xie, K.F.; Fang, S.M. Dendritic branching Z-scheme Cu_2O/TiO_2 heterostructure photocatalysts for boosting H2 production. *J. Phys. Chem. Solids* **2021**, *152*, 109948. [CrossRef]

21. Chang, Y.C.; Chiao, Y.C.; Fun, Y.X. $Cu_2O/CuS/ZnS$ nanocomposite boosts blue LED-light-driven photocatalytic hydrogen evolution. *Catalysts* **2022**, *12*, 1035. [CrossRef]

22. Pu, S.; Wang, H.; Zhu, J.; Li, L.; Long, D.; Jian, Y.; Zeng, Y. Heterostructure $Cu_2O/(001)TiO_2$ effected on photocatalytic degradation of ammonia of livestock houses. *Catalysts* **2019**, *9*, 267. [CrossRef]

23. Marcolongo, D.M.S.; Nocito, F.; Ditaranto, N.; Comparelli, R.; Aresta, M.; Dibenedetto, A. Opto-electronic characterization of photocatalysts based on p, n-junction ternary and quaternary mixed oxides semiconductors (Cu_2O-In2O3 and Cu_2O-In_2O_3-TiO_2). *Catalysts* **2022**, *12*, 153. [CrossRef]

24. Zhang, Q.; Zhang, K.; Xu, D.; Yang, G.; Huang, H.; Nie, F.; Liu, C.; Yang, S. CuO nanostructures: Synthesis, characterization, growth mechanisms, fundamental properties, and applications. *Prog. Mater. Sci.* **2014**, *60*, 208–337. [CrossRef]

25. Li, Z.; Chen, H.; Liu, W. Full-spectrum photocatalytic activity of $ZnO/CuO/ZnFe_2O_4$ nanocomposite as a photofenton-like catalyst. *Catalysts* **2018**, *8*, 557. [CrossRef]

26. Chang, C.J.; Kao, Y.C.; Lin, K.S.; Chen, C.Y.; Kang, C.W.; Yang, T.H. Carbon fiber cloth@BiOBr/CuO as immobilized membrane-shaped photocatalysts with enhanced photocatalytic H_2 production activity. *J. Taiwan Inst. Chem. Eng.* **2023**, *149*, 104998. [CrossRef]

27. Maraj, M.; Raza, A.; Wang, X.; Chen, J.; Riaz, K.N.; Sun, W. Mo-doped CuO nanomaterial for photocatalytic degradation of water pollutants under visible light. *Catalysts* **2021**, *11*, 1198. [CrossRef]

28. Hanif, M.A.; Akter, J.; Kim, Y.S.; Kim, H.G.; Hahn, J.R.; Kwac, L.K. Highly efficient and sustainable ZnO/CuO/g-C3N4 photocatalyst for wastewater treatment under visible light through heterojunction development. *Catalysts* **2022**, *12*, 151. [CrossRef]

29. Wang, N.; Tao, W.; Gong, X.; Zhao, L.; Wang, T.; Zhao, L.; Liu, F.; Liu, X.; Sun, P.; Lu, G. Highly sensitive and selective NO_2 gas sensor fabricated from Cu_2O-CuO microflowers. *Sens. Actuators B Chem.* **2022**, *362*, 131803. [CrossRef]

30. Roy, A.; Jadhav, H.S.; Gil Seo, J. Cu_2O/CuO Electrocatalyst for Electrochemical Reduction of Carbon Dioxide to Methanol. *Electroanalysis* **2022**, *33*, 705–712. [CrossRef]

31. Bayat, F.; Sheibani, S. Enhancement of photocatalytic activity of CuO-Cu_2O heterostructures through the controlled content of Cu_2O. *Mater. Res. Bull.* **2022**, *145*, 111561. [CrossRef]

32. Yang, Y.; Xu, D.; Wu, Q.; Diao, P. Cu_2O/CuO bilayered composite as a high-efficiency photocathode for photoelectrochemical hydrogen evolution reaction. *Sci. Rep.* **2016**, *6*, 35158. [CrossRef]

33. Seo, Y.J.; Arunachalam, M.; Ahn, K.S.; Kang, S.H. Integrating heteromixtured Cu_2O/CuO photocathode interface through a hydrogen treatment for photoelectrochemical hydrogen evolution reaction. *Appl. Surf. Sci.* **2021**, *551*, 149375.

34. Alqahtani, M.S.; Mohamed, S.H.; Hadia, N.M.A.; Rabia, M.; Awad, M.A. Some characteristics of Cu/Cu_2O/CuO nanostructure heterojunctions and their applications in hydrogen generation from seawater: Effect of surface roughening. *Phys. Scr.* **2024**, *99*, 045939. [CrossRef]

35. Son, H.; Lee, J.H.; Uthirakumar, P.; Van Dao, D.; Soon, A.; Lee, I.H. Facile synthesis of flexible and scalable Cu/Cu_2O/CuO nanoleaves photoelectrodes with oxidation-induced self-initiated charge-transporting platform for photoelectrochemical water splitting enhancement. *J. Alloys Compd.* **2023**, *942*, 169094. [CrossRef]

36. Kim, S.; Kim, H.; Hong, S.K.; Kim, D. Fabrication and Photoelectrochemical Properties of a Cu_2O/CuO Heterojunction Photoelectrode for Hydrogen Production from Solar Water Splitting. *Korean J. Mater. Res.* **2016**, *26*, 604–610. [CrossRef]

37. Jeong, D.; Jo, W.; Jeong, J.; Kim, T.; Han, S.; Son, M.K.; Jung, H. Characterization of Cu_2O/CuO heterostructure photocathode by tailoring CuO thickness for photoelectrochemical water splitting. *RSC Adv.* **2022**, *12*, 2632–2640. [CrossRef] [PubMed]

38. Luévano-Hipólito, E.; Torres-Martínez, L.M.; Alejandro Avila-Lopez, M. Visible-light-driven CO_2 reduction and H_2 evolution boosted by 1D Cu_2O/CuO heterostructures. *J. Phys. Chem. Solids* **2022**, *170*, 110924. [CrossRef]

39. Dubale, A.A.; Ahmed, I.N.; Zhang, Y.J.; Yang, X.L.; Xie, M.H. A facile strategy for fabricating C@Cu_2O/CuO composite for efficient photochemical hydrogen production with high external quantum efficiency. *Appl. Surf. Sci.* **2020**, *534*, 147582. [CrossRef]

40. Yoo, H.; Kahng, S.; Kim, J.H. Z-scheme assisted ZnO/Cu$_2$O-CuO photocatalysts to increase photoactive electrons in hydrogen evolution by water splitting. *Sol. Energy Mater. Sol. Cells* **2020**, *204*, 110211. [CrossRef]

41. Balık, M.; Bulut, V.; Erdogan, I.Y. Optical, structural and phase transition properties of Cu$_2$O, CuO and Cu$_2$O/CuO: Their photoelectrochemical sensor applications. *Int. J. Hydrogen Energy* **2019**, *44*, 18744–18755. [CrossRef]

42. John, K.I.; Adenle, A.A.; Adeleye, A.T.; Onyia, I.P.; Amune-Matthews, C.; Omorogie, M.O. Unravelling the effect of crystal dislocation density and microstrain of titanium dioxide nanoparticles on tetracycline removal performance. *Chem. Phys. Lett.* **2021**, *776*, 138725. [CrossRef]

43. Dasineh Khiavi, N.; Katal, R.; Kholghi Eshkalak, S.; Masudy-Panah, S.; Ramakrishna, S.; Jiangyong, H. Visible Light Driven Heterojunction Photocatalyst of CuO–Cu$_2$O Thin Films for Photocatalytic Degradation of Organic Pollutants. *Nanomaterials* **2019**, *9*, 1011. [CrossRef] [PubMed]

44. Lim, Y.-F.; Chua, C.S.; Lee, C.J.J.; Chi, D. Sol–gel deposited Cu$_2$O and CuO thin films for photocatalytic water splitting. *Phys. Chem. Chem. Phys.* **2014**, *16*, 25928–25934. [CrossRef] [PubMed]

45. Li, P.; Zhang, X.; Wang, J.; Xu, B.; Zhang, X.; Fan, G.; Zhou, L.; Liu, X.; Zhang, K.; Jiang, W. Binary CuO/TiO$_2$ nanocomposites as high-performance catalysts for tandem hydrogenation of nitroaromatics. *Colloids Surf. A* **2021**, *629*, 127383. [CrossRef]

46. Xu, H.; Li, X.; Kang, S.Z.; Qin, L.; Li, G.; Mu, J. Noble metal-free cuprous oxide/reduced graphene oxide for enhanced photocatalytic hydrogen evolution from water reduction. *Int. J. Hydrogen Energy* **2014**, *39*, 11578–11582. [CrossRef]

47. Zhang, L.; Jing, D.; Guo, L.; Yao, X. In situ photochemical synthesis of Zn-doped Cu$_2$O hollow microcubes for high efficient photocatalytic H$_2$ production. *ACS Sustain. Chem. Eng.* **2014**, *2*, 1446–1452. [CrossRef]

48. Laskova, B.; Moehl, T.; Kavan, L.; Zukalova, M.; Liu, X.; Yella, A.; Comte, P.; Zukal, A.; Nazeeruddin, M.K.; Graetzel, M. Electron Kinetics in Dye Sensitized Solar Cells Employing Anatase with (1 0 1) and (0 0 1) Facets. *Electrochim. Acta* **2015**, *160*, 296–305. [CrossRef]

49. Thu, P.T.; Thinh, V.D.; Lam, V.D.; Bach, T.N.; Phong, L.T.H.; Tung, D.H.; Manh, D.H.; Khien, N.V.; Anh, T.X.; Le, N.T.H. Decorating of Ag and CuO on ZnO nanowires by plasma electrolyte oxidation method for enhanced photocatalytic efficiency. *Catalysts* **2022**, *12*, 801. [CrossRef]

50. Luo, L.; Zhang, T.; Zhang, X.; Yun, R.; Lin, Y.; Zhang, B.; Xiang, X. Enhanced hydrogen production from ethanol photoreforming by site-specific deposition of Au on Cu$_2$O/TiO$_2$ pn junction. *Catalysts* **2020**, *10*, 539. [CrossRef]

51. Yadav, A.A.; Hunge, Y.M.; Kang, S.W. Visible light-responsive CeO$_2$/MoS$_2$ composite for photocatalytic hydrogen production. *Catalysts* **2022**, *12*, 1185. [CrossRef]

52. Chang, C.J.; Tsai, M.H.; Hsu, Y.H.; Tuan, C.S. Morphology and optoelectronic properties of ZnO rod array/conjugated polymer hybrid films. *Thin Solid Films* **2008**, *516*, 5523–5526. [CrossRef]

53. Tien, T.M.; Chung, Y.J.; Huang, C.T.; Chen, E.L. Fabrication of WS$_2$/WSe$_2$ Z-scheme nano-heterostructure for efficient photocatalytic hydrogen production and removal of congo red under visible light. *Catalysts* **2022**, *12*, 852. [CrossRef]

54. Chang, C.J.; Hung, S.T. Electrochemical deposition of ZnO pore-array structures and photoconductivity of ZnO/polymer hybrid films. *Thin Solid Films* **2008**, *517*, 1279–1283. [CrossRef]

55. Peng, C.; Xu, W.; Wei, P.; Liu, M.; Guo, L.; Wu, P.; Zhang, K.; Cao, Y.; Wang, H.; Yu, H.; et al. Manipulating photocatalytic pathway and activity of ternary Cu$_2$O/(001)TiO$_2$@Ti$_3$C$_2$Tx catalysts for H$_2$ evolution: Effect of surface coverage. *Int. J. Hydrogen Energy* **2019**, *44*, 29975–29985. [CrossRef]

56. Chang, C.J.; Hung, C.Y.; Chen, J.; Li, C.T.; Yu, Y.H. Effects of surface property on loading octaaza bis-α-diimine Ni complex, charge separation, and H$_2$ production activity of MnS@ZnS photocatalysts. *Surf. Interfaces* **2024**, *48*, 104253. [CrossRef]

57. Gopannagari, M.; Kumar, D.P.; Reddy, D.A.; Hong, S.; Song, M.I.; Kim, T.K. In situ preparation of few-layered WS$_2$ nanosheets and exfoliation into bilayers on CdS nanorods for ultrafast charge carrier migrations toward enhanced photocatalytic hydrogen production. *J. Catal.* **2017**, *351*, 153–160. [CrossRef]

58. Yao, Z.; Nie, J.; Hassan, Q.U.; Li, G.; Liao, J.; Zhang, W.; Zhu, L.; Shi, X.; Rao, F.; Chang, J.; et al. Efficient charge separation of a Z-scheme Bi$_5$O$_{7-\delta}$I/ CeO$_2$ − δ heterojunction with enhanced visible light photocatalytic activity for NO removal. *Inorg. Chem. Front.* **2022**, *9*, 2832–2844. [CrossRef]

59. Martínez-Arias, A.; Fernández-García, M.; Gálvez, O.; Coronado, J.M.; Anderson, J.A.; Conesa, J.C.; Soria, J.; Munuera, G. Comparative study on redox properties and catalytic behavior for CO oxidation of CuO/CeO$_2$ and CuO/ZrCeO$_4$ catalysts. *J. Catal.* **2000**, *195*, 207–216. [CrossRef]

60. Wei, B.; Yang, N.; Pang, F.; Ge, J. Cu$_2$O-CuO Hollow Nanospheres as a Heterogeneous Catalyst for Synergetic Oxidation of CO. *J. Phys. Chem. C* **2018**, *122*, 19524–19531. [CrossRef]

61. Chang, C.J.; Chao, P.Y. Efficient photocatalytic hydrogen production by doped ZnS grown on Ni foam as porous immobilized photocatalysts. *Int. J. Hydrogen Energy* **2019**, *44*, 20805–20814. [CrossRef]

 *catalysts*

Article

Hydrothermal Synthesis of La-MoS$_2$ and Its Catalytic Activity for Improved Hydrogen Evolution Reaction

Archana Chaudhary [1], Rais Ahmad Khan [2], Sultan Saad Almadhhi [2], Ali Alsulmi [2], Khursheed Ahmad [3],* and Tae Hwan Oh [3],*

1   Department of Chemistry, Medi-Caps University, Indore 453331, India
2   Department of Chemistry, College of Science, King Saud University, Riyadh 11451, Saudi Arabia
3   School of Chemical Engineering, Yeungnam University, Gyeongsan 38541, Republic of Korea
*   Correspondence: khursheed.energy@gmail.com (K.A.); taehwanoh@ynu.ac.kr (T.H.O.)

Abstract: Herein, we report the synthesis and characterization of lanthanum-doped MoS$_2$ (La-MoS$_2$) via a hydrothermal route. The synthesized La-MoS$_2$ was characterized using powder X-ray diffraction (PXRD), scanning electron microscopy (SEM), and energy-dispersive X-ray (EDX) techniques. The band gap of La-MoS$_2$ was observed to be 1.68 eV, compared to 1.80 eV for synthesized MoS$_2$. In the photoluminescence (PL) spectra, a decrease in the intensity was observed for La-MoS$_2$ compared to MoS$_2$, which suggests that due to doping with charged La^{3+}, separation increases. The as-synthesized MoS$_2$ and La-MoS$_2$ were used for photocatalytic hydrogen evolution reactions (HERs), exhibiting 928 µmol·g^{-1} evolution of H$_2$ in five hours for a 10 mg dose of La-MoS$_2$, compared to 612 µmol·g^{-1} for MoS$_2$. A 50 mg mass of the catalyst (La-MoS$_2$) exhibited enhanced H$_2$ production of 1670 µmol·g^{-1} after five hours. The higher rate of the HER for La-MoS$_2$ is because of doping with La^{3+}. The photocatalytic hydrogen evolution performance of La-MoS$_2$ was also evaluated for different doses of La-MoS$_2$ exhibiting reusability up to the fourth cycle, showing potential applications of La-MoS$_2$ in hydrogen evolution reactions. Mechanistic aspects of the HER on the surface of La-MoS$_2$ have also been discussed.

Keywords: La-doped MoS$_2$; hydrogen production; photocatalysis

Citation: Chaudhary, A.; Khan, R.A.; Almadhhi, S.S.; Alsulmi, A.; Ahmad, K.; Oh, T.H. Hydrothermal Synthesis of La-MoS$_2$ and Its Catalytic Activity for Improved Hydrogen Evolution Reaction. *Catalysts* **2024**, *14*, 893. https://doi.org/10.3390/catal14120893

Academic Editors: Georgios Bampos, Paraskevi Panagiotopoulou and Eleni A. Kyriakidou

Received: 31 October 2024
Revised: 27 November 2024
Accepted: 3 December 2024
Published: 5 December 2024

1. Introduction

In the past few decades, rapid expansion of industries and a growing global population have posed significant challenges for humanity in meeting its energy needs [1]. Over the past few years, energy demand has been risen exponentially [2]. At present, the primary source of energy is fossil fuels, which are rapidly depleting and have harmful environmental effects when utilized [2]. To overcome this challenge, researchers and scientists are working on the development of sustainable technologies designed to reduce environmental impact. Solar energy has been proved to be one of the most efficient and green of all energy sources. The sun provides an enormous amount of solar energy, around 10^{22} joules, which can contribute to the development of next generation energy technologies [3]. Thus, it is essential to utilize solar energy for the development of green and renewable energy technology for the future. In connection with this, hydrogen energy is considered one of the most efficient energy technologies to fulfill energy requirements [4,5]. Photocatalytic hydrogen generation by the use of solar energy on the surface of a catalyst (semiconductor) is a very propitious alternative for resolving the present energy and environmental crisis [6–8]. Hydrogen fuel provides a high yield of energy upon combustion, i.e., 122 kJ/mol, which is far better than any other fossil fuel (gasoline, coal, etc.) [9,10]. The process of hydrogen generation by photocatalysis is environmentally benign, producing no harmful byproducts [11]. The use of photocatalytic materials for the generation of hydrogen was demonstrated for the first time in 1972 [12]. Thereafter, a great number of semiconductor materials have been designed and demonstrated for the efficient generation of hydrogen [13–15]. A series of MoS_2-TiO_2 photocatalysts have been designed by Zhu et al. [16], using a ball-milling process. Due to the introduction of MoS_2, a decrease in the recombination of photogenerated charge carriers was observed, enhancing the photocatalytic performance compared to pure TiO_2. The rate of the HER was enhanced to 150.7 µmol/h for the 4.0 wt% MoS_2-TiO_2 catalyst. Kumar et al. designed defect-rich MoS_2 nanosheets and decorated them with nitrogen-doped ZnO nanorod composites [17]. For this photocatalyst, optimal hydrogen evolution (17.3 mmol h^{-1} g^{-1}) was recorded for 15 wt% defect-rich MoS_2 nanosheets coated on N-ZnO. The results also demonstrated that a defect-induced interfacial contact region developed on encapsulation of defect-rich MoS_2 across N-ZnO, resulting in a composite that further increased the photocatalytic activity of the designed semiconductor system.

Doping of MoS_2 with a metal or nonmetal may generates defects along with alteration of the optical bandgap. For a Co-doped MoS_2/g-C_3N_4 composite, a very good photocatalytic reduction rate of water has been reported by Hu et al. [18] with a 0.31 mmol $g^{-1}h^{-1}$ rate of H_2 evolution due to the generation of an edge-enriched 1T phase as a result of Co doping. Additionally, the rate of photocatalysis can be enhanced by depositing a metal on MoS_2. With metal deposition on MoS_2, the interfacial charge transfer rate is enhanced along with the generation of a local electric field via the Schottky junction. It may also broaden the absorption spectrum to the near-infrared region. Scientists have studied the deposition of several metals, viz., copper [19], silver [20] gold [21], platinum [22], and palladium [23], on MoS_2, showing the creation of plasmonic photocatalysts absorbing strong visible light along with size- and shape-dependent surface plasmon resonance (SPR) effects.

Metal dichalcogenide compounds have received a great deal of attention because of their excellent optoelectronic properties. Metal dichalcogenide compounds including metal sulfides (such as MoS_2), selenides, and tellurides exhibit tunable band gaps and high surface activity, making them suitable for semiconductors, photocatalysis, and energy storage. Metal chalcogenides, especially in two-dimensional forms, have gained attention for hydrogen evolution reactions (HERs) and other electrochemical applications, as they offer excellent electron mobility and catalytic efficiency. MoS_2 is a 2D layered material characterized by a large surface area, tunable electronic properties, and structural flexibility, which make it suitable for catalytic processes. Its catalytic activity in HERs is primarily driven by the active edge sites of the MoS_2 layers, where sulfur vacancies and exposed Mo atoms enhance electron transfer and improve reaction kinetics. One key property of MoS_2 is its relatively low overpotential, especially when engineered with additional defects or doped with metals such as Ni or Co, which introduce more active sites and increase conductivity. The basal plane, which is generally inert, can also be activated through exfoliation, creating more active edge sites. Additionally, MoS_2 has excellent stability at acidic and neutral pH values, making it a durable option for water-splitting applications. Recent research focuses on synthesizing MoS_2 heterostructures with materials such as graphene or carbon nanotubes, further enhancing conductivity and stability. These modifications make MoS_2-based catalysts a promising alternative to noble metals in efficient and cost-effective H_2 production. MoS_2, a layered transition metal dichalcogenide, is generated by the stacking of S–Mo–S atomic layers that are held together by van der Waals forces present among them [24]. When MoS_2 is changed from bulk to the nanoscale, it exhibits unique physiochemical and optical properties and becomes a potential candidate for photocatalysis [25]. Bulk MoS_2 is optically inactive with an indirect bandgap of 1.29 eV; it also demonstrates a low photoluminescence response [26]. However, when the thickness of bulk MoS_2 is reduced to a single layer or a few layers, it exhibits a peak at 1.8–1.9 eV for the direct band gap [27]. In MoS_2, speedy/fast transfer of charge carriers occurs when visible light strikes its surface because of the narrow band gap; therefore, it is a promising and valuable photocatalyst candidate. However, the fast rate of recombination of charge carriers limits it applications in photocatalysis reactions [27,28].

Herein, we report the fabrication of a lanthanum-doped MoS_2 photocatalyst for a hydrogen evolution reaction. La-MoS_2 was prepared using a hydrothermal method, and its photocatalytic activity was examined for a hydrogen evolution reaction under visible light irradiation. The proposed photocatalyst La-MoS_2 also demonstrated decent stability for hydrogen evolution under visible light irradiation.

2. Results and Discussion

2.1. Material Characterization

To characterize the generated phase, the powder X-ray diffraction patterns of pure MoS_2 and La-MoS_2 were recorded and are shown in Figure 1. Four distinctive peaks at 2θ values of 13.1°, 32.4°, 35.8°, and 57.9° were observed in the samples, corresponding to the (002), (100), (103), and (110) planes of MoS_2. The PXRD patterns confirm the presence of a hexagonal crystal structure (2H-MoS_2), which is in accordance with the existing literature (JCPDS No.: 0037-1492). On comparing the PXRD patterns of both the samples of pure MoS_2 and La-MoS_2, it was confirmed that almost the same peak position could be observed in both cases, with a slight shift in La-MoS_2. The slight shift in the peak position after doping may be due to the larger ionic radius of lanthanum (La^{3+}) as compared to molybdenum (Mo^{4+}), which results in the expansion of the crystalline lattice and is consistent with Bragg's equation [29]. In the PXRD pattern of La-MoS_2, a higher peak intensity could be observed as compared to MoS_2; this increase in the peak intensity may be ascribed to the different electron densities of La^{3+} (dopant) and MoS_2 (host), which primarily relate to the scattering factor, structure factor, etc. In the case of La-MoS_2, the increase in the structure factor may be attributed to increased crystallite size, which results in an increase in the intensity of PXRD peaks on doping with La^{3+} [30]. The absence of peaks for elemental La^{3+} confirmed that the crystal structure of MoS_2 had not been changed due to doping. To calculate the particle sizes of both MoS_2 and La-MoS_2, the Debye–Scherrer formula ($D = K$ λ / β cosθ, where $K = 0.89$, $\lambda = 0.154$, D is the average crystalline size, is β is full width half maxima (FWHM) and θ is the Bragg angle) has been used in a previous study. The average crystallite sizes of MoS_2 and La-Mos_2 samples have been calculated as 17.5 and 21.2 nm, respectively, showing an increase in the crystallite size of MoS_2 on doping with La^{3+} [31].

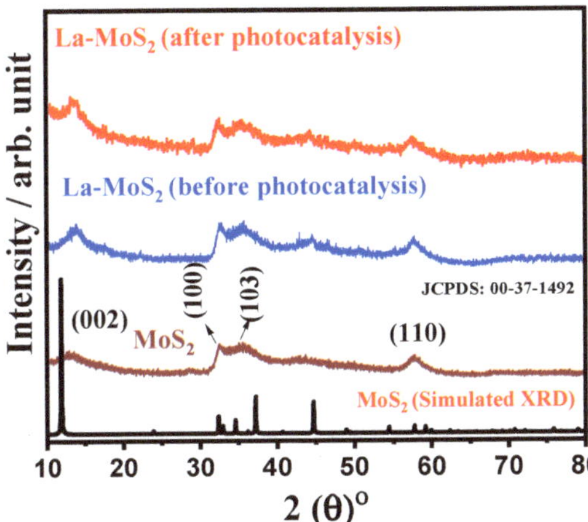

Figure 1. PXRD patterns of as-obtained MoS_2, La-MoS_2, and La-MoS_2 after stability test.

Figure 2 presents surface morphology of synthesized MoS_2 (Figure 2a) and La-MoS_2 (Figure 2c). For both of the samples, spherical morphologies were observed, with an average diameter of approximately one nanometer (Figure 2b,d). The average sizes of the MoS_2 and La-MoS_2 particles were found to be 648.26 nm and 676.01 nm, respectively, using particle size distribution curves (Figure 2b,d).

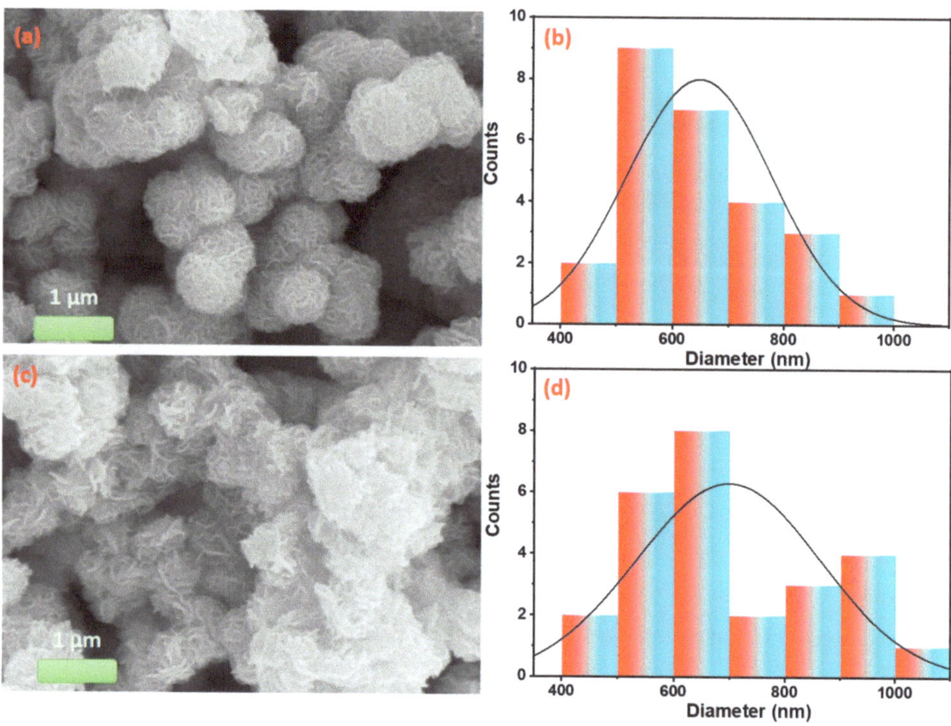

Figure 2. SEM image (**a**) and particle size distribution (**b**) of MoS_2. SEM image (**c**) and particle size distribution (**d**) of La-MoS_2.

From the FESEM images, it appeared that the surfaces of the MoS_2 and La-MoS_2 microspheres were made up of nanoflakes. Hollow microspheres of MoS_2 have also been reported by Afanasiev and Bezverkhy [32], who produced them via a heat treatment method, and by Chen et al. [33], who produced them via a direct sulfidization route; these results are similar to ours. Moreover, in these synthesized microspheres, numerous thin-stretched, folding flakes have also been observed, which are helpful in hydrogen evolution reactions.

When these active sites are exposed to light, they assist in the transportation of charge carriers along with participating in oxidation and reduction reactions in the course of the photocatalytic procedure [34]. EDX spectra of MoS_2 and La-MoS_2 sub-microspheres are shown in Figure S1. As shown in the EDX spectrum of MoS_2, two signals indicating the elements Mo and S in the elemental composition were observed, while in the EDX spectrum of La-MoS_2, signals for the element La were also detected. The obtained results again indicated the phase purity of the synthesized samples.

In order to check the distribution of the elements in the samples, elemental mapping was also performed. Images of element mapping of MoS_2 revealed that the sample contained Mo and S as the main elements, and both elements were uniformly distributed in the sample (Figure 3a–c). Images of elemental mapping along with the corresponding overlay and FESEM images for La-MoS_2 sample are presented in Figure 3d–g, showing the presence of the element La in the MoS_2 sample. It is evident from the images that the element La was distributed in lower amounts than Mo and S.

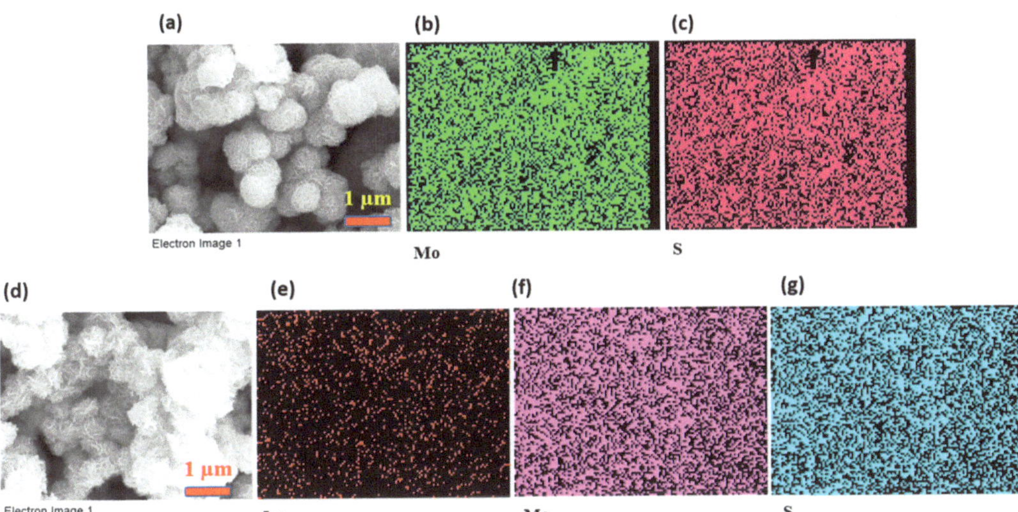

Figure 3. (**a**) Electron micrograph and mapping images of the elements (**b**) Mo and (**c**) S in the prepared MoS_2. (**d**) Electron micrograph and mapping images of the elements (**e**) La, (**f**) Mo, and (**g**) S in the prepared La-MoS_2.

To calculate the optical bandgaps of the prepared samples, UV–visible spectra were been recorded for both the as-prepared MoS_2 and La-MoS_2 samples. Figure 4a depicts the UV–visible spectra of both the samples. The results of UV–visible spectroscopy indicated that there was a significant shift in the absorption band towards visible region in the case of La-MoS_2. To calculate the band gap energy of both the samples, a Tauc plot was created using the formula $(\alpha h\upsilon)^2 = A\,(h\upsilon - Eg)$. A plot of $(\alpha h\upsilon)^2$ against $h\upsilon$ is shown in Figure 4b. The band gap of as-prepared MoS_2 was calculated as 1.80 eV, similar to the band gap of MoS_2 monolayer material [35], whereas a significant reduction in the band gap (1.68 eV) was noticed after doping with La^{3+}, which further affirms that upon doping with La^{3+}, the visible light absorption property of the synthesized photocatalyst was enhanced.

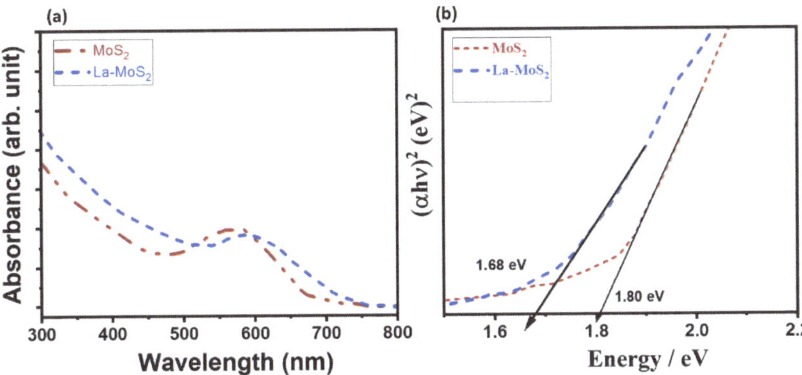

Figure 4. (**a**) UV-vis spectra and (**b**) Tauc plots of MoS_2 and La-MoS_2.

PL spectra of MoS_2 and La-MoS_2 are shown in Figure S2. The main role of photoluminescence (PL) spectroscopy is to investigate migration, electron transfer efficiency, and electron trapping in semiconductor materials [36]. When the rate of recombination of electron–hole pairs is high, the peaks in the PL spectrum are intense. Thus, it is clear from the recorded PL spectra of MoS_2 and $La\text{-}MoS_2$ that on doping with La^{3+}, the rate of recombination of electron–hole pairs decreases, as the intensity $La\text{-}MoS_2$ is less than that of pure MoS_2. The decrease in the recombination rate is due to the introduction of a discrete energy level due to doping with La^{3+}.

2.2. Photocatalytic Performance of La-MoS$_2$

To analyze the photocatalytic efficiency, the synthesized MoS_2 and $La\text{-}MoS_2$ samples were suspended in water using lactic acid as a sacrificial agent, and the whole suspensions were stimulated with solar irradiation. Figure 5a shows the results of hydrogen production using 10 mg of photocatalyst. With increasing time, the photocatalytic efficiency increases for both MoS_2 and $La\text{-}MoS_2$, but for $La\text{-}MoS_2$, the rate of the photocatalytic hydrogen evolution rection (HER) was much higher than for MoS_2. After 5 h, the amount of H_2 was found to be 612 $\mu mol \cdot g^{-1}$ for MoS_2 and 918 $\mu mol \cdot g^{-1}$ for $La\text{-}MoS_2$. Thus, it may be concluded that $La\text{-}MoS_2$ works better for the HER as compared to MoS_2 due to charge separation. Hence, for further experiments, $La\text{-}MoS_2$ was taken into consideration. The effects of different doses of $La\text{-}MoS_2$ are shown in Figure 5b. The doses of $La\text{-}MoS_2$ were varied between 10–50 mg. The results were collected at intervals of one hour for a total of five hours. As shown in Figure 5b, with an increasing dose of the photocatalyst, the amount of H_2 production also increased.

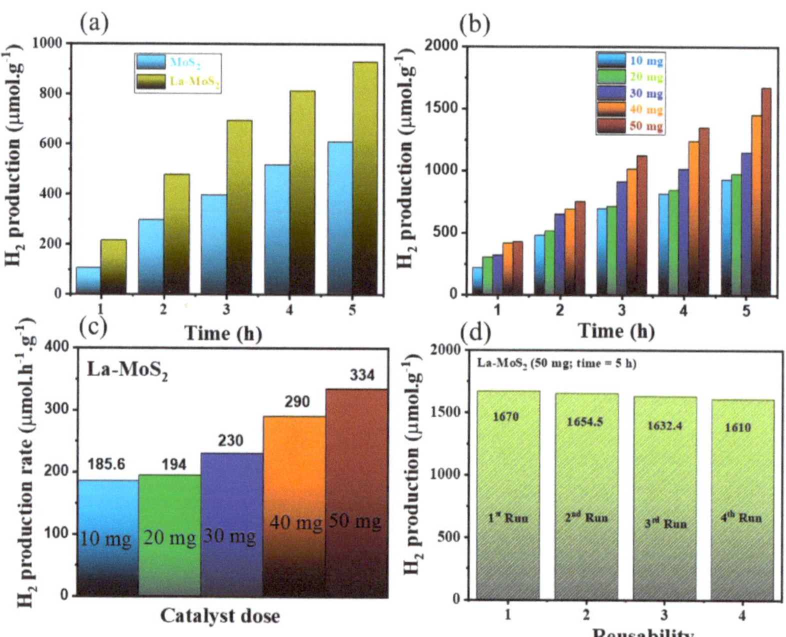

Figure 5. (**a**) H_2 evolution activity of MoS_2 and $La\text{-}MoS_2$ (10 mg catalyst) in lactic acid. (**b**) Effect of different doses of catalyst ($La\text{-}MoS_2$; 10–50 mg) in lactic acid. (**c**) H_2 evolution rates for different catalyst doses. (**d**) Reusability test.

The highest amount of H_2, 1670 $\mu mol \cdot g^{-1}$, was obtained for 50 mg catalyst. The rates of H_2 evolution for different catalyst doses are summarized in Figure 5c. Figure 5c shows that when the dose of $La\text{-}MoS_2$ was increased, the rate of H_2 production increased more than 1.5-folds. It reached from 185.6 $\mu mol \cdot h^{-1} \cdot g^{-1}$ to 334 $\mu mol \cdot h^{-1} \cdot g^{-1}$. In order to use a catalyst in real-time applications, it is essential to test the stability and reusability of the photocatalyst; accordingly, for $La\text{-}MoS_2$, a reusability test was performed, and it showed almost consistent results up to four consecutive cycles (Figure 5d). The XRD results after the stability test also reflected the acceptable stability of the La-MoS$_2$ (Figure 1).

A potential mechanism for the hydrogen evolution reaction on the surface of La-MoS$_2$ is shown in Scheme 1. The H$_2$ evolution is initiated when visible light strikes the surface of La-MoS$_2$ and it absorbs photons with energy (hυ) either equal to or greater than its energy band gap [1]. On absorbing the energy from solar radiation, electrons jump from the valence band to the conduction band, creating electron–hole pairs and leaving the holes in the valence band. The photogenerated electrons that are present in the conduction band reduce H$^+$ into H$_2$, whereas the holes present in the valence band combine with H$_2$O and break it down into O$_2$ and H$^+$ [37]. These holes also react with the scavenger lactic acid, changing it to pyruvic acid. The enhanced rate of photocatalytic production over the surface of La-MoS$_2$ may be attributed to the narrow band gap and synergistic interactions. This may improve the electron transport process, and enhanced H$_2$ evolution activity can be observed. Therefore, it can be concluded that controlling the transfer and migration of charge carriers may enhance the efficiency of photocatalysts, such as the spatial separation of carriers, elongating their lifetime and thus increasing their photocatalytic performance.

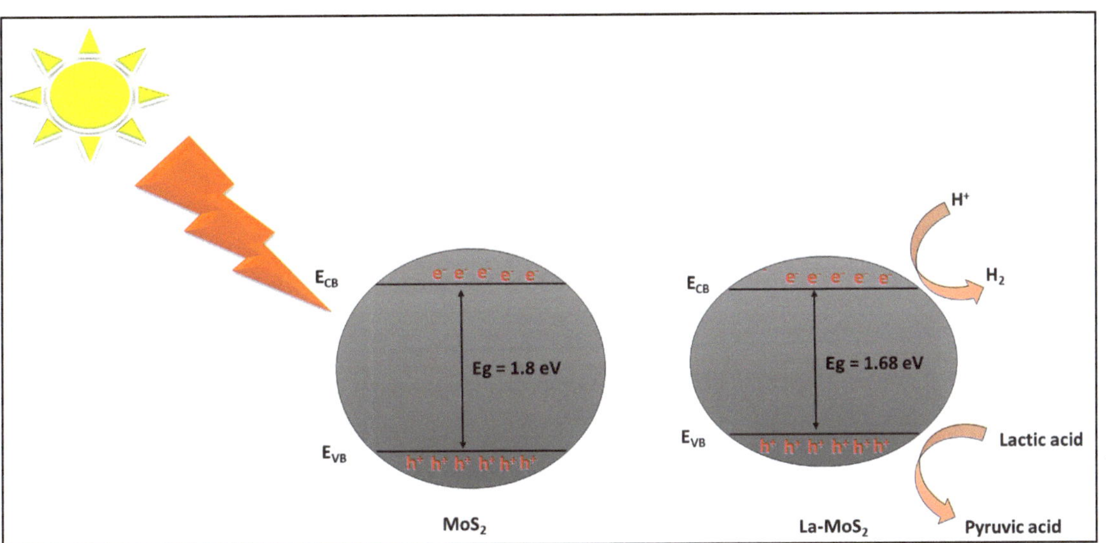

Scheme 1. Potential mechanism of H$_2$ generation.

In previous years, various photocatalysts have been reported for photocatalytic H$_2$ evolution. In particular, MoS$_2$ exhibited acceptable performance for the generation of H$_2$ under visible light irradiation. Xin et al. [38] reported the use of MoS$_2$ as a photocatalyst; it achieved a demonstrated H$_2$ evolution rate of 99.4 μmol·h^{-1}·g^{-1} with lactic acid as a sacrificial reagent. The authors further adopted P-doped MoS$_2$ as a photocatalyst and explored it for H$_2$ evolution. An enhanced H$_2$ evolution rate of 278.8 μmol·h^{-1}·g^{-1} was observed with a similar environment and conditions. The authors stated that doping with the element P significantly improved the catalytic activity of P-MoS$_2$. In another published article, MoS$_2$ was used as a photocatalyst, and its photocatalytic activity was evaluated in presence of SO$_3^{2-}$ sacrificial reagent [39]. The authors found that pristine MoS$_2$ had lower photocatalytic activity for H$_2$ evolution and a low H$_2$ evolution rate of 39 μmol·h^{-1}·g^{-1}. The authors also used pristine ZnO as a photocatalyst, which achieved a demonstrated H$_2$ evolution rate of 22 μmol·h^{-1}·g^{-1}. However, an MoS$_2$/ZnO composite demonstrated a significant improvement in its H$_2$ evolution rate, and a decent H$_2$ evolution rate of 235 μmol·h^{-1}·g^{-1} was obtained. Pristine MoS$_2$ also exhibited H$_2$ evolution at a rate of 185 $\mu mol \cdot h^{-1} \cdot g^{-1}$ in the presence of methanol as a sacrificial reagent [40]. In another work, $MoS_2/g\text{-}C_3N_4$ composite-based investigations revealed that H_2 evolution at a rate of 441.3 $\mu mol \cdot h^{-1} \cdot g^{-1}$ could be observed in presence of triethanolamine as a sacrificial reagent [41]. Wei et al. [42] also proposed the synthesis of $MoN_{1\cdot2x}S_{2-1\cdot2x}@g\text{-}C_3N_4$ as a photocatalyst for H_2 production applications. An interesting H_2 evolution rate of 360.4 $\mu mol \cdot h^{-1} \cdot g^{-1}$ was observed for the photocatalyst $MoN_{1\cdot2x}S_{2-1\cdot2x}@g\text{-}C_3N_4$ in the presence of triethanolamine as a sacrificial reagent. In 2023, Wang et al. [43] proposed the hydrothermal synthesis of O, $P\text{–}MoS_2$ for H_2 production application under visible light. The authors used triethanolamine/acetonitrile as sacrificial reagents and reported that the H_2 evolution rate was 339.3 $\mu mol \cdot h^{-1} \cdot g^{-1}$. The aforementioned studies showed that MoS_2-based photocatalysts have a significant role in photocatalytic H_2 evolution studies. The results obtained in the present study are compared with previously reported articles in Table 1. The comparison of results showed that $La\text{-}MoS_2$ showed a decent response for hydrogen production compared to the many previously reported photocatalysts (Table 1).

Table 1. Comparison of H_2 evolution rate of $La\text{-}MoS_2$ with published articles.

Photocatalysts	H_2 Evolution Rate ($\mu mol \cdot h^{-1} \cdot g^{-1}$)	Light Source	Sacrificial Agent	Reference
$La\text{-}MoS_2$	334	300 W; Xe lamp (λ = 420 nm)	Lactic acid	This study
$P\text{-}MoS_2$	278.8	300 W; Xe lamp (λ = 420 nm)	Lactic acid	38
MoS_2	99.4	300 W; Xe lamp (λ = 420 nm)	Lactic acid	38
MoS_2	39	300 W; Xe lamp (λ = 420 nm)	SO_3^{2-}	39
MoS_2/ZnO	235	300 W; Xe lamp (λ = 420 nm)	SO_3^{2-}	39
ZnO	22	300 W; Xe lamp (λ = 420 nm)	SO_3^{2-}	39
$g\text{-}C_3N_4$	54	300 W; Xe lamp (λ = 420 nm)	Methanol	40
MoS_2	185	300 W; Xe lamp (λ = 420 nm)	Methanol	40
$MoS_2/g\text{-}C_3N_4$	441.3	300 W; Xe lamp (λ = 420 nm)	Triethanolamine	41
$MoN_{1\cdot2x}S_{2-1\cdot2x}@g\text{-}C_3N_4$	360.4	300 W; simulated solar light	Triethanolamine	42
O, $P\text{–}MoS_2$	339.3	LED light; λ = 420 nm	Triethanolamine/acetonitrile	43

3. Materials and Methods

3.1. Materials

Ammonium molybdate tetrahydrate $(NH_4)_6Mo_7O_{24}\cdot4H_2O$; 99.98% trace metal basis), thiourea (ACS reagent, $\geq$99.0%), and lanthanum (III) nitrate hexahydrate (99.999% trace metal basis) were purchased from Merck (Mumbai, India). All the other used chemicals and materials were of analytical grade and used as received from Sigma (Mumbai, India), Alfa Aesar (Indore, India), and Merck (Mumbai, India). No further treatment or purification was carried out.

3.2. Synthesis of La-MoS$_2$

In this study, we adopted hydrothermal synthesis method for the preparation of pristine MoS_2 and $La\text{-}MoS_2$ materials. In brief, hydrothermal treatment of the reaction solution of $(NH_4)_6Mo_7O_{24}\cdot4H_2O$ and thiourea yielded MoS_2. In brief, 0.6 g of thiourea was slowly added to an aqueous solution of 0.5 g ammonium molybdate (30 mL), and this reaction solution was stirred for 25–30 min at room temperature. A clear solution was obtained, which was then poured into a Teflon reactor and sealed tightly in a stainless steel autoclave (the autoclave needed to be carefully tightened to prevent the melting of the Teflon cup). This autoclave was heated at 180 °C for 18 h in a muffle furnace. After cooling down the furnace, the autoclave was opened; the MoS_2 was then collected, washed with DI water and ethanol, and dried at 70 °C for 8 h in a vacuum oven. In further investigations, La-MoS$_2$ was also prepared using a hydrothermal method as illustrated in Scheme 2.

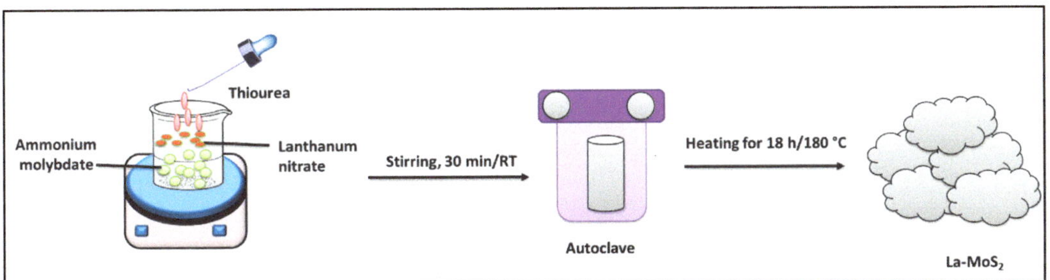

Scheme 2. Representation of the synthetic process for the preparation of La-MoS$_2$.

Typically, 0.6 g of thiourea was added to an aqueous solution of 0.5 g ammonium molybdate. Furthermore, 5 wt% (0.055 g) lanthanum nitrate was added to the above reaction solution and stirred for 25–30 min at room temperature. After stirring, this solution was transferred to a Teflon reactor, sealed tightly in a stainless steel autoclave, and heated at 180 °C for 18 h in a muffle furnace (Scheme 2). After cooling down the furnace, the autoclave was opened, and La-MoS$_2$ was collected using centrifugation, washed with DI water and ethanol, and dried at 70 °C for 8 h in a vacuum oven.

3.3. Instruments

A Rigaku powder X-ray diffractometer (model Rint 2500, manufactured in Rigaku, Tokyo, Japan) with wavelength of 1.5406 Å and Cu/Kα radiation was used to record the PXRD results of the synthesized samples. A Supra Zeiss-55 field-emission scanning electron microscope was used to record the surface morphological images of the synthesized samples (Zeiss, Jena, Germany). The energy-dispersive X-ray spectroscopy (EDX) results were obtained on a Horiba EDX spectroscope (Horiba, Kyoto, Japan). An ultraviolet–visible spectrophotometer (UV-vis spectrophotometer model Cary 100 (Varian, Palo Alto, CA, USA) was used to capture the UV-vis spectra of the synthesized samples. The photoluminescence (PL) spectra of the samples were obtained on a Dongwoo Optron PL spectrometer. The hydrogen evolution results were obtained on a gas chromatograph with a thermal conductivity detector (TCD).

3.4. Photocatalytic H$_2$ Evolution

We used a quartz tube reactor as the photocatalytic H$_2$ production set-up. A mass of 10 mg of the catalyst (La-MoS$_2$ or MoS$_2$) was added to a mixture of 80 mL DI water and 20 mL lactic acid. The reaction solution was purged with nitrogen gas for 40 min to remove unnecessary gases or oxygen. The quartz tube containing the catalyst and reaction solution was closed with an airtight seal and used for photocatalytic H$_2$ evolution processes under visible light irradiation (Figure S3). A 300 W xenon lamp with wavelength = 420 nm was used as the light source. Different doses of the catalyst (La-MoS$_2$; 10, 20, 30, 40, and 50 mg) were used to optimize the performance of the photocatalyst for improved H$_2$ evolution. The generated H$_2$ was taken out using a syringe and measured by employing a gas chromatograph (in combination with a TCD).

4. Conclusions

In conclusion, lanthanum-doped molybdenum disulfide (La-MoS$_2$) was synthesized via a hydrothermal route. To characterize La-MoS$_2$, powder X-ray diffraction (PXRD), scanning electron microscopy (SEM), and energy-dispersive X-ray (EDX) techniques were used. The band gap of La-MoS$_2$ was calculated as 1.68 eV, which is less than the band gap of synthesized MoS$_2$ (i.e., 1.80 eV). In the photoluminescence (PL) spectra, a decrease in intensity was observed for La-MoS$_2$ compared to MoS$_2$, indicating increased charge separation for La-MoS$_2$. The as-synthesized MoS$_2$ and La-MoS$_2$ were used for photocatalytic hydrogen evolution reactions (HERs), and La-MoS$_2$ exhibited an improved H$_2$ production rate compared to the MoS$_2$. The increased photocatalytic performance may be because of the generation of another discrete energy level due to doping with La^{3+}. Mechanistic aspects of the HER on the surface of La-MoS$_2$ have also been discussed.

Supplementary Materials: The following supporting information can be downloaded at: https://www.mdpi.com/article/10.3390/catal14120893/s1, Figure S1: EDX spectra of MoS$_2$ and La-MoS$_2$ sub-microspheres; Figure S2. PL spectra of MoS$_2$ and La-MoS$_2$ sub-microspheres; Figure S3: Schematic illustration of H$_2$ evolution reaction.

Author Contributions: Conceptualization: A.C. and R.A.K.; methodology: K.A. and A.C.; validation: R.A.K., S.S.A. and A.A.; formal analysis: A.C.; investigation: K.A.; resources, R.A.K.; writing—original draft preparation: K.A. and A.C.; writing—review and editing: T.H.O. All authors have read and agreed to the published version of the manuscript.

Funding: Researchers Supporting Project number (RSP2024R400), King Saud University, Riyadh, Saudi Arabia.

Data Availability Statement: The authors elect not to share the data.

Acknowledgments: Authors gratefully acknowledged Researchers Supporting Project number (RSP2024R400), King Saud University, Riyadh, Saudi Arabia.

Conflicts of Interest: The authors declare no conflicts of interest.

References

1. Rahman, A.; Jennings, J.R.; Tan, A.L.; Khan, M.M. Molybdenum Disulfide-Based Nanomaterials for Visible-Light-Induced Photocatalysis. *ACS Omega* **2022**, *7*, 22089–22110. [CrossRef] [PubMed]
2. Christoforidis, K.C.; Fornasiero, P. Photocatalytic Hydrogen Production: A Rift into the Future Energy Supply. *Chem. Cat. Chem.* **2017**, *9*, 1523–1544. [CrossRef]
3. Lewis, N.S.; Nocera, D.G. Powering the planet: Chemical challenges in solar energy utilization. *Proc. Natl. Acad. Sci. USA* **2006**, *103*, 15729–15735. [CrossRef] [PubMed]
4. Furukawa, H.; Yaghi, O.M. Storage of hydrogen, methane, and carbon dioxide in highly porous covalent organic frameworks for clean energy applications. *J. Am. Chem. Soc.* **2009**, *131*, 8875–8883. [CrossRef]
5. Ganguly, P.; Harb, M.; Cao, Z.; Cavallo, L.; Breen, A.; Dervin, S.; Dionysiou, D.D. 2D Nanomaterials for Photocatalytic Hydrogen Production. *ACS Energy Lett.* **2019**, *4*, 1687–1709. [CrossRef]
6. Gupta, U.; Rao, C. Hydrogen generation by water splitting using MoS$_2$ and other transition metal dichalcogenides. *Nano Energy* **2017**, *41*, 49–65. [CrossRef]
7. Teets, T.S.; Nocera, D.G. Photocatalytic hydrogen production. *Chem. Commun.* **2011**, *47*, 9268–9274. [CrossRef]
8. Kumaravel, V.; Mathew, S.; Bartlett, J.; Pillai, S.C. Photocatalytic hydrogen production using metal doped TiO$_2$: A review of recent advances. *Appl. Catal. B* **2019**, *244*, 1021–1064. [CrossRef]
9. Noyan, O.F.; Hasan, M.M.; Pala, N. A Global Review of the Hydrogen Energy Eco-System. *Energies* **2023**, *16*, 1484. [CrossRef]
10. Li, Y.; Somorjai, G.A. Nanoscale advances in catalysis and energy applications. *Nano Lett.* **2010**, *10*, 2289–2295. [CrossRef]
11. Omer, A.M. Energy, environment and sustainable development. *Renew. Sustain. Energy Rev.* **2008**, *12*, 2265–2300. [CrossRef]
12. Fujishima, A.; Honda, K. Electrochemical photolysis of water at a semiconductor electrode. *Nature* **1972**, *238*, 37–38. [CrossRef] [PubMed]
13. Grätzel, M. Photoelectrochemical cells. *Nature* **2001**, *414*, 338. [CrossRef] [PubMed]
14. Wang, X.; Maeda, K.; Thomas, A.; Takanabe, K.; Xin, G.; Carlsson, J.M.; Domen, K.; Antonietti, M. A metal-free polymeric photocatalyst for hydrogen production from water under visible light. *Nat. Mater.* **2009**, *8*, 76–80. [CrossRef]
15. Zhou, X.; Liu, N.; Schmidt, J.; Kahnt, A.; Osvet, A.; Romeis, S.; Zolnhofer, E.M.; Marthala, V.R.R.; Guldi, D.M.; Peukert, W. Noble-Metal-Free Photocatalytic Hydrogen Evolution Activity: The Impact of Ball Milling Anatase Nanopowders with TiH$_2$. *Adv. Mater.* **2017**, *29*, 1604747. [CrossRef]
16. Zhu, Y.; Ling, Q.; Liu, Y.; Wang, H.; Zhu, Y. Photocatalytic H$_2$ evolution on MoS$_2$–TiO$_2$ catalysts synthesized via mechanochemistry. *Phys. Chem. Chem. Phys.* **2015**, *17*, 933–940. [CrossRef]
17. Kumar, S.; Kumar, A.; Rao, V.N.; Kumar, A.; Shankar, M.V.; Krishnan, V. Defect-Rich MoS2 Ultrathin Nanosheets-Coated Nitrogen-Doped ZnO Nanorod Heterostructures: An Insight into in-Situ-Generated ZnS for Enhanced Photocatalytic Hydrogen Evolution. *ACS Appl. Energy Mater.* **2019**, *2*, 5622–5634. [CrossRef]
18. Hu, Y.; Yu, X.; Liu, Q.; Wang, L.; Tang, H. Highly Metallic Co-Doped MoS$_2$ Nanosheets as an Efficient Cocatalyst to Boost Photoredox Dual Reaction for H$_2$ Production and Benzyl Alcohol Oxidation. *Carbon* **2022**, *188*, 70–80. [CrossRef]

19. Yusuf, M.; Song, S.; Park, S.; Park, K.H. Multifunctional Core-Shell Pd@Cu on MoS2 as a Visible Light-Harvesting Photocatalyst for Synthesis of Disulfide by S S Coupling. *Appl. Catal. A* **2021**, *613*, 118025. [CrossRef]

20. Hu, J.; Zhang, C.; Li, X.; Du, X. An Electrochemical Sensor Based on Chalcogenide Molybdenum Disulfide-Gold-Silver Nanocomposite for Detection of Hydrogen Peroxide Released by Cancer Cells. *Sensors* **2020**, *20*, 6817. [CrossRef]

21. Lai, H.; Ma, G.; Shang, W.; Chen, D.; Yun, Y.; Peng, X.; Xu, F. Multifunctional Magnetic Sphere-MoS$_2$@Au Hybrid for Surface-Enhanced Raman Scattering Detection and Visible Light Photo-Fenton Degradation of Aromatic Dyes. *Chemosphere* **2019**, *223*, 465–473. [CrossRef] [PubMed]

22. Gottam, S.R.; Tsai, C.-T.; Wang, L.-W.; Lin, C.-C.; Chu, S.-Y. Highly Sensitive Hydrogen Gas Sensor Based on a MoS2-Pt Nanoparticle Composite. *Appl. Surf. Sci.* **2020**, *506*, 144981. [CrossRef]

23. Chi, M.; Zhu, Y.; Jing, L.; Wang, C.; Lu, X. Fabrication of Ternary MoS$_2$-Polypyrrole-Pd Nanotubes as Peroxidase Mimics with a Synergistic Effect and Their Sensitive Colorimetric Detection of L-Cysteine. *Anal. Chim. Acta* **2018**, *1035*, 146–153. [CrossRef] [PubMed]

24. Lin, T.; Zhong, L.; Guo, L.; Fu, F.; Chen, G. Seeing Diabetes: Visual Detection of Glucose Based on the Intrinsic Peroxidase-like Activity of MoS2 Nanosheets. *Nanoscale* **2014**, *6*, 11856–11862. [CrossRef] [PubMed]

25. Liu, Y.; Xie, Y.; Liu, L.; Jiao, J. Sulfur vacancy induced high performance for photocatalytic H$_2$ production over 1T@2H phase MoS$_2$ nanolayers. *Catal. Sci. Technol.* **2017**, *7*, 5635–5643. [CrossRef]

26. Okuno, Y.; Lancry, O.; Tempez, A.; Cairone, C.; Bosi, M.; Fabbri, F.; Chaigneau, M. Probing the nanoscale light emission properties of a CVD-grown MoS2 monolayer by tip-enhanced photoluminescence. *Nanoscale* **2018**, *10*, 14055–14059. [CrossRef]

27. Deokar, G.; Rajput, N.S.; Vancsó, P.; Ravaux, F.; Jouiad, M.; Vignaud, D.; Cecchet, F.; Colomer, J.F. Large area growth of vertically aligned luminescent MoS$_2$ nanosheets. *Nanoscale* **2017**, *9*, 277–287. [CrossRef]

28. Pradhan, G.; Sharma, A.K. Anomalous Raman and photoluminescence blue shift in mono- and a few layered pulsed laser deposited MoS$_2$ thin films. *Mater. Res. Bull.* **2018**, *102*, 406–411. [CrossRef]

29. Marin, R.; Back, M.; Mazzucco, N.; Enrichi, F.; Frattini, R.; Benedetti, A.; Riello, P. Unexpected optical activity of cerium in Y$_2$O$_3$:Ce^{3+}, Yb^{3+}, Er^{3+} up and down-conversion system. *Dalton Trans.* **2013**, *42*, 16837. [CrossRef]

30. Revathy, J.S.; Priya, N.S.C.; Sandhya, K.; Rajendran, D.N. Structural and optical studies of cerium doped gadolinium oxide phosphor. *Bull. Mater. Sci.* **2021**, *44*, 13. [CrossRef]

31. Kumar, M.J.K.; Kalathi, J.T. Low-temperature sonochemical synthesis of high dielectric Lanthanum doped Cerium oxide nanopowder. *J. Alloy. Compd.* **2018**, *748*, 348–354. [CrossRef]

32. Afanasiev, P.; Bezverkhy, I. Synthesis of MoS^x (5 > x > 6) Amorphous Sulfides and Their Use for Preparation of MoS$_2$ Monodispersed. *Chem. Mater.* **2002**, *14*, 2826–2830. [CrossRef]

33. Chen, X.; Wang, X.; Wang, Z.; Yu, W.; Qian, Y. Direct sulfidization synthesis of high-quality binary sulfides (WS$_2$, MoS$_2$, and V$_5$S$_8$) from the respective oxides. *Mater. Chem. Phys.* **2004**, *87*, 327–331. [CrossRef]

34. Riaz, K.N.; Yousaf, N.; Tahir, M.B.; Israr, Z.; Iqbal, T. Facile hydrothermal synthesis of 3D flower-like La-MoS$_2$ nanostructure for photocatalytic hydrogen energy production. *Int. J. Energy Res.* **2019**, *43*, 491–499. [CrossRef]

35. Radisavljevic, B.; Radenovic, A.; Brivio, J. Single-layer MoS$_2$ transistors. *Nat. Nanotechnol.* **2011**, *6*, 147–150. [CrossRef]

36. Schneider, J.; Matsuoka, M. Takeuchi Understanding TiO$_2$ photocatalysis: Mechanisms and materials. *Chem. Rev.* **2014**, *114*, 9919–9986. [CrossRef]

37. Galińska, A.; Walendziewski, J. Photocatalytic water splitting over Pt–TiO$_2$ in the presence of sacrificial reagents. *Energy Fuels* **2005**, *19*, 1143–1147. [CrossRef]

38. Xin, X.; Song, Y.; Guo, S.; Zhang, Y.; Wang, B.; Wang, Y.; Li, X. One-step synthesis of P-doped MoS$_2$ for efficient photocatalytic hydrogen production. *J. Alloys Compd.* **2020**, *829*, 154635. [CrossRef]

39. Hunge, Y.M.; Yadav, A.A.; Kang, S.-W.; Lim, S.J.; Kim, H. Visible light activated MoS$_2$/ZnO composites for photocatalytic degradation of ciprofloxacin antibiotic and hydrogen production. *J. Photochem. Photobiol. A Chem.* **2023**, *434*, 114250. [CrossRef]

40. Li, D.; Zeng, G.; Wu, Y.; Zhou, Z.; Guo, W.; Li, C. Au induced in-situ formation of ultra-stable 1T-MoS$_2$ on polymeric carbon nitride toward promoted photocatalytic hydrogen production. *Int. J. Hydrogen Energy* **2023**, *48*, 34363–34369. [CrossRef]

41. Xu, Y.; Ouyang, J.; Zhang, L.; Long, H.; Song, Y.; Cui, Y. Embedded 1T-rich MoS$_2$ into C$_3$N$_4$ hollow microspheres for effective photocatalytic hydrogen production. *Chem. Phys. Lett.* **2023**, *814*, 140331. [CrossRef]

42. Wei, X.; Wang, M.; Ali, S.; Wang, J.; Zhou, Y.; Zuo, R.; Zhong, Q.; Zhan, C. Enhanced photocatalytic H$_2$ evolution on g-C$_3$N$_4$ nanosheets loaded with nitrogen-doped MoS$_2$ as cocatalysts. *Int. J. Hydrogen Energy* **2024**, *89*, 691–702. [CrossRef]

43. Wang, L.; Hu, Y.; Xu, J.; Huang, Z.; Lao, H.; Xu, X.; Xu, J.; Tang, H.; Yuan, R.; Wang, Z.; et al. Dual non-metal atom doping enabled 2D 1T-MoS$_2$ cocatalyst with abundant edge-S active sites for efficient photocatalytic H$_2$ evolution. *Int. J. Hydrogen Energy* **2023**, *48*, 16987–16999. [CrossRef]

 *catalysts*

Article

Aqueous Phase Reforming by Platinum Catalysts: Effect of Particle Size and Carbon Support

Xuan Trung Nguyen [1], Ella Kitching [2], Thomas Slater [2], Emanuela Pitzalis [1], Jonathan Filippi [3], Werner Oberhauser [3] and Claudio Evangelisti [1,*]

[1] Institute of Chemistry of OrganoMetallic Compounds (ICCOM), National Research Council (CNR), Via G. Moruzzi 1, 56124 Pisa, Italy; trung.nguyen@pi.iccom.cnr.it (X.T.N.); emanuela.pitzalis@cnr.it (E.P.)

[2] Cardiff Catalysis Institute (CCI), Cardiff University, Maindy Road, Cardiff CF24 4HQ, UK; kitchingea@cardiff.ac.uk (E.K.); slatert2@cardiff.ac.uk (T.S.)

[3] Institute of Chemistry of OrganoMetallic Compounds (ICCOM), National Research Council (CNR), Via Madonna del Piano 10, 50019 Sesto Fiorentino, Italy; jonathan.filippi@iccom.cnr.it (J.F.); werner.oberhauser@cnr.it (W.O.)

* Correspondence: claudio.evangelisti@cnr.it

Abstract: Aqueous phase reforming (APR) is a promising method for producing hydrogen from biomass-derived feedstocks. In this study, carbon-supported Pt catalysts containing particles of different sizes (below 3 nm) were deposited on different commercially available carbons (i.e., Vulcan XC72 and Ketjenblack EC-600JD) using the metal vapor synthesis approach, and their catalytic efficiency and stability were evaluated in the aqueous phase reforming of ethylene glycol, the simplest polyol containing both C–C and C–O bonds. High-surface-area carbon supports were found to stabilize Pt nanoparticles with a mean diameter of 1.5 nm, preventing metal sintering. In contrast, Pt single atoms and clusters (below 0.5 nm) were not stable under the reaction conditions, contributing minimally to catalytic activity and promoting particle growth. The most effective catalyst Pt_A/C^K, containing a mean Pt NP size of 1.5 nm and highly dispersed on Ketjenblack carbon, demonstrated high hydrogen site time yield (8.92 min^{-1} at 220 °C) and high stability under both high-temperature treatment conditions and over several recycling runs. The catalyst was also successfully applied to the APR of polyethylene terephthalate (PET), showing potential for hydrogen production from plastic waste.

Keywords: aqueous phase reforming; ethylene glycol; Pt catalyst; hydrogen production; nanoparticles; PET

Citation: Nguyen, X.T.; Kitching, E.; Slater, T.; Pitzalis, E.; Filippi, J.; Oberhauser, W.; Evangelisti, C. Aqueous Phase Reforming by Platinum Catalysts: Effect of Particle Size and Carbon Support. *Catalysts* **2024**, *14*, 798. https://doi.org/10.3390/catal14110798

Academic Editors: Georgios Bampos, Paraskevi Panagiotopoulou and Eleni A. Kyriakidou

Received: 10 October 2024
Revised: 25 October 2024
Accepted: 30 October 2024
Published: 7 November 2024

1. Introduction

In the search for sustainable energy solutions, renewable hydrogen (H$_2$) production has emerged as an important element in transitioning toward a low-carbon energy system. Traditionally, H$_2$ has been primarily produced from fossil fuels, such as natural gas, naphtha, or coal [1]. With the increasing environmental concerns related to non-renewable resource consumption, the focus has shifted toward exploring biomass as a H$_2$ production source, thanks to its CO$_2$ neutrality. Aqueous phase reforming (APR) has garnered significant attention as an effective method for converting biomass-derived feedstocks into hydrogen [2]. APR stands out due to its ability to process a wide range of organic sources, including mixed polyol streams, glycerol from biodiesel production, and waste streams from food industries and biorefineries [3]. These feedstocks can be valorized through APR to produce valuable hydrogen streams, thus addressing both waste management and energy production challenges. Furthermore, APR can be utilized as a solution to emerging waste problems; for example, increasing plastic waste is projected to reach 12 billion tons by 2050, demanding solutions for sustainability [4]. The valorization of plastic waste has attracted considerable interest as a sustainable alternative to traditional disposal methods. In addition to the widely explored technologies, such as catalytic pyrolysis or steam reforming [5], APR remains relatively under-explored for hydrogen production from plastic waste.

The APR process is typically carried out at moderate temperatures, ranging from 220 to 265 °C, and pressures between 15 and 50 bar [6]. These conditions facilitate the conversion of alcohols and sugars into dihydrogen and carbon dioxide, thanks to the thermodynamically favorable water–gas shift reaction. The process's ability to operate at lower temperatures and higher pressures compared to conventional steam reforming eliminates the need to vaporize the feed stream, offering significant energy-saving advantages [7]. However, achieving high hydrogen selectivity in APR is challenging due to side reactions, such as methanation and dehydration/hydrogenation, which consume hydrogen to produce alkanes, alcohols, and organic acids [6,8].

Typically, APR is performed over heterogenous catalysts. Extensive research has identified supported metal catalysts, particularly those based on Pt, as highly effective for APR due to their high activity and selectivity toward hydrogen production [9,10]. The nature of the catalyst, including the active sites, support materials, and their interactions, significantly influences the product distribution and overall efficiency of the process.

The particle size of the NPs plays a critical role in the catalytic performance of the catalyst, as it affects the number of exposed active sites and also their electronic configuration [11]. Previous studies have reported the effect of Pt particle sizes in the aqueous phase reforming process [8]. However, different conclusions were drawn from the literature. The earliest result reported by Lehnert and Claus indicated that larger particles improved H_2 selectivity thanks to the abundance of terrace Pt sites [12]. The opposite conclusion was drawn by Wawrzet et al. when they studied glycerol APR with a Pt/Al_2O_3 catalyst [13]. They found that large particles slightly improved the conversion but reduced H_2 selectivity. Later, additional reports on the particle size effect on APR supported the result of Wawrzet et al. [14–17]. Generally, it is believed that smaller Pt particles (of approximately 1.5 nm) improve H_2 productivity. It must be noted that the particle sizes in the mentioned studies ranged from 1.1 nm upwards, while, to the best of our knowledge, no study has pointed out the role of sub-nano-Pt clusters in this reaction. Single-atom and cluster sites behave particularly differently from nanoparticles (NPs), prompting us to investigate this matter for the optimization of Pt catalysts [18,19].

In addition to the active metal, the support also plays an important role, particularly in catalyst activity and stability [20,21]. Compared to the commonly studied Al_2O_3 and SiO_2, which often suffer from low durability under APR conditions, carbonaceous supports have shown to be effective and stable for APR [22,23]. Moreover, the textural, morphological, and chemical properties of carbonaceous materials can be tuned widely to accommodate the objective. The interaction of the carbon support and the active metal can significantly influence catalyst performance. Therefore, the investigation of the effect of support on catalyst activity and stability is desirable to develop efficient catalysts for APR.

Metal vapor synthesis (MVS) is an advanced method that enables the formation of ultra-small NPs that can be immobilized on different supports without significant changes in particle size distribution [24]. This approach is particularly advantageous to decouple the effect of the particle size and role of support, in addition to its ability to generate an effective highly dispersed supported catalyst. In this work, we report for the first time the use of the MVS approach to make efficient Pt catalysts for APR of ethylene glycol (EG) as a model substrate. This substrate was chosen as it is the smallest polyol containing both C-O and C-C bonds. The simplicity of EG allows for easier assessment of the catalyst's activity, while still incorporating the key chemical structure found in the target substrates used in APR, such as polyols and carbohydrates. The effects of support and particle sizes (in a range less than 2 nm) were also investigated. Furthermore, the best-performing catalysts were used in the application of APR using polyethylene terephthalate (PET) as a substrate to make H_2, exhibiting the feasibility of the catalyst in providing an alternative for waste plastic valorization.

2. Results and Discussion

2.1. Synthesis and Characterization of Carbon-Supported Pt Catalysts

Pt-based catalysts were synthesized by the MVS method, as depicted in Scheme 1 [25]. This approach guarantees the formation of heterogeneous catalysts containing small Pt NPs (<5 nm), with narrow size distribution, in their reduced form without residues (such as halides derived from metal salt precursors, or stabilizing agents). Moreover, the metal NPs' size can be controlled by exploiting the coordinative properties of the solvent used. In particular, we recently reported that carbon-supported Pt catalysts with a very high metal dispersion and sub-nano-Pt clusters, regardless of the support employed, were obtained by the co-condensation of Pt atoms and mesitylene solvent vapor [26]. Unlike mesitylene, the less-coordinating property of acetone was exploited to prepare Pt-NPs of slightly larger sizes [27]. As supports, we selected two commercially available carbonaceous supports, largely used for (electro)catalytic applications: Vulcan XC72 (C^V) and Ketjenblack EC-600JD (C^K) (see Table S1 for their textural and structural properties). The simple addition of MVS-derived acetone or mesitylene-solvated Pt nanoclusters (Pt_A and Pt_M, respectively) to the desired support (i.e., C^V or C^K) gave the corresponding heterogeneous supported catalysts Pt_A/C^V, Pt_AC^K, and Pt_M/C^K as in Scheme 1. Additionally, each catalyst was subjected to heat treatment at 500 °C under an inert atmosphere for 2 h, to evaluate the stability and possible restructuring of the Pt NPs under these conditions. As a result, six different Pt-based catalysts were obtained, all samples containing a Pt loading of 2.0 wt.%, as confirmed by ICP-OES.

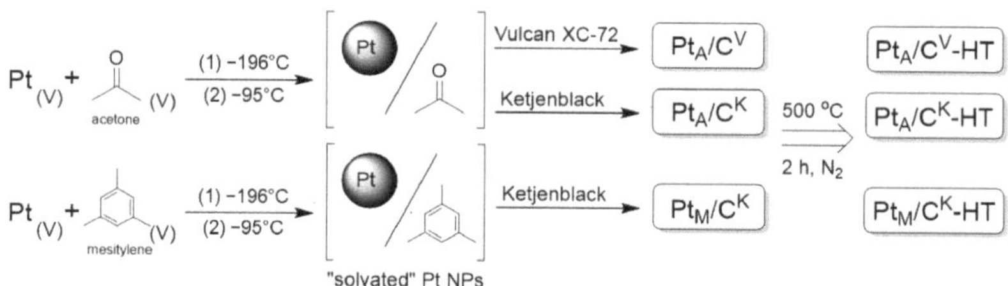

Scheme 1. Schematic synthesis of Pt-based catalysts supported on carbon by MVS method.

An HRTEM analysis of both the freshly prepared catalysts derived from Pt–acetone nanoclusters (i.e., Pt_A/C^K and Pt_A/C^V) showed the presence of Pt-NPs having a mean diameter of 1.5 nm (standard deviation, SD = 0.3 nm), highly dispersed onto the carbon supports (Figure 1). On the other hand, the Pt_M/C^K sample obtained from Pt–mesitylene nanoclusters (Figure S1, Supplementary Information) exhibited, as expected, Pt NPs < 1.5 nm in size, suggesting the presence of sub-nano-atoms/clusters that were not detected by a conventional TEM analysis.

In order to gain further information about the structure of the catalysts, high-angle annular dark field (HAADF) imaging with an aberration-corrected scanning transmission electron microscope (STEM) was conducted. To obtain deeper information about the structure of the catalysts in the sub-nanometer range, HAADF imaging with aberration-corrected STEM was carried out on both Pt–acetone- and Pt–mesitylene-derived catalysts dispersed on C^K carbon. The analysis revealed a strong difference between the two samples that showed a completely diverse arrangement of the Pt atoms in the sub-nanometer range (Figure 2). Indeed, the Pt_M/C^K catalyst showed a relatively high proportion of single atoms (6.2 at.%) and clusters below 1 nm (mean diameter of 0.4 nm, single atoms excluded from calculation) (Figure 2, right side). On the other hand, in the analogous samples prepared from Pt–acetone, no Pt single atoms or small clusters (<0.2 at.%) were observed, but only well-formed NPs centered at 1.5 nm (Figure 2, left side), confirming the above discussed particle size distribution obtained by conventional TEM. An analysis of the Fourier transforms of atomic resolution HAADF-STEM images was conducted to identify crystallographic directions of the nanoparticle facets in samples Pt_M/C^K and Pt_A/C^K. Where NPs formed in the Pt_M/C^K sample, the majority lacked distinct facets, and instead appeared to be amorphous, with structures ranging from circular and ovoid to chain-like branches of atoms. In Pt_A/C^K samples, the majority of the NPs that formed appeared to be circular, amorphous, and therefore non-faceted. However, in the 1.0–1.5 nm diameter range, a minority of FCC NPs with {111} facets were observed.

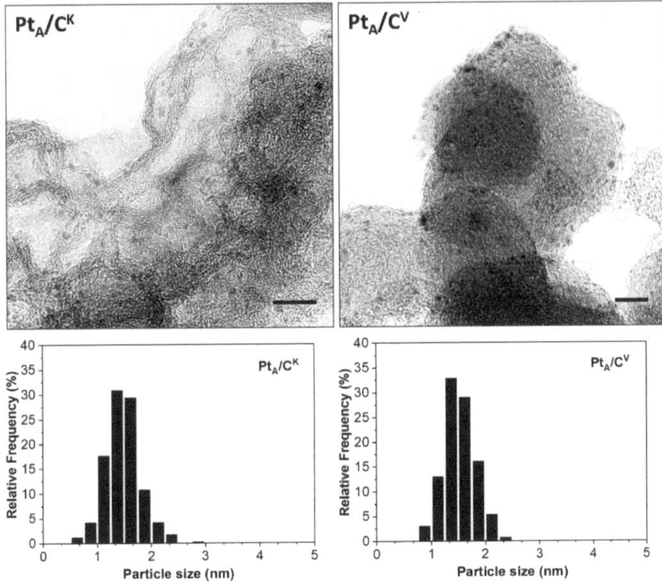

Figure 1. Representative TEM images at high magnification (the bar in the images corresponds to 10 nm) and particle size distributions of Pt_A/C^K (**left side**) and Pt_A/C^V (**right side**).

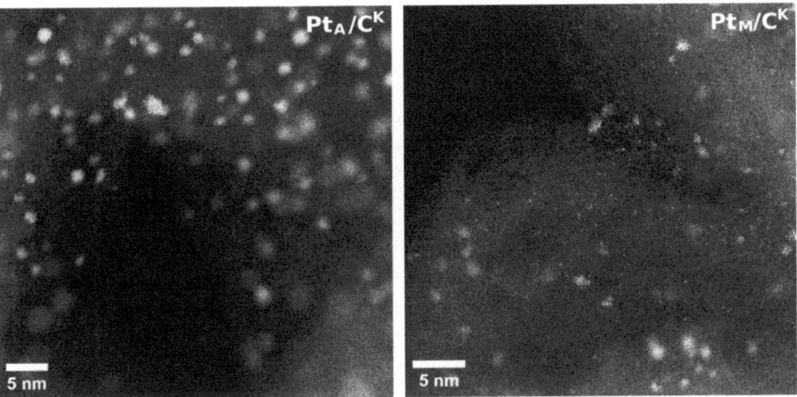

Figure 2. Representative HAADF STEM images of Pt_A/C^K (**left side**) and Pt_M/C^K (**right side**).

The TEM analysis of the heat-treated samples (500 °C, 2 h) demonstrated the crucial role of both the starting NP sizes and the support on the stability of the catalyst against the sintering of the Pt active phase. Interestingly, the best result in terms of stability was observed with the Pt_A/C^K system, where the Pt NPs retained the initial sizes after thermal treatment (d_m = 1.5 nm; SD = 0.3) (Figure 3, left side). The Pt–acetone-derived sample supported on Vulcan carbon showed a slight increase in the NP sizes (d_m = 1.6 nm; SD = 0.6 nm) (Figure 3, right side), demonstrating the high thermal stability of the acetone-derived NPs towards the metal sintering, regardless of the carbon support (i.e., C^K or C^V) used. On the other hand, the mesitylene-derived samples revealed a lower thermal stability. Indeed, for the Pt_M/C^K-HT, a significant increase over the mean NP size (3.6 nm; SD = 7.6) was observed (Figure S2, Supplementary Information). Pt_M/C^K-HT NPs typically fell into one of two size categories, a 1–3 nm diameter or greater than a 10 nm diameter, resulting in a high standard deviation of particle size. While some of the smaller NPs in the 1–3 nm diameter range remained amorphous, the majority crystallized into FCC structures, alongside the larger NPs. These crystalline FCC NPs displayed both {110} and {111} facets, with the {111} facets observed approximately three times more frequently than the {110} facets.

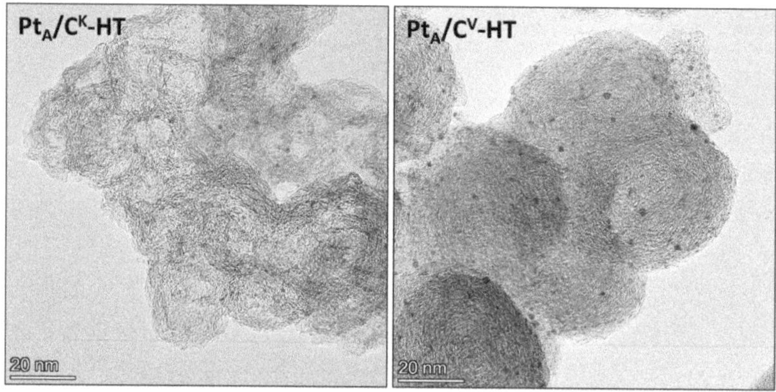

Figure 3. Representative TEM images of Pt_A/C^K-HT (**left side**) and Pt_A/C^V-HT (**right side**).

The mean particle diameters observed for the different carbon-supported Pt catalysts are summarized in Table 1, together with the corresponding Pt dispersions determined by estimating the total surface atoms and bulk atoms in each observed NP [28]. As a result, the freshly prepared Pt–acetone-derived catalysts (i.e., Pt_A/C^K and Pt_A/C^V) give, as expected, very similar metal dispersions of 63% and 62%, respectively (entries 1 and 2, Table 1), whereas the Pt–mesitylene-derived catalyst (entry 3) exhibited a higher dispersion (73%). The results acquired on the heat-treated catalysts revealed a high stability of the NPs in the Pt–acetone-derived samples, which contrasts with the Pt_M/C^K, where due to the NP aggregation, an inhomogeneous size distribution was registered; hence, the estimation of the Pt dispersion based on TEM measurements was not reliable.

Table 1. Characterization of MVS-derived carbon-supported Pt catalysts.

Entry	Catalyst [a]	Mean Pt NP Diameter/Standard Deviation	Dispersion (Calculated)
1	Pt_A/C^K	1.5/0.31 nm	63%
2	Pt_A/C^V	1.5/0.34 nm	62%
3	Pt_M/C^K	0.35/0.35 nm [b]	73% [c]
4	Pt_A/C^K-HT	1.5/0.29 nm	63%
5	Pt_A/C^V-HT	1.6/0.54 nm	50%
6	Pt_M/C^K-HT	3.3/7.38 nm	n.d. [d]

[a] The Pt loading was fixed to 2.0 wt.% for all the catalysts; [b] single-site atoms included; [c] single-site atoms not included; [d] not determined.

The reported results above reveal that the solvent used for NP synthesis by the MVS method plays a crucial role not only in the final Pt NPs' size, but also in the thermal stability of the particles against the sintering process. In the MVS procedure, when the Pt–solvent solid matrix is warmed, metal cluster/particle nucleation takes place and the interaction between the metal atoms and the solvent drives the formation of the final NPs [29]. As discussed above, the interaction of acetone with Pt atoms led to the formation of NPs with a size distribution centered on a diameter of 1.5 nm, with a high stability against sintering at a high temperature. This mean diameter was previously reported to correspond to the formation of metastable Pt NPs containing 147 Pt atoms in both cuboctahedral and icosahedral geometries by DFT [30]. On the other hand, the single atoms and the very small clusters observed in the Pt_M/C^K had the tendency to form larger NPs after heat treatment, suggesting a lower stability to elevated temperature (see Table 1) [31,32].

To further characterize the surface properties of Pt nanoparticles (NPs) in the MVS-derived Pt catalysts, CO-stripping experiments, known for their high sensitivity and reliability for the determination of the surface area of Pt NP surfaces, were conducted [33–35]. The oxidation potential, at which the reaction between surface-adsorbed CO and OH^- groups occurs via a Langmuir–Hinshelwood mechanism, is significantly influenced by the size, morphology, and degree of agglomeration of the NPs. The CV profiles of each catalyst resemble the typical features of platinum (Figure S3, Supplementary Information), with hydrogen adsorption/desorption peaks in the 0–0.1 V vs. RHE range, surface Pt-OH formation around 0.6 V vs. RHE, and the platinum surface oxidation wave at potentials higher than 1 V, followed by the Pt-O reduction in the reverse scan around 0.6 V. The samples using C^K as support show higher differential background capacitance due to the higher surface area of the support itself. CO-stripping experiments (Figure 4) show onset, peak potential, and measured electrochemical accessible surface area (EASA) that are very close for acetone–Pt samples, suggesting that the nature of platinum in each sample is similar. An increase in EASA when heating Pt_A/C^K at 500 °C for 2 h was observed, suggesting that the heat treatment is effective to remove acetone from the micropores of the carbon support, leading to an increase in the exposed metal surface. On the contrary, a shift to a lower onset and high EASA reported on Pt_M/C^K aligned well with the presence of the smallest Pt NPs, Pt clusters, and atoms recorded by the TEM analysis. Additionally, heat treatment of Pt_M/C^K negatively impacted EASA, a reduction of about 60% (Figure 4B). Pt_M/C^K-HT also exhibited a broad and poorly defined J/E curve, consistent with the presence of NPs having a wide size distribution, confirming that a noticeable sintering phenomenon occurred during the heat treatment, as observed with the TEM analysis.

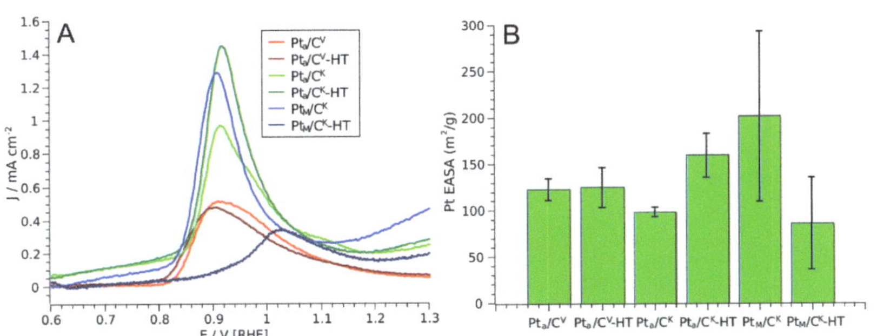

Figure 4. CO-stripping voltammetry (baseline-corrected) of the MVS-derived Pt/C catalysts (**A**); CO stripping-derived Platinum Electrochemical Accessible Surface Area (EASA) of the MVS-derived Pt/C catalysts with error bars (standard deviation) (**B**).

2.2. Catalytic Performance of MVS-Derived Supported Pt Catalysts

2.2.1. Aqueous Phase Reforming of Ethylene Glycol

Ethylene glycol (EG) was selected as a model compound to evaluate the catalytic performance of the carbon-supported Pt catalysts in the APR process. EG is an attractive platform molecule, not only due to its potential as a secondary raw material sourced from biomass, sugars, and alcohols [9], but also because it is the simplest polyol containing both C-C and C-O bonds. Thus, EG reforming can serve as a model process for the direct production of hydrogen from biomass-derived sugars, such as glucose, and sugar alcohols like sorbitol.

The mechanism of hydrogen production from EG was previously extensively investigated by Dumesic et al. [36] and is summarized in Scheme 2. The formation of H_2 and CO_2 takes place by means of C-C and C-O bond cleavage and a further water gas shift reaction (WGS). Meanwhile, side reactions leading to liquid products mainly include (a) dehydration followed by hydrogenation to form ethanol and methanol by further hydrogenolysis; (b) dehydrogenation producing acetic acid. Finally, methane and high-carbon alkanes can be formed via Fischer–Tropsch and methanation reactions in the gas phase.

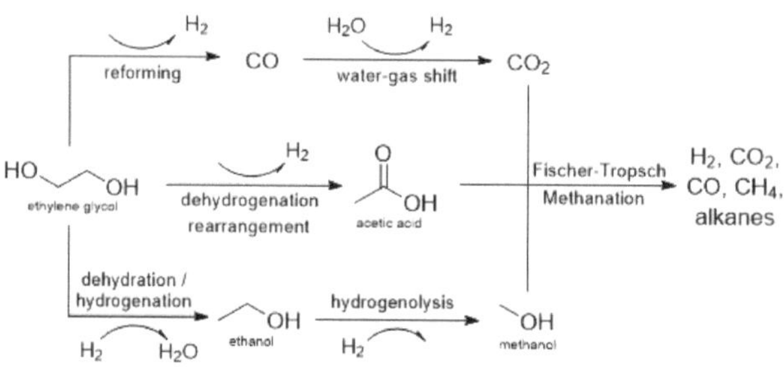

Scheme 2. Reaction pathway of APR of ethylene glycol.

The catalytic results obtained with the Pt-based catalysts are reported in Figure 5 and in Table 2. The turnover frequency (TOF) of the catalyst was calculated as the moles of ethylene glycol (EG) converted per minute per exposed Pt atom, using the measured dispersion. Likewise, the hydrogen site time yield (H_2 STY) was determined as the amount of hydrogen produced per minute per exposed Pt atom. The reactions were performed in mild conditions (T = 220 °C for 4 h) compared to conventional APR conditions in order to properly compare and point out the catalytic properties of the different Pt catalysts.

Generally, a negligible amount of CO, CH_4, and other higher alkanes were found in the gas phase of all catalytic runs. This is expected as Pt-based catalysts are not strong methanation catalysts, while a low temperature favors the WGS, minimizing the formation of CO. Very high selectivity (ranging from 81% to 90%) towards gaseous products (i.e., H_2 and CO_2) was observed for all the catalysts, along with a small amount of liquid products, primarily ethanol and methanol (Figure 5B). No formation of acetic acid was observed. It has been reported that acetic acid could be formed via a bifunctional route: dehydrogenation by the metal phase and rearrangement by an acidic support [36]. In our case, since the supports used (i.e., C^V and C^K) are slightly basic [37], it is likely that the rearrangement step was suppressed, resulting in a negligible amount of acetic acid formed. This is particularly beneficial as carboxylic acid products are commonly undesirable, offering low-value chemicals compared to the feed.

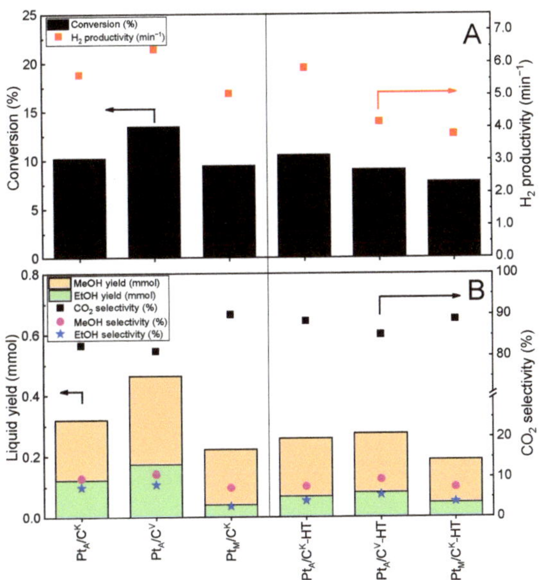

Figure 5. (**A**) EG conversion and hydrogen productivity of supported Pt catalysts. (**B**) Liquid yield and CO_2 selectivity of supported Pt catalysts. Reaction conditions: 220 °C, 4 h, 50 mg Pt/C (2.0 Pt wt.%, 5.1 μmol Pt), 20 mL H_2O, EG 5 wt.%.

Table 2. Numerical data of APR of EG by supported MVS-derived Pt catalysts [a].

Entry	Catalyst	EG Conv. (%)	H_2 Sel. (%)	TOF (min⁻¹)	H_2 STY (min⁻¹)
		Freshly prepared catalyst			
1	Pt_A/C^K	10.2%	83%	2.14 min⁻¹	8.92 min⁻¹
2	Pt_A/C^V	13.9%	77%	2.67 min⁻¹	10.32 min⁻¹
3	Pt_M/C^K	9.5%	81%	1.70 min⁻¹	6.91 min⁻¹
		Heat-treated catalysts (500 °C, 2 h, N_2)			
4	Pt_A/C^K—HT	10.5%	89%	2.06 min⁻¹	9.25 min⁻¹
5	Pt_A/C^V—HT	9.0%	79%	2.07 min⁻¹	8.31 min⁻¹
6	Pt_M/C^K—HT	7.8%	73%	n.d. [b]	n.d.

[a] Reaction conditions: T = 220 °C, t = 4 h, 50 mg Pt/C (2.0 Pt wt.%, 5.1 μmol Pt), 20 mL H_2O, EG 5 wt.%; [b] not determined.

The comparison among the freshly prepared Pt catalysts (left side of Figure 5 and entries 1–3 of Table 2) points out the crucial role of the initial size of the Pt active phase, which was controlled by the solvent used for the catalyst's synthesis (i.e., acetone or mesitylene), and the kind of carbon support (i.e., Vulcan or Ketjenblack) on the activity and stability of the catalyst. The stabilizing effect of the carbonaceous support was clearly observed in the results, particularly when acetone was used as the coordinating solvent. While freshly prepared Pt_A/C^V showed higher activity compared to Pt_A/C^K, it experienced significant deactivation after heat treatment, with about a 30% drop in both ethylene glycol conversion and hydrogen productivity. The TEM analysis revealed that heat treatment caused a larger NP size distribution in Pt_A/C^V (i.e., lower Pt dispersion, see Table 1), likely explaining its lower performance. In contrast, Pt_A/C^K demonstrated exceptional stability up to 500 °C with virtually no change in EG conversion (from 10.2% to 10.5%), H_2 productivity (from 5.62 min⁻¹ to 5.83 min⁻¹), or liquid product distribution. This finding is supported by the unchanged morphology and size of Pt_A/C^K found by TEM characterization. When examining the dispersion of Pt NPs, the calculated

TOF and H_2 site time yield (STY) highlighted the effect of particle size. Pt_A/C^K and Pt_A/C^K-HT shared similar particle size distributions and showed comparable values (8.92 min^{-1} and 9.25 min^{-1}, respectively). The heat-treated Pt_A/C^V-HT, despite showing only a slight increase in particle size and distribution, suffered a significant drop in both TOF and H_2 STY (a reduction of nearly one-fifth in both values) compared to the pristine Pt_A/C^V. This considerable decrease in performance cannot be attributed solely to the loss of the Pt surface due to sintering. Instead, it aligns with previous studies that suggest that quantum effects become significant when NP sizes fall below 1.6 nm [13,15–17]. For example, Yamaguchi and Tai reported a 36% decrease in TOF when Pt particle size increased from 1.5 nm to 1.6 nm. The effect of particle size upon heat treatment was further corroborated by the case of Pt_A/C^K. Thus, the results showed that C^K provides greater stability for supported Pt NPs, particularly those containing NPs of 1.5 nm in size. The enhanced stability of the C^K-supported system could be attributed to a higher specific area and pore volume, which facilitates higher dispersion of Pt NPs [38,39]. Interestingly, hydrogen selectivity for both Pt_A/C^K and Pt_A/C^V slightly improved after heat treatment (from 83% to 89%, and from 77% to 79%, respectively). This could be due to increased particle crystallinity, as well-ordered Pt facets may facilitate C–C bond cleavage. Indeed, heat-treated samples showed higher crystallinity than that observed in the freshly prepared samples (vide supra). This improvement in crystallinity, and consequently in hydrogen selectivity, could also result from exposure to reaction conditions, as discussed in the recycling tests reported below.

Surprisingly, when mesitylene was used as the coordinating solvent, despite the abundance of sub-nanometer Pt sites (i.e., single atoms or clusters), much lower TOF and H_2 STY values were observed (entry 3, Table 2) with respect to the analogous system derived from acetone containing NPs. Moreover, it must be considered that for the calculation of the TOF of the Pt_M/C^K, the contribution of the single Pt atoms was excluded (see dispersion value reported in Table 1). These results suggest that the presence of sub-nanometer Pt atoms and clusters does not boost the catalytic efficiency of this process. This observation can be tentatively explained by the reaction mechanism of APR. Cleavage of C–C and C–O bonds occurs through the adsorption of reactants and intermediates on Pt surfaces. Density Functional Theory (DFT) calculations and microkinetic studies suggest that the most stable binding modes of these intermediates are bidentate or tridentate configurations, which require adjacent Pt atoms or Pt planes to facilitate dissociation reactions [9,40–42].

In contrast to Pt_A catalysts, upon heating to 500 °C, the mesitylene–Pt-based catalyst (i.e., Pt_M/C^K-HT) showed a significant drop in catalytic activity (a reduction of about 25% in both conversion and H_2 productivity) (Figure 5). This strong decrease in performance of the system likely reflects the severe sintering that occurred, as observed by the TEM analysis. Indeed, the high mobility of Pt single atoms and clusters onto the carbon support led to particle size growth in the heat-treated catalysts (see Table 1), which negatively impacted the catalyst's performance. Overall, acetone-derived Pt NPs dispersed on C^K support proved to be the most effective catalyst for the APR of EG, in terms of both activity and stability under heat treatment. Additionally, a comparison between commercial Pt/Al_2O_3 (Sigma-Aldrich, St. Louis, MO, USA) containing Pt NPs with a mean diameter of 4.7 nm (Figure S5), serving as a benchmark catalyst, showed a remarkable improvement in Pt_A/C^K in EG conversion and H_2 productivity (10.2% vs. 5.5% and 5.62 min^{-1} vs. 3.04 min^{-1}), demonstrating the superior efficiency of our MVS-derived catalyst (Figure S4).

2.2.2. Recycling Tests

Since Pt_A/C^K was the best-performing catalyst among those screened, we proceeded to investigate the stability of this catalyst with recycling tests. In the recycling test, a higher amount of catalyst was used to minimize the effect of material loss during recovery. The reaction was performed at 240 °C for 2 h under 7 bar N_2. After collecting the gas and liquid samples, we filtered and dried the catalyst, followed by running three more catalytic reactions under the same experimental conditions. The results are shown in Figure 6. It can be clearly seen that the catalyst possesses good stability with a slight decrease in EG conversion only after the first run (from 30.9% to 22.2%), then stabilizing in the consecutive runs. On the contrary, the H_2 productivity did not experience any significant change (from 4.96 min^{-1} Pt to 4.82 min^{-1}). In the liquid phase, after the first run, where ethanol and methanol selectivity were high, the following runs showed consistent liquid composition. It is likely that the catalyst underwent restructuring in the first run to be stabilized by the reaction environment. In fact, the observed decrease in the yield of the liquid products along with an increased H_2 selectivity was similar to the heat treatment case. Afterwards, the catalyst performed consistently, indicating that it reached a stable configuration. The TEM analysis of the catalyst after four runs (Figure S6) shows only a slight increase in the mean NP size (dm = 1.7 nm). Moreover, the ICP-OES analysis of the used catalyst showed negligible leaching of Pt into the solution, confirming the stability of our catalyst. In conclusion, our Pt_A/C^K is a promising catalyst for APR thanks to its high stability and activity.

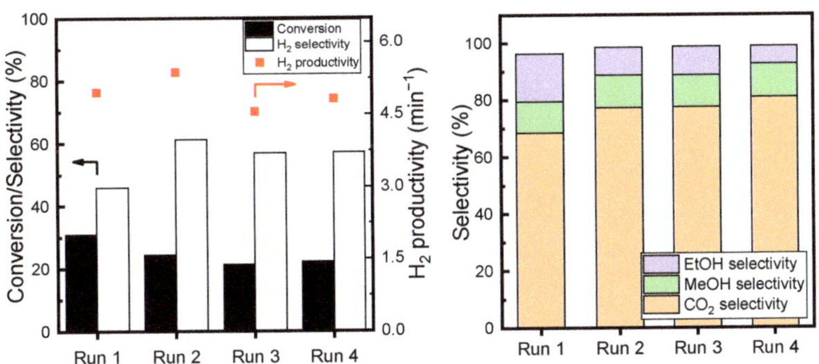

Figure 6. Recycling test of Pt_A/C^K in APR of EG. Reaction conditions: 240 °C, 2 h, 200 mg of catalyst (2.0 Pt wt.%, 20.5 µmol Pt), 20 mL H_2O, EG 5 wt.%.

2.2.3. Aqueous Phase Reforming of PET

In this study, we utilized the highest-performing MVS catalyst (Pt_A/C^K) to explore hydrogen generation through the APR of polyethylene terephthalate (PET), a primary plastic accounting for 8% by weight of solid waste worldwide [43]. Our data are shown in Table 3. A blank test shows that depolymerization reaches completion with 100% recovery of EG in 2 h at 240 °C under 7 bar N_2. From the results, H_2 productivity of Pt_A/C^K was found to be 1.14 min^{-1} (entry 1, Table 3). Increasing the reaction time and the amount of the catalyst resulted in lower productivity (entry 2, Table 3). This is likely due to a negative effect of the formation of products that slowed down the reaction rate. Indeed, the formation of gaseous products, resulting in an increase in pressure in the batch condition, was reported to have a negative effect on the APR reaction [8].

Table 3. Aqueous Phase Reforming of PET under neutral condition [a].

Entry	Catalyst	PET (g) /Metal (mol)	Time	EG Conv. (%)	H$_2$ Prod. (min^{-1})	Ref.
1	Pt_A/C^K	136,500	2 h	12.8%	1.14	This work
2	Pt_A/C^K	68,250	4 h	38.5%	0.87	This work
3 [b]	Pt_A/C^K	68,250	2 h	0%	0	This work
4 [c]	Pt/Al$_2$O$_3$ com	68,250	4 h	10.3%	0.13	This work
5	Ru/MEC	10,100	-	-	0.67	[44]
6	Ru/ZnO/MEC	10,100	-	-	1.26	[45]
7	Pt/ENS	4333	-	-	0.9	[5]

[a] Reaction conditions: 240 °C, Pt catalyst, 50 mL H_2O, 700 mg of crushed PET; [b] NaOH 4 wt.%; [c] commercial Pt/Al$_2$O$_3$ 1 wt.%, purchased from Sigma-Aldrich (Milwaukee, WI, USA).

The productivity of the Pt_A/C^K for APR of PET was lower than that obtained in the APR process of EG, carried out in analogous reaction conditions (i.e., 4.96 min^{-1}, see Figure 5, Run 1). In addition to the fact that the depolymerization reaction happened quantitatively, it is likely that the presence of produced terephthalic acid (TPA) has a negative effect on APR of EG. We observed a similar phenomenon with the commercial Pt/Al_2O_3 catalyst, which exhibited much lower productivity in the APR of PET than in the case of pure EG (entry 4, Table 3 and Figure S4, Supplementary Information). This behavior is distinctively different from the Ru-based catalyst reported by Su et al., where TPA did not influence the activity of the catalyst [45]. Furthermore, the authors reported that basic conditions are beneficial for the process. On the other hand, in our case, the use of basic conditions showed a detrimental effect on the efficiency of the Pt-based catalyst (entry 3, Table 3). Increasing the pH of the solution resulted in negatively charged terephthalates, which might strongly coordinate to Pt atoms of the electron-deficient Pt surface, preventing the absorption of EG and thus deactivating the catalyst [46]. In fact, several studies have reported the inhibition effect of aromatic carboxylic acid on noble metal particle growth and stabilization [47,48].

Literature examples concerning the usage of heterogeneous catalysts for APR of PET plastic are still limited. In Table 3, we compared our catalyst's productivity with all reported studies on this subject to date. As all of the studies were performed at different conditions and reactor configurations, a direct comparison of the catalysts is challenging. Nevertheless, we chose H_2 productivity as a representative indicator as it could somewhat reflect the intrinsic catalytic activity of the materials. As shown, Pt_A/C^K was among the highest H_2 productivities among examined catalysts under neutral conditions. Our catalyst can perform efficiently without the assistance of a base (i.e., corrosiveness of the base is avoided), in contrast to Ru-based catalysts [44]. A further optimization of operating parameters for the best-performing Pt catalyst will be investigated in a forthcoming study.

3. Materials and Methods

3.1. Catalyst Synthesis

The Pt nanoparticles were generated by metal vapor synthesis (MVS), which was reported previously [26]. To prepare the supported Pt catalysts, platinum vapor was generated at 10^{-5} mbar by resistive heating of a tungsten wire coated with approximately 100 mg of electrodeposited platinum. This metal vapor was co-condensed with 100 mL of solvent (either acetone or mesitylene) vapor at $-196\ ^{\circ}C$ (liquid nitrogen) in the MVS reactor. The reactor was then warmed to the melting point of the solid matrix ($-95\ ^{\circ}C$ for Pt–acetone matrix and $-40\ ^{\circ}C$ for Pt–mesitylene), resulting in a dark brown solution (95 mL). This solvated Pt solution was siphoned at $-40\ ^{\circ}C$ into a Schlenk tube and stored at $-20\ ^{\circ}C$. The ICP-OES analysis showed that the Pt solution contained approximately 0.5 mg/mL Pt. To obtain the supported catalysts, the solutions were added to the suspension of carbonaceous supports (Vulcan XC-72 and/or Ketjenblack EC600J) in the matching solvent. The mixtures were stirred at 25 $^{\circ}C$ for 20 h to ensure complete deposition. The solids were collected after washing three times with 50 mL of *n*-pentane each. Finally, the catalysts were dried under reduced pressure at room temperature.

For heat treatment of catalysts, a set amount of a catalyst (150 mg) was loaded into a ceramic boat and heated under N_2 flow (100 mL/min) in a tubular furnace to a designated temperature. The heating rate was 10 $^{\circ}C$/min and the dwell time was 2 h at 500 $^{\circ}C$. The heat-treated catalysts were labeled with "HT" (for heat-treated) as a suffix.

3.2. Catalyst Characterization

TEM imaging was performed using a Talos™ F200X G2 TEM (Thermo Scientific, Waltham, MA, USA). Samples for the analysis were prepared by their suspension in isopropyl alcohol and were ultrasonically dispersed followed by the deposition of a drop of the suspension on a holey carbon-coated copper grid (300 mesh). The histograms of the metal particle size distribution were obtained by counting at least 300 particles in the micrographs. The mean particle diameter (d_m) was calculated by using the formula $d_m = \sum d_i n_i / \sum n_i$, where ni is the number of particles with diameter d_i. Metal dispersion was calculated based on the work of Borodziński et al. [28].

High-angle annular dark field (HAADF) STEM images were acquired on a JEOL ARM200F microscope and a JEOL ARM300F (JEOL, Tokyo, Japan) at the electron Physical Science Imaging Centre (ePSIC) at Diamond Light Source. The JEOL ARM200F was operated at an acceleration voltage of 200 kV, with a convergence semi-angle of 23 mrad and an inner collection angle of 79 mrad. The JEOL ARM300F was operated at an acceleration voltage of 300 kV, a convergence semi-angle of 26.2 mrad, a beam current of 25 pA, and an approximate inner collection angle of 92.6 mrad. Further HAADF-STEM images were acquired on a Thermo Fisher Spectra 200 (Thermo Fischer, Waltham, MA, USA) at Cardiff University, which was operated at an accelerating voltage of 200 kV, a beam current of 60 pA, and a convergence semi-angle of 29.5 mrad. The HAADF images were collected with an inner collection angle of 56 mrad, and an outer collection angle of 200 mrad. A 5–15 min beam shower was completed prior to imaging on all microscopes to reduce contamination. Samples for electron microscopy were prepared by their suspension in isopropyl alcohol and ultrasonic dispersion, followed by a deposition of a drop of the suspension onto a 300-mesh holey carbon-coated grid of either molybdenum, copper, or gold. The particle size distribution analysis was completed via the ParticleSpy Python package (reference to https://zenodo.org/records/5094360, accessed on 30 July 2023). Automated particle detection and measurement were performed using Otsu filtering, with the manual counting verification of single atoms to ensure accuracy.

ICP-OES analyses were carried out with an ICP-Optical emission dual-view Perkin Elmer OPTIMA 8000 apparatus (Perkin Elmer, Waltham, MA, USA). For analyzing the MVS solution, 0.5 mL of a Pt-SMA solution was heated over a heating plate in a porcelain crucible to remove the solvent, followed by dissolving the solid residue in 2 mL of aqua regia in 6 h. The solution was then diluted with 50 mL of HCl 0.5 M. For catalyst loading measurements, approximately 10 mg of a catalyst was dissolved in 4 mL of aqua regia and fluxed for 6 h. The solution was then diluted with 50 mL of HCl 0.5 M. The quantification of Pt was performed with calibration with Pt solutions of known concentrations. The limit of detection (l.o.d) calculated for platinum was 2 ppb. In all cases, the results of MVS solution and catalyst loading were highly consistent. A quantitative deposition of NPs onto the support was obtained, leading to a final 2.0 wt.% Pt loading for all the catalysts.

The CO-stripping analysis was performed as follows: each catalyst sample powder was dispersed (10 mg) in a mixture of 100 μL ultrapure water and 1 g 2-propanol by using an ultrasonic bath for 60 min. The homogeneous ink was then cast onto a glassy carbon disk (Pine™ Electrodes, A = 0.1963 cm^2) in two steps: 5 μL of ink was added in 2.5 μL steps using a 2–20 μL micro-pipette and dried after each step; the ink quantity was determined by weight before drying. After complete drying, 1.6 μL of 0.5% Nafion in a 2-propanol solution was cast on top of the catalyst layer. Three replicates were prepared for each sample. After that, the electrode was used for cyclic voltammetry experiments in a 3-electrode cell, using a platinum gauze as a counter-electrode and Ag | AgCl | KCl^{sat} as a reference electrode. The electrolyte (0.5 M H$_2$SO$_4$, pH 0.3) was purged with nitrogen for 30 min prior to the experiments. The electrode was first scanned for 8 cycles between −0.2 V and 1.2 V vs. Ag | AgCl | KCl^{sat}; after that, the solution was saturated with pure CO by bubbling for 30 min. Then, a potential of 0 V vs. Ag | AgCl | KCl^{sat} was applied for 30 min. Then, still applying the potential, the solution was purged with N$_2$ for 30 min and then the potential was scanned between 0 V and 1.2 V vs. Ag | AgCl | KCl^{sat} for three scans. The first scan showed an oxidation peak around 0.7 V vs. Ag | AgCl | KCl^{sat}, corresponding to the oxidation of chemisorbed CO. The area was determined by integrating the charge of the peak and dividing by the CO charge for cm^2 (0.420 mC/cm^2, for linearly bond CO) and then dividing by the metal loading.

3.3. Catalytic Tests

Aqueous phase reforming of ethylene glycol was carried out in a 100 mL stainless steel autoclave equipped with a mechanical stirrer and a temperature controller. The catalyst (equivalent to 5.1 μmol Pt) and 20 mL of a 5 wt.% aqueous EG solution were loaded into the vessel. The reactor was thoroughly purged with nitrogen and pressurized to an initial pressure of 10 bar, which served as an internal standard for the final quantification of the products present in the gas phase. The mixture was subsequently stirred at 1000 rpm and heated to 220 °C. Heating time was approximately 10 min. After 4 h at 220 °C, the reactor was quenched to room temperature in an ice water bath. Gaseous products were collected in a multilayer foil gas sampling bag (Supelco, Bellefonte, PA, USA). After depressurization, the liquid products were collected and filtered. The spent catalyst was recovered by filtration, washed with a copious amount of DI water, and dried overnight at 100 °C. Similar procedures were performed at 240 °C, 7 bar of N_2, and for 2 h.

Gas and liquid phase products were analyzed by gas chromatography (Shimadzu 2010Pro; Shimadzu, Kyoto, Japan) equipped with a thermal conductivity detector (TCD) and a flame ionization detector (FID). H_2, CO, CO_2, N_2, and CH_4 were detected by TCD using a Carboxen 1010 (Supelco, Bellefonte, PA, USA). Nitrogen was used as an internal standard for the quantification of permanent gasses and methane. Volatile liquid phase products (methanol, ethanol, acetic acid, unreacted ethylene glycol) were also analyzed by gas chromatography equipped with a CP-Wax 52 CB GC column (Agilent, Santa Clara, CA, USA) and a flame ionization detector (FID). The quantification of the products was performed with a calibration curve using 1-propanol as an internal standard.

Recycling tests were performed as follows: 200 mg of a catalyst was loaded into the reactor and the reaction was performed at 240 °C for 2 h. After the catalyst was filtered, and washed with a copious amount of DI water, it was dried in an oven at 105 °C overnight. Then, the resulting solid was loaded back into an autoclave to perform the next cycle. Gas and liquid products were analyzed as previously mentioned. The procedure was repeated 4 times. Loss of the catalyst during handling was taken into account in the next cycle.

Aqueous phase reforming of PET was performed in a 100 mL stainless steel autoclave equipped with a mechanical stirrer and a temperature controller. Typically, 100 mg of a catalyst was charged with 700 mg of crushed PET and 20 mL of water in the reactor. The reactor was purged with N_2 several times and charged with 7 bar of N_2, followed by heating up to 240 °C within 15 min. After 4 h, the reactor was cooled with an ice bath. Gas was collected with a gas bag and the solid was separated by filtration. The gas phase and liquid phase products were analyzed by GC-FID. The solid phase was washed with concentrated NaOH to dissolve terephthalic acid (TPA), followed by neutralization to recover TPA crystals.

4. Conclusions

To summarize, Pt NPs of different size distributions (mean diameter below 2 nm) were deposited on commercially available carbons (i.e., Vulcan XC72, C^V, and/or Ketjenblack EC-600JD, C^K) following different synthesis protocols by the metal vapor synthesis method. Their catalytic efficiency was evaluated and compared for the aqueous phase reforming of ethylene glycol. The most active and stable catalyst, Pt_A/C^K, displayed a mean Pt NP size of 1.5 nm, was obtained from Pt–acetone, and was highly dispersed on the carbon support with the highest surface area (i.e., C^K). The system exhibits very high H_2 site time yield (8.92 min^{-1} at 220 °C) compared to a commercially available alumina-supported Pt catalyst. High stability towards metal sintering was proven for both high-temperature treatment (i.e., 500 °C) and several batch recycling runs under the reaction conditions (240 °C) with negligible H_2 yield loss and/or metal leaching. In contrast, an analogous system supported on the same carbon but containing a major fraction of sub-nanometer Pt single atoms and clusters showed an inferior catalytic efficiency together with a low stability towards the formation of large particles (>3 nm). Our study suggests an optimal Pt particle size of around 1.5 nm for this process as a good compromise between Pt dispersion and stability.

The MVS-derived catalyst was also effectively applied to the APR process of polyethylene terephthalate (PET), showing potential for H_2 production from recycled plastics. This work highlights the fundamental role of the use of advanced synthesis approaches that are able to strictly control metal nanoparticle size and the kind of support materials in optimizing APR processes for sustainable hydrogen production.

Supplementary Materials: The following supporting information can be downloaded at: https://www.mdpi.com/article/10.3390/catal14110798/s1, Section S1: Parameter for Evaluation of APR process; Table S1. Morphological properties of CV and CK; Figure S1. Representative TEM image at high magnification (bar in images corresponds to 10 nm) of PtM/CK; Figure S2. Representative HAADF-ac-STEM images of PtM/CK-HT; Figure S3: Blank voltammetry (8th scan) of the various Pt/C samples; Figure S4. EG conversion and hydrogen productivity of PtA/CK and commercial Pt/Al2O3; Figure S5. Representative TEM images and Pt particle size distribution of commercial Pt/Al2O3 1 wt.%; Figure S6. Representative TEM image and Pt particle size distribution of PtA/CK after 4 catalytic runs; Table S2. Gas composition of APR of PET with Pt-based catalyst.

Author Contributions: Conceptualization, X.T.N. and C.E.; methodology, X.T.N. and C.E.; validation, X.T.N., E.P. and C.E.; formal analysis, X.T.N., E.K. and J.F.; data curation, X.T.N., E.K. and C.E.; writing—original draft preparation, X.T.N.; writing—review and editing, X.T.N., E.K., T.S., E.P., J.F., W.O. and C.E.; visualization, X.T.N., E.K. and C.E.; supervision, T.S. and C.E.; project administration, C.E.; funding acquisition, C.E. All authors have read and agreed to the published version of the manuscript.

Funding: This research was funded by the European Union—NextGeneration EU through the Italian Ministry of Environment and Energy Security POR H_2 AdP MMES/ENEA with involvement of CNR and RSE, PNRR—Mission 2, Component 2, Investment 3.5 "Ricerca e sviluppo sull'idrogeno", CUP: B93C22000630006. XT.N. and C. E. would also like to thank the European Union's Horizon 2020 Research and Innovation Program under the Marie Skłodowska-Curie Actions—Innovative Training Networks (MSCA-ITN) BIKE project, Grant Agreement 813748.

Data Availability Statement: All data generated or analyzed during this study are included in this published article and its Supplementary Information Files.

Acknowledgments: The authors thank Mohsen Danaie from ePSIC, Diamond Light Source, Harwell Science & Innovation Campus, UK, for his contribution to HAADF STEM images. We also thank Carlo Bartoli from ICCOM-CNR—Florence for his assistance in the catalytic experiment. We also thank Diamond Light Source for access and support in use of the electron Physical Science Imaging Centre (Instrument E01 and proposal number MG33438) that contributed to the results presented here. We would like to thank the ERDF (European Regional Development Fund) and Wolfson Foundation for funding the CCI (Cardiff Catalysis Institute) Electron Microscopy Facility.

Conflicts of Interest: The authors declare no conflicts of interest.

References

1. Dunn, S. Hydrogen Futures: Toward a Sustainable Energy System. *Int. J. Hydrogen Energy* **2002**, *27*, 235–264. [CrossRef]
2. Cortright, R.D.; Davda, R.R.; Dumesic, J.A. Hydrogen from Catalytic Reforming of Biomass-Derived Hydrocarbons in Liquid Water. *Nature* **2002**, *418*, 964–967. [CrossRef] [PubMed]
3. González-Arias, J.; Zhang, Z.; Reina, T.R.; Odriozola, J.A. Hydrogen Production by Catalytic Aqueous-Phase Reforming of Waste Biomass: A Review. *Environ. Chem. Lett.* **2023**, *21*, 3089–3104. [CrossRef]
4. Geyer, R.; Jambeck, J.R.; Law, K.L. Production, Use, and Fate of All Plastics Ever Made. *Sci. Adv.* **2017**, *3*, e1700782. [CrossRef]
5. Ruiz-Garcia, C.; Baeza, J.A.; Oliveira, A.S.; Roldán, S.; Calvo, L.; Gilarranz, M.A. Exploration of Operating Conditions in the Direct Aqueous-Phase Reforming of Plastics. *Fuel* **2024**, *374*, 132446. [CrossRef]
6. Davda, R.R.; Shabaker, J.W.; Huber, G.W.; Cortright, R.D.; Dumesic, J.A. A Review of Catalytic Issues and Process Conditions for Renewable Hydrogen and Alkanes by Aqueous-Phase Reforming of Oxygenated Hydrocarbons over Supported Metal Catalysts. *Appl. Catal. B Environ.* **2005**, *56*, 171–186. [CrossRef]
7. Chen, G.; Li, W.; Chen, H.; Yan, B. Progress in the Aqueous-Phase Reforming of Different Biomass-Derived Alcohols for Hydrogen Production. *J. Zhejiang Univ.-Sci. A* **2015**, *16*, 491–506. [CrossRef]
8. Pipitone, G.; Zoppi, G.; Pirone, R.; Bensaid, S. A Critical Review on Catalyst Design for Aqueous Phase Reforming. *Int. J. Hydrogen Energy* **2022**, *47*, 151–180. [CrossRef]
9. Davda, R.R.; Shabaker, J.W.; Huber, G.W.; Cortright, R.D.; Dumesic, J.A. Aqueous-Phase Reforming of Ethylene Glycol on Silica-Supported Metal Catalysts. *Appl. Catal. B Environ.* **2003**, *43*, 13–26. [CrossRef]

10. Huber, G.W.; Shabaker, J.W.; Evans, S.T.; Dumesic, J.A. Aqueous-Phase Reforming of Ethylene Glycol over Supported Pt and Pd Bimetallic Catalysts. *Appl. Catal. B Environ.* **2006**, *62*, 226–235. [CrossRef]

11. Boudart, M. Catalysis by Supported Metals**This Work Was Sponsored by Grant GK 648 of the National Science Foundation. Its Continuity over the Years Has Been Assured by a Generous Grant from the Petroleum Research Fund of the American Chemical Society. In *Advances in Catalysis*; Eley, D.D., Pines, H., Weisz, P.B., Eds.; Academic Press: Cambridge, MA, USA, 1969; Volume 20, pp. 153–166. ISBN 0360-0564.

12. Lehnert, K.; Claus, P. Influence of Pt Particle Size and Support Type on the Aqueous-Phase Reforming of Glycerol. *Catal. Commun.* **2008**, *9*, 2543–2546. [CrossRef]

13. Wawrzetz, A.; Peng, B.; Hrabar, A.; Jentys, A.; Lemonidou, A.A.; Lercher, J.A. Towards Understanding the Bifunctional Hydrodeoxygenation and Aqueous Phase Reforming of Glycerol. *J. Catal.* **2010**, *269*, 411–420. [CrossRef]

14. Barbelli, M.L.; Pompeo, F.; Santori, G.F.; Nichio, N.N. Pt Catalyst Supported on α-Al$_2$O$_3$ Modified with CeO$_2$ and ZrO$_2$ for Aqueous-Phase-Reforming of Glycerol. *Catal. Today* **2013**, *213*, 58–64. [CrossRef]

15. Kim, Y.; Kim, M.; Jeong, H.; Kim, Y.; Choi, S.H.; Ham, H.C.; Lee, S.W.; Kim, J.Y.; Song, K.H.; Yoon, C.W.; et al. High Purity Hydrogen Production via Aqueous Phase Reforming of Xylose over Small Pt Nanoparticles on a γ-Al$_2$O$_3$ Support. *Int. J. Hydrogen Energy* **2020**, *45*, 13848–13861. [CrossRef]

16. Chen, A.; Chen, P.; Cao, D.; Lou, H. Aqueous-Phase Reforming of the Low-Boiling Fraction of Bio-Oil for Hydrogen Production: The Size Effect of Pt/Al$_2$O$_3$. *Int. J. Hydrogen Energy* **2015**, *40*, 14798–14805. [CrossRef]

17. Yamaguchi, W.; Tai, Y. Size-Dependent Catalytic Activity of Platinum Nanoparticles for Aqueous-Phase Reforming of Glycerol. *Chem. Lett.* **2014**, *43*, 313–315. [CrossRef]

18. Cui, X.; Li, W.; Ryabchuk, P.; Junge, K.; Beller, M. Bridging Homogeneous and Heterogeneous Catalysis by Heterogeneous Single-Metal-Site Catalysts. *Nat. Catal.* **2018**, *1*, 385–397. [CrossRef]

19. Li, Z.; Ji, S.; Liu, Y.; Cao, X.; Tian, S.; Chen, Y.; Niu, Z.; Li, Y. Well-Defined Materials for Heterogeneous Catalysis: From Nanoparticles to Isolated Single-Atom Sites. *Chem. Rev.* **2020**, *120*, 623–682. [CrossRef]

20. Shabaker, J.W.; Davda, R.R.; Huber, G.W.; Cortright, R.D.; Dumesic, J.A. Aqueous-Phase Reforming of Methanol and Ethylene Glycol over Alumina-Supported Platinum Catalysts. *J. Catal.* **2003**, *215*, 344–352. [CrossRef]

21. Koichumanova, K.; Vikla, A.K.K.; de Vlieger, D.J.M.; Seshan, K.; Mojet, B.L.; Lefferts, L. Towards Stable Catalysts for Aqueous Phase Conversion of Ethylene Glycol for Renewable Hydrogen. *ChemSusChem* **2013**, *6*, 1717–1723. [CrossRef]

22. Antal, M.J., Jr.; Allen, S.G.; Schulman, D.; Xu, X.; Divilio, R.J. Biomass Gasification in Supercritical Water. *Ind. Eng. Chem. Res.* **2000**, *39*, 4040–4053. [CrossRef]

23. Vikla, A.K.K.; Simakova, I.; Demidova, Y.; Keim, E.G.; Calvo, L.; Gilarranz, M.A.; He, S.; Seshan, K. Tuning Pt Characteristics on Pt/C Catalyst for Aqueous-Phase Reforming of Biomass-Derived Oxygenates to Bio-H$_2$. *Appl. Catal. A Gen.* **2021**, *610*, 117963. [CrossRef]

24. Pitzalis, E.; Psaro, R.; Evangelisti, C. From Metal Vapor to Supported Single Atoms, Clusters and Nanoparticles: Recent Advances to Heterogeneous Catalysts. *Inorganica Chim. Acta* **2022**, *533*, 120782. [CrossRef]

25. Oberhauser, W.; Evangelisti, C.; Tiozzo, C.; Bartoli, M.; Frediani, M.; Passaglia, E.; Rosi, L. Platinum Nanoparticles onto Pegylated Poly(Lactic Acid) Stereocomplex for Highly Selective Hydrogenation of Aromatic Nitrocompounds to Anilines. *Appl. Catal. A Gen.* **2017**, *537*, 50–58. [CrossRef]

26. Oberhauser, W.; Evangelisti, C.; Jumde, R.P.; Psaro, R.; Vizza, F.; Bevilacqua, M.; Filippi, J.; Machado, B.F.; Serp, P. Platinum on Carbonaceous Supports for Glycerol Hydrogenolysis: Support Effect. *J. Catal.* **2015**, *325*, 111–117. [CrossRef]

27. Oberhauser, W.; Evangelisti, C.; Tiozzo, C.; Vizza, F.; Psaro, R. Lactic Acid from Glycerol by Ethylene-Stabilized Platinum-Nanoparticles. *ACS Catal.* **2016**, *6*, 1671–1674. [CrossRef]

28. Borodziński, A.; Bonarowska, M. Relation between Crystallite Size and Dispersion on Supported Metal Catalysts. *Langmuir* **1997**, *13*, 5613–5620. [CrossRef]

29. Klabunde, K.J.; Li, Y.X.; Tan, B.J. Solvated Metal Atom Dispersed Catalysts. *Chem. Mater.* **1991**, *3*, 30–39. [CrossRef]

30. Al-Shareef, R.; Harb, M.; Saih, Y.; Ould-Chikh, S.; Anjum, D.H.; Candy, J.-P.; Basset, J.-M. Precise Control of Pt Particle Size for Surface Structure–Reaction Activity Relationship. *J. Phys. Chem. C* **2018**, *122*, 23451–23459. [CrossRef]

31. Chen, J.; Chan, K.-Y. Size-Dependent Mobility of Platinum Cluster on a Graphite Surface. *Mol. Simul.* **2005**, *31*, 527–533. [CrossRef]

32. Martin, T.E.; Mitchell, R.W.; Boyes, E.D.; Gai, P.L. Atom-by-Atom Analysis of Sintering Dynamics and Stability of Pt Nanoparticle Catalysts in Chemical Reactions. *Philos. Trans. R. Soc. A Math. Phys. Eng. Sci.* **2020**, *378*, 20190597. [CrossRef] [PubMed]

33. Maillard, F.; Savinova, E.R.; Simonov, P.A.; Zaikovskii, V.I.; Stimming, U. Infrared Spectroscopic Study of CO Adsorption and Electro-Oxidation on Carbon-Supported Pt Nanoparticles: Interparticle versus Intraparticle Heterogeneity. *J. Phys. Chem. B* **2004**, *108*, 17893–17904. [CrossRef]

34. Maillard, F.; Schreier, S.; Hanzlik, M.; Savinova, E.R.; Weinkauf, S.; Stimming, U. Influence of Particle Agglomeration on the Catalytic Activity of Carbon-Supported Pt Nanoparticles in CO Monolayer Oxidation. *Phys. Chem. Chem. Phys.* **2005**, *7*, 385–393. [CrossRef]

35. Ciapina, E.G.; Santos, S.F.; Gonzalez, E.R. Electrochemical CO Stripping on Nanosized Pt Surfaces in Acid Media: A Review on the Issue of Peak Multiplicity. *J. Electroanal. Chem.* **2018**, *815*, 47–60. [CrossRef]

36. Shabaker, J.W.; Dumesic, J.A. Kinetics of Aqueous-Phase Reforming of Oxygenated Hydrocarbons: Pt/Al$_2$O$_3$ and Sn-Modified Ni Catalysts. *Ind. Eng. Chem. Res.* **2004**, *43*, 3105–3112. [CrossRef]

37. Amin, A.S.; Caidi, A.; Lange, T.; Radev, I.; Sandbeck, D.J.S.; Philippi, W.; Kräenbring, M.-A.; Öztürk, M.; Peinecke, V.; Lerche, D.; et al. Key Control Characteristics of Carbon Black Materials for Fuel Cells and Batteries for a Standardized Characterization of Surface Properties. *Part. Part. Syst. Charact.* **2024**, 2400069. [CrossRef]
38. Bevilacqua, M.; Bianchini, C.; Marchionni, A.; Filippi, J.; Lavacchi, A.; Miller, H.; Oberhauser, W.; Vizza, F.; Granozzi, G.; Artiglia, L.; et al. Improvement in the Efficiency of an OrganoMetallic Fuel Cell by Tuning the Molecular Architecture of the Anode Electrocatalyst and the Nature of the Carbon Support. *Energy Environ. Sci.* **2012**, *5*, 8608–8620. [CrossRef]
39. Yang, Z.; Kim, C.; Hirata, S.; Fujigaya, T.; Nakashima, N. Facile Enhancement in CO-Tolerance of a Polymer-Coated Pt Electrocatalyst Supported on Carbon Black: Comparison between Vulcan and Ketjenblack. *ACS Appl. Mater. Interfaces* **2015**, *7*, 15885–15891. [CrossRef]
40. Kandoi, S.; Greeley, J.; Simonetti, D.; Shabaker, J.; Dumesic, J.A.; Mavrikakis, M. Reaction Kinetics of Ethylene Glycol Reforming over Platinum in the Vapor versus Aqueous Phases. *J. Phys. Chem. C* **2011**, *115*, 961–971. [CrossRef]
41. Faheem, M.; Saleheen, M.; Lu, J.; Heyden, A. Ethylene Glycol Reforming on Pt(111): First-Principles Microkinetic Modeling in Vapor and Aqueous Phases. *Catal. Sci. Technol.* **2016**, *6*, 8242–8256. [CrossRef]
42. Gu, G.H.; Wittreich, G.R.; Vlachos, D.G. Microkinetic Modeling of Aqueous Phase Biomass Conversion: Application to Ethylene Glycol Reforming. *Chem. Eng. Sci.* **2019**, *197*, 415–418. [CrossRef]
43. George, N.; Kurian, T. Recent Developments in the Chemical Recycling of Postconsumer Poly(Ethylene Terephthalate) Waste. *Ind. Eng. Chem. Res.* **2014**, *53*, 14185–14198. [CrossRef]
44. Su, H.; Li, T.; Wang, S.; Zhu, L.; Hu, Y. Low-Temperature Upcycling of PET Waste into High-Purity H_2 Fuel in a One-Pot Hydrothermal System with in Situ CO_2 Capture. *J. Hazard. Mater.* **2023**, *443*, 130120. [CrossRef] [PubMed]
45. Su, H.; Hu, Y.; Feng, H.; Zhu, L.; Wang, S. Efficient H_2 Production from PET Plastic Wastes over Mesoporous Carbon-Supported Ru-ZnO Catalysts in a Mild Pure-Water System. *ACS Sustain. Chem. Eng.* **2023**, *11*, 578–586. [CrossRef]
46. Horányi, G.; Nagy, F. Investigation of Adsorption Phenomena on Platinized Platinum Electrodes by Tracer Methods: IV. Adsorption of Aromatic Compounds. *J. Electroanal. Chem. Interfacial Electrochem.* **1971**, *32*, 275–282. [CrossRef]
47. Wu, Y.; Cai, S.; Wang, D.; He, W.; Li, Y. Syntheses of Water-Soluble Octahedral, Truncated Octahedral, and Cubic Pt–Ni Nanocrystals and Their Structure–Activity Study in Model Hydrogenation Reactions. *J. Am. Chem. Soc.* **2012**, *134*, 8975–8981. [CrossRef]
48. Shiraishi, Y.; Tanaka, H.; Sakamoto, H.; Hayashi, N.; Kofuji, Y.; Ichikawa, S.; Hirai, T. Synthesis of Au Nanoparticles with Benzoic Acid as Reductant and Surface Stabilizer Promoted Solely by UV Light. *Langmuir* **2017**, *33*, 13797–13804. [CrossRef]

Article

Hydrogen and CO Over-Equilibria in Catalytic Reactions of Methane Reforming

Vitaliy R. Trishch [1], Mykhailo O. Vilboi [1], Gregory S. Yablonsky [2,*] and Dmytro O. Kovaliuk [1]

[1] Igor Sikorsky Kyiv Polytechnic Institute, National Technical University of Ukraine, 03056 Kyiv, Ukraine; tvr-ihf@lll.kpi.ua (V.R.T.); vmo-ihf@lll.kpi.ua (M.O.V.); d.kovalyuk@kpi.ua (D.O.K.)

[2] Department of Energy, Environmental and Chemical Engineering, McKelvey School of Engineering, Washington University in St. Louis, St. Louis, MO 63130, USA

* Correspondence: gy@wustl.edu

Abstract: Hydrogen and carbon monoxide over-equilibria have been found computationally in kinetic dependencies of methane-reforming catalytic reactions (steam and dry reforming) using the conditions of the conservatively perturbed equilibrium (CPE) phenomenon, i.e., at the initial equilibrium concentration of hydrogen or carbon monoxide. The influence of the pressure, temperature, flow rate and composition of the initial mixture on the position of the CPE point (the extremum point) was investigated over a wide domain of parameters. The CPE phenomenon significantly increases the product concentration (H_2 and CO) at the reactor length, which is significantly less than the reactor length required to reach equilibrium. The CPE point is interpreted as the "turning point" in kinetic behaviour. Recommendations on temperature and pressure regimes are different from the traditional ones related to Le Chatelier's law. The obtained results provide valuable information on optimal reaction conditions for complex reversible chemical transformations, offering potential applications in chemical engineering processes.

Keywords: modelling; optimization; methane reforming; chemical equilibrium; conservatively perturbed equilibrium; extremum; CPE point

Citation: Trishch, V.R.; Vilboi, M.O.; Yablonsky, G.S.; Kovaliuk, D.O. Hydrogen and CO Over-Equilibria in Catalytic Reactions of Methane Reforming. *Catalysts* **2024**, *14*, 773. https://doi.org/10.3390/catal14110773

Academic Editor: Guido Busca

Received: 12 September 2024
Revised: 19 October 2024
Accepted: 29 October 2024
Published: 31 October 2024

1. Introduction

Increasing the intensity of a complex catalytic reaction is a fundamental challenge in chemical technology, particularly when it comes to achieving the kinetic characteristics above equilibrium (such as rate, concentration and selectivity) under transient non-steady-state conditions. Chemical equilibrium represents the final state of a complex chemical reaction for both closed systems and open systems of infinite length.

The equilibrium composition is characterized by its uniqueness and stability. In a closed chemical system, this implies that for a fixed quantity of chemical elements at a given temperature, the system will always reach the same chemical composition from any initial state, ensuring that the equilibrium composition is both unique and stable. Zel'dovich [1,2] qualitatively demonstrated these properties in 1938. Since the 1960s, numerous researchers, including Shapiro, Shapley [3], Aris [4,5], Horn, Jackson [6], Volpert, Khudyaev [7,8] and Gorban' [9], have examined this problem more rigorously. Their detailed findings are documented in various monographs, e.g., [10].

Calculating the equilibrium composition has become essential for addressing the numerous issues in chemical and biochemical engineering. These calculations are typically based on a procedure that employs a list of components with known chemical potentials minimizing the Gibbs free energy. The system temperature and total amount of any particular chemical element are assumed to remain constant throughout the process.

2. Analysis of Previous Studies

Conservatively perturbed equilibrium (CPE). Definition and properties.

In [11–15], a new phenomenon in chemical kinetics, conservatively perturbed equilibrium (CPE), was described and analyzed. To achieve this phenomenon, some initial concentrations of components are replaced by their corresponding equilibrium concentrations while keeping the total amount of each chemical element constant. The temperature of the system should be held constant too.

The computational phenomenon CPE is carried out as follows:

1. The values of the equilibrium concentrations of all components are determined.
2. Some components are selected so that their initial concentrations differ from the equilibrium concentrations.
3. At least one component is selected such that its initial concentration is equal to the equilibrium value.
4. Perturbations introduced before (see item 2) must comply with all conservation laws applicable to this reaction system.
5. The evolution of all concentrations is observed as they tend toward equilibrium.

The CPE phenomenon is illustrated based on consecutive reactions [16]:

$$A \underset{k_1^+,k_1^-}{\overset{}{\rightleftarrows}} B \underset{k_2^+,k_2^-}{\overset{}{\rightleftarrows}} C, \tag{1}$$

In accordance with the mass–action law, the steady-state kinetic model for the plug flow reactor (PFR) is expressed as follows:

$$\frac{dC_A}{d\tau} = -k_1^+ C_A + k_1^- C_B,$$

$$\frac{dC_B}{d\tau} = k_1^+ C_A - k_1^- C_B - k_2^+ C_B + k_2^- C_C, \tag{2}$$

$$\frac{dC_C}{d\tau} = k_2^+ C_B - k_2^- C_C,$$

where C_A, C_B, and C_C are the concentrations of substances A, B, and C, respectively; τ is the residence time. k_1^+, k_1^-, k_2^+, and k_2^- are the rate coefficients of the corresponding forward and reverse reactions.

Obviously, $C_A + C_B + C_C = C_{A,0} + C_{B,0} + C_{C,0} = 1$, where $C_{A,0}$, $C_{B,0}$, and $C_{C,0}$ are the initial concentrations of substances A, B, and C, respectively. The following kinetic parameters and equilibrium concentrations are used:

$$k_1^+ = 10, k_1^- = 2, k_2^+ = 6, k_2^- = 2$$
$$C_{A,\,eq.} = 0.714, C_{B,\,eq.} = 0.238, C_{C,eq.} = 0.048, \tag{3}$$

where $C_{A,eq.}$, $C_{B,eq.}$, and $C_{C,eq.}$ are the equilibrium concentrations of substances A, B, and C, respectively.

The concentration dependencies for reactions (1) (Figure 1) performed in the PFR with the CPE phenomenon represent the properties of the CPE phenomenon.

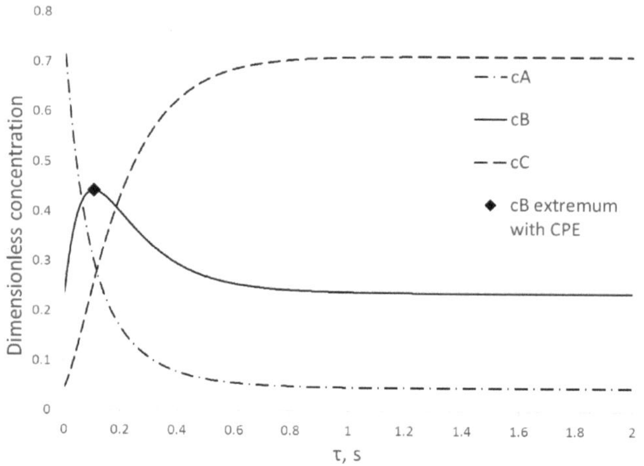

Figure 1. Dimensionless concentration dependencies for reactions (1).

The concentration of component B passes through the unavoidable extremum reaches its equilibrium value. The initial concentrations (compositions) are given in Table 1.

Table 1. Selected initial concentrations (compositions).

	Composition 1	**Composition 2**	**Composition 3**
C_A	0.818	0.727	0.545
C_B	**0.136**	**0.136**	**0.136**
C_C	0.045	0.136	0.318

Three concentration dependencies for different initial compositions of the mixture are shown (Figure 2). In this linear case, two important features are demonstrated: (1) the CPE point does not change its position when the initial composition of the mixture changes; (2) the greater the deviation of the concentration of disturbed substances within balance limitations, the higher the magnitude of the CPE extremum. For the general linear case, the rigorous theoretical proof of these two findings is still lacking. However, for typical examples of simple linear mechanisms, these properties are demonstrated computationally. This extremum's value offers insights into the reaction mechanism and kinetic parameters, helping optimize conditions for complex reversible chemical processes. The CPE method was recently tested in a batch reactor during an esterification reaction involving ethanol, benzyl alcohol, and acetic acid, yielding two esters and water [17,18].

The goal of this paper is a continuation of our previous work [19] in which the CPE phenomenon was just demonstrated computationally for the reforming of methane (RM) reactions using a limited number of parameters. In this paper, we are going to study this phenomenon systematically, presenting the effect of different process parameters over a wide range, i.e., pressure, initial mixture composition, and flow rate, as well as the kinetic dependencies, especially on CPE point characteristics. Also, the temperature domain will be modified. In this paper, the physicochemical meaning of the CPE point will be interpreted.

The obtained CPE characteristics will be compared with cases without CPE and recommendations which correspond to the known Le Chatelier's law.

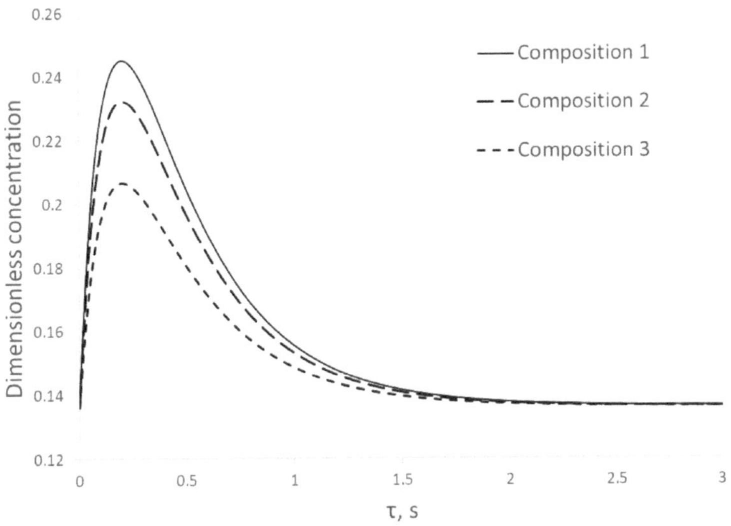

Figure 2. Dimensionless concentration dependencies for different initial compositions, A and C ($C_{B,0} = C_{B,eq.}$).

Steam- and dry-reforming overall reactions are considered in the reforming of methane (RM) (see Table 2).

Table 2. Selected reactions for methane reforming.

Reaction	ΔH^0 (kJ/mol)	Link
$CH_4 + CO_2 \leftrightarrow 2CO + 2H_2$	+247	
$CH_4 + H_2O \leftrightarrow CO + 3H_2$	+206	(4)

3. Results

Figure 3 shows the effect of the CPE phenomenon on the concentration profiles of H_2 and CO in comparison with the no-CPE case.

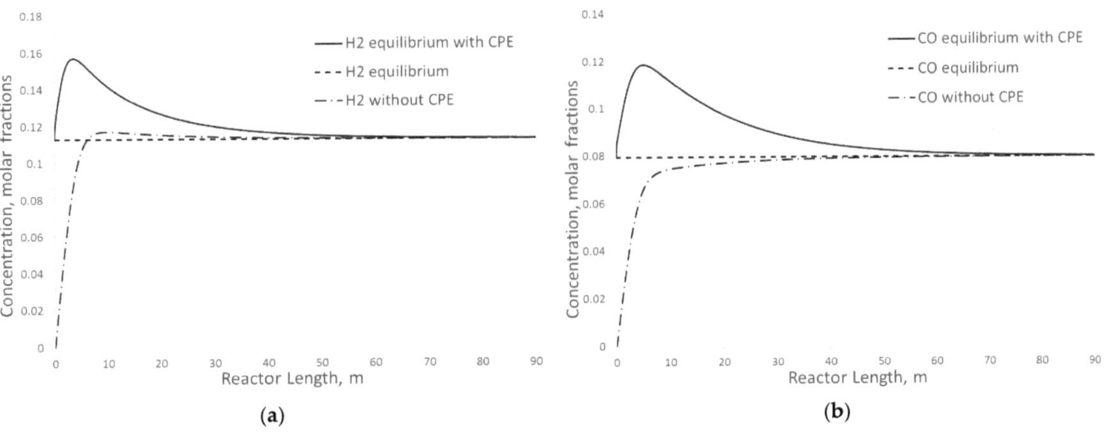

Figure 3. Longitudinal profile of mole fractions. Cases with and without CPE phenomenon: (**a**) hydrogen; (**b**) carbon monoxide.

The system is characterized by the transition from the initial equilibrium value to the final one via an extremum, which is unavoidable and well distinguished. Computer calculations show that the system with no CPE either has no extrema or has a very slightly pronounced one. Under given pressure and temperature conditions, the concentration of products can be increased using the CPE phenomenon by 39% and 49% for H_2 and CO, respectively. Additionally, utilizing the CPE phenomenon allows for a reduction in reactor length due to the faster occurrence of the extremum (CPE point) compared to the case without the CPE phenomenon. Later, the kinetic dependencies with the CPE phenomenon and without it will be compared in more detail.

For the CPE hydrogen case, by definition, the initial hydrogen concentration equals its equilibrium concentration. Five concentration dependencies are obtained (H_2 and H_2O, CO and CO_2, and CH_4 as well). The H_2 and CO dependencies are characterized by extrema, hydrogen by the maximum and water by the minimum, respectively. The CO and CO_2 dependencies are monotone: the CO and CO_2 concentrations are increased and decreased, respectively. As for the methane concentration, it reaches its equilibrium value at the CPE point. Similarly, dependencies which correspond to the CPE CO case can be interpreted. Figures 4 and 5 demonstrate the effect of pressure and temperature on the kinetic dependencies, especially on the position of the CPE point regarding the products, H_2 and CO, respectively.

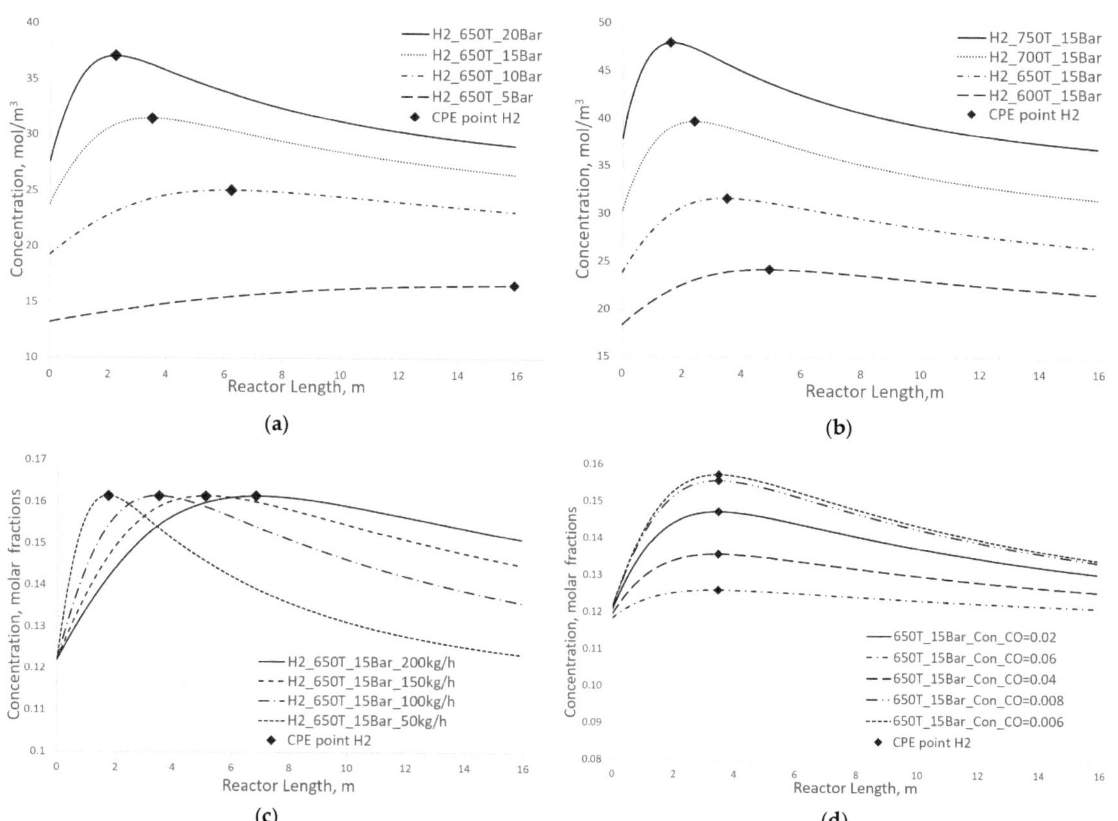

Figure 4. Longitudinal profiles of the H_2 concentration: (**a**) with pressure change, (**b**) with temperature change, (**c**) with flow rate change, and (**d**) with changes in the initial concentration indicated.

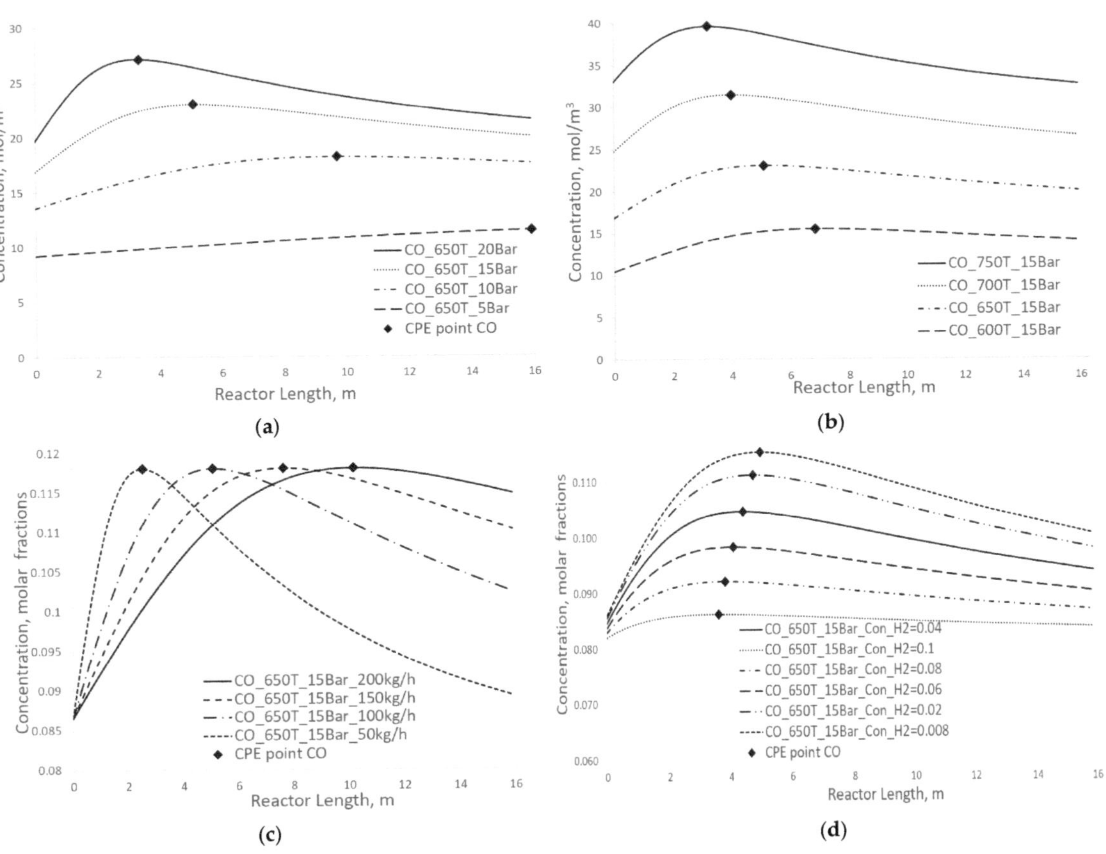

Figure 5. Longitudinal profiles of CO concentration: (**a**) with pressure change, (**b**) with temperature change, (**c**) with flow rate change, and (**d**) with changes in the initial concentration indicated.

As can be seen from Figures 4 and 5, pressure, temperature, and flow rate all affect the position of the CPE point. An increase in temperature or pressure has a positive effect on the value of the CPE point, causing the magnitude of this point to increase. Additionally, higher temperatures or pressures result in a shorter reactor length required to reach the CPE point. With the increase in the flow rate, the magnitude of the CPE point does not change. However, the length of the reactor at which the extremum occurs is increased in such a way that the residence time of the CPE point remains almost constant. These findings are consistent with previous results obtained for simple linear mechanisms [11,12]. When temperature and pressure are increased simultaneously, the CPE point is reached faster than when these parameters are changed separately (for example, increasing temperature while maintaining constant pressure).

The kinetic behaviour of CPE extrema varies significantly depending on the initial concentrations of H_2 and CO. For hydrogen, the position of the CPE point changes insignificantly. The special theoretical analysis, which is in the stage of preparation, shows that for linear models, the position of the CPE point does not depend on initial concentrations. For our model, which is non-linear, the position of the CPE point, generally, is not the same for different initial concentrations. The obtained difference is a result of computer calculations using our non-linear model.

CPE phenomenon vs. no-CPE cases. CPE phenomenon and Le Chatelier's law

For a more detailed understanding, computer simulations without CPE were systematically performed.

Specifically, we analyzed the effect of pressure and temperature changes for dry-reforming and steam-reforming reactions for two cases:

(a)   These reactions are performed simultaneously (see Figures 6 and 7);
(b)   Reactions of steam and dry reforming are performed separately. See Figures 8 and 9 for hydrogen concentrations. Similar results can be obtained for CO concentrations.

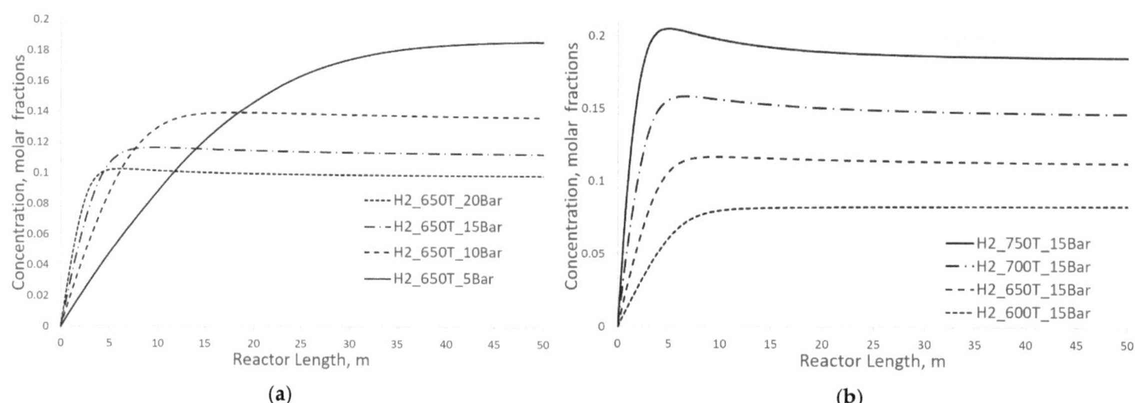

(a)

(b)

Figure 6. Two reforming reactions with no CPE. Longitudinal molar fraction profiles of hydrogen: (a) pressure change; (b) temperature change.

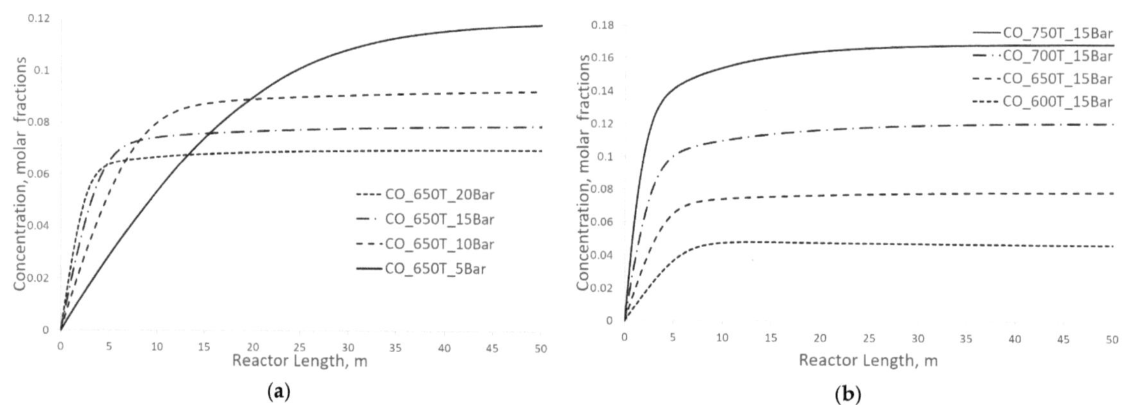

(a)

(b)

Figure 7. Two reforming reactions with no CPE. Longitudinal molar fraction profiles of carbon monoxide: (a) pressure change; (b) temperature change.

Both steam- and dry-reforming reactions are endothermic and both result in an increase in the number of molecules, i.e., four molecules of the products compared to two molecules of the reactants. In accordance with Le Chatelier's law, the optimal equilibrium regimes are the following ones: the temperature is increased and the pressure is decreased.

Our computer calculations demonstrate similar behaviour for both cases regardless of whether methane reforming has occurred for two processes together of for a separate process. All kinetic dependencies exhibit a plateau, which reflects the near-equilibrium regime. This plateau increases as the temperature rises and decreases as the pressure rises.

Beyond the equilibrium regime, both temperature and pressure affect the concentration of the products, H_2 and CO, in proportion to the kinetic term of Equation (5), $k_1 P_{CH_4}$ or $k_2 P_{CH_4}$.

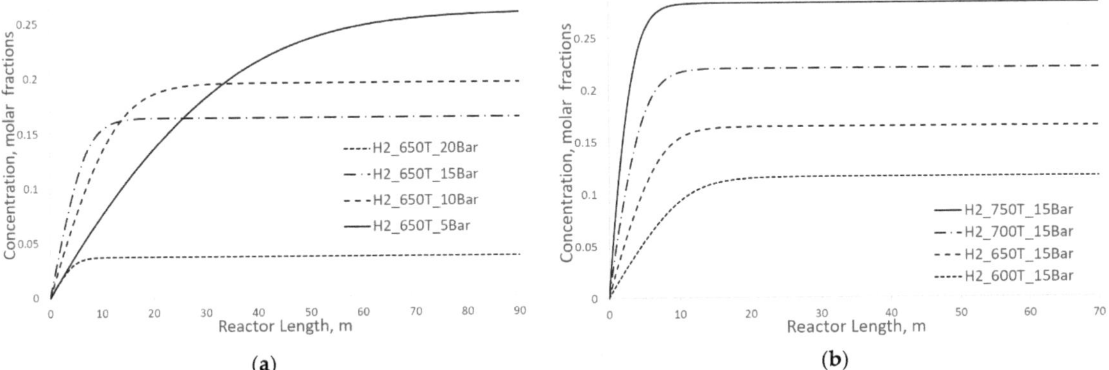

(a)

(b)

Figure 8. Steam reforming. Longitudinal molar fraction profiles of hydrogen: (**a**) pressure change; (**b**) temperature change.

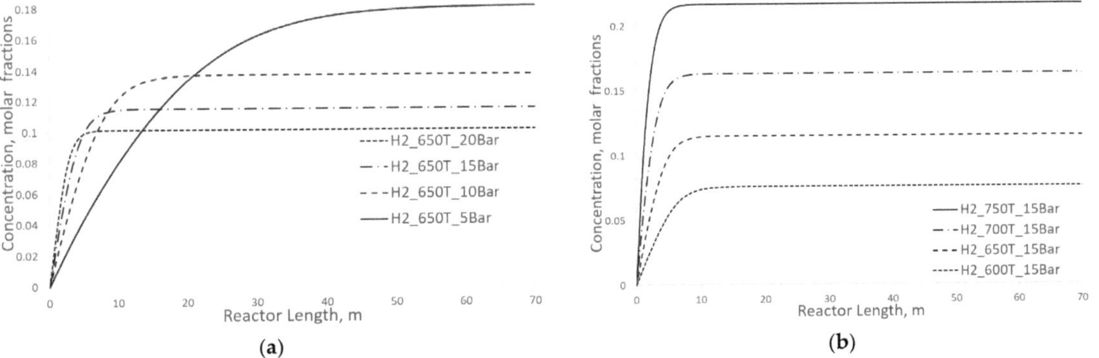

(a)

(b)

Figure 9. Dry reforming. Longitudinal molar fraction profiles of hydrogen: (**a**) pressure change; (**b**) temperature change.

The CPE phenomenon, where both temperature and pressure increase the concentration of H_2 and CO, is different from the recommendations of Le Chatelier's law and similar to kinetic behaviour beyond the domain of Le Chatelier's law. However, as mentioned, the concentrations of products for the CPE regimes are significantly higher than those for no-CPE cases.

4. Methods

The mathematical modelling of reforming reactions was based on kinetic equations of dry and steam reforming on a nickel catalyst, which were taken from the literature [20].

$$\begin{aligned}
Dry\ reforming: \quad & r_1 = k_1 P_{CH_4}\left(1 - \frac{P_{CO}^2\, P_{H_2}^2}{P_{CH_4}\, P_{CO_2}\, K_{dr}}\right),\ (s^{-1}) \\
Steam\ reforming: \quad & r_2 = k_2 P_{CH_4}\left(1 - \frac{P_{CO}\, P_{H_2}^3}{P_{CH_4}\, P_{H_2O}\, K_{sr}}\right),\ (s^{-1})
\end{aligned} \tag{5}$$

Generally, these equations are non-linear. However, in both cases, the kinetic terms $k_1 P_{CH_4}$ and $k_2 P_{CH_4}$ are linear. In accordance with the CPE requirements, in any experiment, the total amount of any element (C, H, and O) must be conserved.

$$C_{CH_4} + C_{CO_2} + C_{CO} = C_{CH_4,in} + C_{CO_2,in} + C_{CO,in} = N_C,$$
$$4C_{CH_4} + 2C_{H_2O} + 2C_{H_2} = 4C_{CH_4,in} + 2C_{H_2O,in} + 2C_{H_2,in} = N_H, \qquad (6)$$
$$2C_{CO_2} + C_{H_2O} + C_{CO} = 2C_{CO_2,in} + C_{H_2O,in} + C_{CO,in} = N_O,$$

where N is the total number of atoms of a specific element (C, H, or O) in all the respective compounds.

Industrial steam and dry reformers usually operate under pressures between 15–20 bar and 1–10 bar and temperatures between 700–1100 °C and 800–1100 °C, respectively. For safety reasons, operation at pressures above 25 bar should be avoided, as this is considered a critical threshold for many industrial reactors.

For industrial use, an operating temperature of 800–850 °C is generally recommended. Operating above 850 °C can lead to catalyst deactivation due to sintering and increases the risk of reactor tube failure. The simulations covered a temperature range from 873.15 K (600 °C) to 1023.15 K (750 °C). The parameters of the system are listed in Table 3.

Table 3. Parameters of the system.

Object	Parameters
Input	Temperature: 873.15–1023.15 K with 50 K intervals. Pressure: 5–20 bar Mass flow: 50–200 kg/h
Reactor	Reaction set: dry and steam reforming Calculation mode: isothermic Property package: Peng–Robinson Reactive volume: 0.5; 3 m^3 Tube diameter: 200 mm

In this case, we consider that the concentrations of the catalytic intermediates are negligible in comparison with the concentrations of the reactants C_{CH_4}, C_{CO_2}, C_{H_2O}, C_{CO}, and C_{H_2}.

Table 4 has three columns: one column for equilibrium concentrations and two columns for the initial concentrations for the CPE hydrogen and carbon monoxide cases, respectively.

Table 4. Concentrations (mole fractions) and balances.

	Equilibrium	Concentrations Initial Concentrations CPE for H2	CPE for CO	Balance of Components	Equilibrium	Initial Concentrations
C_{CO_2}	0.2101	0.2711	0.1808	C	0.659	0.659
C_{CO}	0.0814	0.0001	**0.0814 eq**	H	2.152	2.152
C_{H_2}	0.1171	**0.1171 eq**	0.0001	O	0.726	0.726
C_{CH_4}	0.3674	0.3877	0.3966			
C_{H_2O}	0.2240	0.1834	0.2825			

5. Discussion

CPE point: analysis and interpretation. CPE as the turning point.

To gain a detailed understanding of the physicochemical significance of the CPE phenomenon, the typical longitudinal profiles of reaction rates are presented. For the CPE case related to hydrogen (Figure 10a), reaction rates for dry and steam reforming are shown. Evidently, the steam-reforming reaction undergoes a reversal, crossing the equilibrium point and changing its direction to become exothermic. The reaction rate reaches zero precisely at the CPE point for hydrogen. Therefore, this CPE point is, at the same time, the turning point for the steam-reforming reaction.

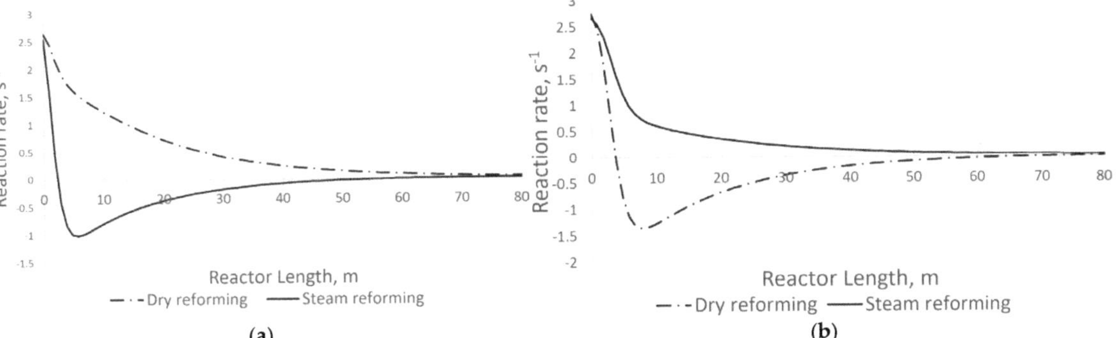

(a)   **(b)**

Figure 10. Reaction rates for the hydrogen CPE case (**a**) and for the carbon monoxide CPE case (**b**).

Similarly, for the CPE case related to CO (Figure 10b), the dry-reforming reaction also reverses, crossing the equilibrium point and becoming exothermic. The rate of dry reforming reaches zero at the CPE point for CO, which occurs at the same time as the turning point for this reaction.

The corresponding longitudinal profiles of component concentrations are presented in Figure 11a,b. In the case of the hydrogen CPE point, the hydrogen concentration maximum coincides with the water concentration minimum. The CO_2 concentration decrease is accompanied by the CO concentration increase. For the carbon monoxide CPE point, the CO maximum coincides with the CO_2 minimum, and the water concentration decrease is accompanied by the hydrogen concentration increase.

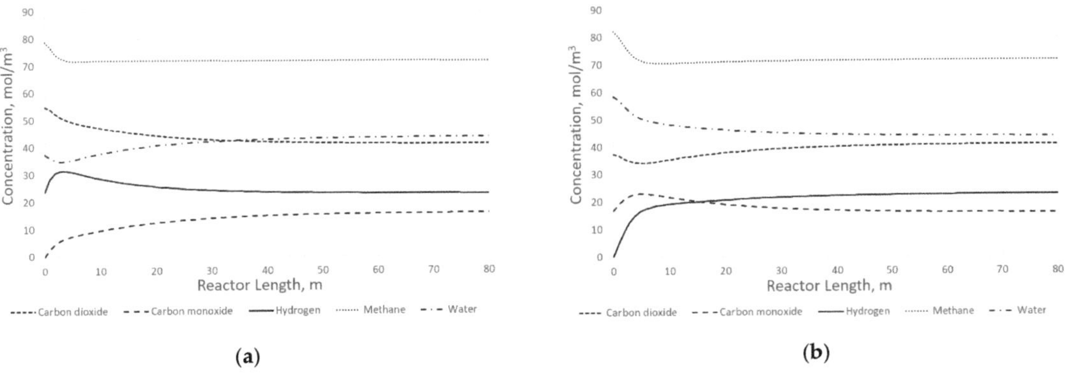

(a)   **(b)**

Figure 11. Longitudinal profiles of concentrations: (**a**) hydrogen CPE case; (**b**) carbon monoxide CPE case.

There is an interesting feature in the kinetic behaviour of our system. Initially, in the hydrogen CPE case, the concentration of hydrogen is equal to the equilibrium concentration. Then, under the conditions that the concentration of CO is zero or small, the hydrogen concentration can only increase, not decrease. This is achieved via forward reactions of dry and steam reforming because reverse reactions are absent or negligible. As H_2 accumulates and reaches its extremum at the CPE point, the reverse steam-reforming reaction starts to play a more significant role, consuming 3 moles of H_2 and 1 mole of CO and decreasing the H_2 concentration. This is one reason for the over-equilibrium, which is also a momentary equilibrium in the steam-reforming reaction. So, at the CPE point, this reaction reverses, and the CPE point becomes the turning point. This is a physicochemical explanation for our over-equilibrium effect.

A similar explanation can be used for the carbon monoxide CPE case, which is observed when the initial concentration of CO is high and H_2 is negligible or absent. Then, the concentration of CO increases due to forward reactions. As CO accumulates and reaches its extremum at the CPE point, the reverse dry-reforming reaction starts to play a more significant role, consuming 2 moles of H_2 and 2 moles of CO and decreasing the concentration of CO. The different behaviour in CPE for hydrogen and carbon monoxide arises because of the difference in the stoichiometry of the products (CO and H_2).

Starting from the CPE point, an almost steady-state concentration of methane is observed (Figure 11a,b).

The reason for this is the following: one of the two reactions, steam or dry reforming, goes in the reverse direction and balances the other reaction, dry or steam reforming, respectively. Formally, starting from the CPE point, the following condition is approximately fulfilled: $r_{dry} = r_{steam} = r$.

Then, in this domain, which is related to the CPE hydrogen case (Figure 11a), the rates of the other substances are as follows:

$$r_{H_2} = 2r_{dry} - 3r_{steam} = -r; \; r_{H_2O} = r_{steam} = r; \; r_{CO_2} = -r_{dry} = -r; \; r_{CO} = 2r_{dry} - r_{steam} = r$$

It is easy to understand that, in this scenario, the combined overall equation is realized, $H_2 + CO_2 = CO + H_2O$, i.e., the reverse water–gas shift reaction (RWGS).

In another domain related to the CPE carbon monoxide case (Figure 11b), the rates are as follows:

$$r_{H_2} = 3r_{steam} - 2r_{dry} = r; \; r_{H_2O} = -r_{steam} = -r; \; r_{CO_2} = r_{dry} = r; \; r_{CO} = r_{steam} - 2r_{dry} = -r$$

The combined overall reaction can be presented as $H_2O + CO = CO_2 + H_2$, i.e., the water–gas shift reaction (WGS).

Evidently, the water–gas shift reaction, forward or reverse, can be obtained as a linear combination of dry- and steam-reforming reactions. To summarize, this reaction can be considered an essential simplification of our chemical process: two reactions of methane reforming 'degenerate' into one water–gas shift reaction, which can be forward or reverse.

6. Conclusions

The kinetic dependencies for the two methane-reforming reactions, dry and steam reforming, were studied systematically for the conservatively perturbed equilibrium (CPE) phenomenon, i.e., at the initial equilibrium concentration of hydrogen or carbon monoxide, over a wide domain of parameters: temperature (873.15–1023.15 K), pressure (5–20 bar), the initial concentrations of other substances, and flow rate (50–200 kg/h). The obtained results confirmed that the changes in pressure and temperature significantly influence the CPE effect, resulting in 'over-equilibrium' for hydrogen and carbon monoxide, which is achieved at the residence time and corresponding reactor length, which are much shorter than the length of a typical industrial reactor. The recommendations for temperature and pressure control based on the CPE phenomenon were compared with recommendations based on Le Chatelier's law related to equilibrium regimes, revealing a notable difference between them.

A physicochemical interpretation of the CPE points, both for the hydrogen and carbon monoxide cases, was given. The over-equilibrium observed under CPE conditions is due to the absence of reverse reaction rates. Furthermore, beyond CPE, the two reforming reactions become opposite and balance each other out. Finally, these two reactions 'degenerate' into the water–gas shift reaction, which can be either forward or reverse.

These findings suggest that the CPE phenomenon could be utilized as an efficient tool for understanding the kinetic dependencies of reforming processes and for optimizing industrial reactors as well. This could lead to the development of new optimal strategies in various industrial applications, particularly in processes where maximizing reaction efficiency and product yield is crucial.

Author Contributions: Conceptualization, G.S.Y. and V.R.T.; methodology, D.O.K.; software, V.R.T.; validation, M.O.V.; resources, M.O.V.; data curation, G.S.Y.; writing—original draft preparation, V.R.T.; writing—review and editing, G.S.Y., D.O.K., and M.O.V.; visualization, V.R.T.; supervision, G.S.Y. All authors have read and agreed to the published version of the manuscript.

Funding: This research received no external funding.

Data Availability Statement: Data are contained within the article.

Conflicts of Interest: The authors declare no conflicts of interest.

References

1. Zel'dovich, Y.B. A proof of the uniqueness of the solution of the equations for the law of mass action. *Sel. Work. Yakov Borisovich Zeldovich* **1938**, *1*, 144–148.
2. Zel'dovich, Y.B. *Selected Works of Yakov Borisovich Zeldovich: Chemical Physics and Hydrodynamics*; Princeton University Press: Princeton, NJ, USA, 2014.
3. Shapiro, N.Z.; Shapley, L.S. Mass action laws and the Gibbs free energy function. *J. Soc. Ind. Appl. Math.* **1965**, *13*, 353–375. [CrossRef]
4. Aris, R. Prolegomena to the rational analysis of systems of chemical reactions. *Arch. Ration. Mech. Anal.* **1965**, *19*, 81–99. [CrossRef]
5. Aris, R. Prolegomena to the rational analysis of systems of chemical reactions II. Some addenda. *Arch. Ration. Mech. Anal.* **1968**, *27*, 356–364. [CrossRef]
6. Horn, F.; Jackson, R. General mass action kinetics. *Arch. Ration. Mech. Anal.* **1972**, *47*, 81–116. [CrossRef]
7. Vol'pert, A.I. Deferential equations on Figures. *Math. USSR Sbornik* **1972**, *17*, 571. [CrossRef]
8. Vol'pert, A.I.; Khudyaev, S.I. *Analysis in Classes of Discontinuous Functions and Equations of Mathematical Physics*; Martinus Nijhous: Dordrecht, The Netherlands, 1985.
9. Gorban, A.N. On the problem of boundary equilibrium points. *React. Kinet. Catal. Lett.* **1980**, *15*, 315–319.
10. Marin, G.B.; Yablonsky, G.S.; Constales, D. *Kinetics of Chemical Reactions: Decoding Complexity*, 2nd ed.; John Wiley–VCH: Weinheim, Germany, 2019.
11. Yablonsky, G.S.; Branco, D.P.; Marin, G.B.; Constales, D. Conservatively Perturbed Equilibrium (CPE) in chemical kinetics. *Chem. Eng. Sci.* **2019**, *196*, 384–390. [CrossRef]
12. Yiming, X.; Xinquan, L.; Constales, D.; Yablonsky, G.S. Perturbed and Unperturbed: Analyzing the Conservatively Perturbed Equilibrium (Linear Case). *Entropy* **2020**, *22*, 1160. [CrossRef] [PubMed]
13. Trishch, V.R.; Beznosyk, Y.O.; Yablonsky, G.S.; Constales, D. The phenomenon of conservative-perturbed equilibrium in conditions different reactors. *Bull. NTUU "Igor Sikorsky Kyiv Polytech. Inst." Ser. Chem. Eng. Ecol. Resour. Conserv.* **2021**, *10*, 2617–9741. [CrossRef]
14. Trishch, V.R.; Beznosyk, Y.O.; Constales, D.; Yablonsky, G.S. Over-Equilibrium as a Result of Conservatively-Perturbed Equilibrium (Acyclic and Cyclic Mechanisms). *J. Non-Equilib. Thermodyn.* **2022**, *47*, 103–110. [CrossRef]
15. Trishch, V.R.; Yablonsky, G.S.; Constales, D.; Beznosyk, Y.O. Conservatively perturbed equilibrium in multi-route catalytic reactions. *J. Non-Equilib. Thermodyn.* **2023**, *48*, 229–241. [CrossRef]
16. Yablonsky, G.S.; Constales, D.; Marin, G.B. Equilibrium relationships for non-equilibrium chemical dependencies. *Chem. Eng. Sci.* **2011**, *66*, 20004. [CrossRef]
17. Gorban, A.N.; Constales, D.; Yablonsky, G.S. Transient Concentration Extremum and Conservatively Perturbed Equilibrium. *Chem. Eng. Sci.* **2020**, *231*, 116295. [CrossRef]
18. Peng, B.; Zhu, X.; Constales, D.; Yablonsky, G.S. Experimental verification of conservatively perturbed equilibrium for a complex non-linear chemical reaction. *Chem. Eng. Sci.* **2020**, *229*, 116008. [CrossRef]
19. Vilboi, M.O.; Trishch, V.R.; Yablonsky, G.S. Conservatively Perturbed Equilibrium (CPE)—Phenomenon as a Tool for Intensifying the Catalytic Process: The Case of Methane Reforming Processes. *J. Catal.* **2024**, *14*, 395. [CrossRef]
20. Wei, J.; Iglesia, E. Isotopic and kinetic assessment of the mechanism of reactions of CH_4 with CO_2 or H_2O to form synthesis gas. *J. Catal.* **2004**, *224*, 370–383. [CrossRef]

catalysts

Article

Kinetic Characterization of Pt/Al$_2$O$_3$ Catalyst for Hydrogen Production via Methanol Aqueous-Phase Reforming

José Sousa [1,2,3], Paranjeet Lakhtaria [1,2], Paulo Ribeirinha [1,2], Werneri Huhtinen [4], Johan Tallgren [4] and Adélio Mendes [1,2,*]

[1] LEPABE—Laboratory for Process Engineering, Environment, Biotechnology and Energy, Faculty of Engineering, University of Porto, Rua Dr. Roberto Frias, 4200-465 Porto, Portugal; jmsousa@fe.up.pt (J.S.)

[2] ALiCE—Associate Laboratory in Chemical Engineering, Faculty of Engineering, University of Porto, Rua Dr. Roberto Frias, 4200-465 Porto, Portugal

[3] Departamento de Química, Escola de Ciências da Vida e'do Ambiente, Universidade de Trás-os-Montes e Alto Douro, Quinta de Prados, 5000-801 Vila Real, Portugal

[4] VTT Technical Research Centre of Finland Ltd., Tietotie 4 C, P.O. Box 1000, FI-02044 VTT Espoo, Finland

* Correspondence: mendes@fe.up.pt

Abstract: Compared to steam reforming, methanol aqueous-phase reforming (APR) converts methanol to hydrogen and carbon dioxide at lower temperatures, but also displays lower conversion rates. Herein, methanol APR is studied over the active Pt/Al$_2$O$_3$ catalyst under different operating conditions. Studies were conducted at different temperatures, pressures, methanol mass fractions, and residence times. APR performance was evaluated in terms of methanol conversion, hydrogen production rate, hydrogen selectivity, and by-product formation. The results revealed that an increase in operating pressure and methanol mass fraction had an adverse effect on the APR performance. Conversely, it was found that hydrogen selectivity was unaffected by the operating pressure and residence time for the methanol feed mass fraction of 5%. For the methanol feed mass fraction of 55%, hydrogen selectivity was improved by operating pressure and residence time. The alumina support phase change to boehmite as well as sintering and leaching of the catalytic particles were observed during catalyst stability experiments. Additionally, a comparison between methanol steam reforming (MSR) and APR was also performed.

Keywords: aqueous-phase reforming; methanol; catalyst; hydrogen production; Pt/Al$_2$O$_3$

Citation: Sousa, J.; Lakhtaria, P.; Ribeirinha, P.; Huhtinen, W.; Tallgren, J.; Mendes, A. Kinetic Characterization of Pt/Al$_2$O$_3$ Catalyst for Hydrogen Production via Methanol Aqueous-Phase Reforming. *Catalysts* **2024**, *14*, 741. https://doi.org/10.3390/catal14100741

Academic Editors: Georgios Bampos, Paraskevi Panagiotopoulou and Eleni A. Kyriakidou

Received: 19 August 2024
Revised: 14 October 2024
Accepted: 15 October 2024
Published: 21 October 2024

1. Introduction

Climate change, greenhouse gas emission, and the energy crisis have recently become major global concerns. The integration of carbon-neutral technologies, renewable energy resources, and reducing the usage of fossil fuels have become a priority [1,2]. For this energy transition, hydrogen has received a lot of attention, interest, and commitment from both the industrial and academic sectors. Hydrogen has been pushed forward for energy decarbonization, to mitigate the energy crisis pressed by geo-political conflicts, and to reach the "Paris agreement" [3,4].

The advantages of hydrogen are linked to its usability in a wide range of applications. For instance, hydrogen can be used directly as a fuel or as a raw material for the production of synthetic fuels and organic substances [5]. Yet, the main issues related to the use of hydrogen are related to its safe and efficient transportation and storage [6]. Even though the gravimetric energy density of hydrogen is excellent, the volumetric density is too low [5]. Hence, in situ hydrogen production has been suggested as an attractive approach, especially for systems with a demand for continuous hydrogen supply.

For in situ hydrogen production, there are different technologies, such as reforming and electrolysis [7,8]. Today, 96% of the world's hydrogen is still produced from fossil fuels, namely natural gas and coal [8,9]. To improve the sustainability of hydrogen production, interest has been focused on the development of more environmentally friendly technologies. Apart from electrolysis, one interesting pathway is the use of renewable liquid fuels as feedstock for hydrogen production [10,11]. Biomass-derived organic waste streams containing alcohols have been investigated as potential bio-based feedstocks [12,13]. Compared to hydrogen, oxygenated hydrocarbons have a higher boiling point and, hence, have higher gravimetric density and are easier to transport and store [14]. In addition to its environmental merits, converting waste streams to value-added products is economically efficient.

There already exist mature reforming technologies of oxygenated hydrocarbons, such as the steam reforming of methanol (MSR) or ethanol [12,15]. The challenges with these technologies are related to the selectivity of the reaction, mainly to avoid the production of CO, in addition to the high amounts of energy necessary for vaporizing the reactants. The aqueous-phase reforming (APR) emerged then to address these challenges as a simpler and more energy-efficient technology [16,17]. The APR allows for converting the feeding fuel to hydrogen and CO_2 in a single-step catalytic process at temperatures lower than the well-established MSR. In APR, reactants are in the liquid phase and the reaction products are in the gas phase [18,19].

The APR of oxygenated hydrocarbons is often performed in a temperature range of 200–250 °C. To improve the overall efficiency of the APR system, the use of waste heat from other processes would be beneficial. The APR system should work with pressures above the bubble point pressure of the feedstock mixture to keep it in the aqueous phase. Bubble point pressure depends on the composition of the feedstock mixture and the selected operating temperature. In general, the working pressure should be between 1.5 and 5.0 MPa [14].

Mainly heterogeneous catalysts have been considered for the APR of oxygenated hydrocarbons [20]. Three phases exist in APR systems: solid (the catalyst), liquid (the feedstock), and gas (the evolved hydrogen and CO_2). This leads to mass transfer limitations, since the evolving gas bubbles may block the active sites of catalysts and decrease the hydrogen production rate [21]. To improve the reaction rate, it has been proposed to use an inert sweeping gas. The sweeping gas drives the evolved gas bubbles of hydrogen and CO_2 out of the catalyst active sites, improving the reaction performance of the APR [13,22].

Methanol has been used as a model compound in APR studies due to the absence of strong C-C bonds [22]. Methanol is a common by-product of chemical and biochemical processes and it has been pointed out as a very attractive green energy vector, which makes it cheap and readily available [23]. The overall APR reaction equation is given below, Equation (1):

$$CH_3OH\ (l) + H_2O\ (l) \rightleftharpoons 3\,H_2\ (g) + CO_2\ (g) \tag{1}$$

The methanol APR is based on two main reactions [24]: methanol dehydrogenation [Equation (2)] and WGS reaction [Equation (3)]. In the methanol dehydrogenation reaction, a methoxy intermediate is formed due to the split of the O-H bond of methanol [22]. Next, hydrogen atoms bonded to carbon in the methoxy intermediate are activated, which causes the decomposition of methoxy into hydrogen and carbon monoxide. Methanol dehydrogenation is followed by WGS, in which cleavage of O-H bonds of water leads to the formation of O^{2-} ion and hydrogen. Extremely active O^{2-} ion reacts with carbon monoxide forming carbon dioxide [22].

$$CH_3OH\ (l) \rightleftharpoons 2\,H_2\ (g) + CO\ (g) \tag{2}$$

$$CO\ (g) + H_2O\ (l) \rightleftharpoons H_2\ (g) + CO_2\ (g) \tag{3}$$

The APR usually displays a high selectivity; only trace amounts of hydrogenated by-products are produced [Equations (4)–(6)]:

$$CO\ (g) + 3\,H_2\ (g) \rightleftharpoons CH_4\ (g) + H_2O\ (g) \tag{4}$$

$$CO_2\ (g) + 4\,H_2\ (g) \rightleftharpoons CH_4\ (g) + 2\,H_2O\ (g) \tag{5}$$

$$2\,CO\,(g) + 5\,H_2\,(g) \rightleftharpoons C_2H_6\,(g) + 2\,H_2O\,(g) \tag{6}$$

With the careful selection of the catalyst and operating conditions, these hydrogenated by-products can be mostly avoided [24].

The methanol APR research has been mostly focused on finding suitable catalysts that promote the reforming reactions at temperatures as low as possible and inhibit undesired reactions [14,25]. Amongst noble (Pt, Ru) and non-noble (Ni, Cu) metals, Pt-based catalysts are the most used for methanol APR. Dumesic's group [17,18] studied the APR of different alcohols (e.g., methanol and ethylene glycol) with Pt supported on alumina (Pt/Al_2O_3). Specifically for methanol, their study was conducted at temperatures/pressures of 210 °C/2.24 MPa and 225 °C/2.96 MPa, on a Pt/Al_2O_3 catalyst with a Pt mass fraction of 0.59%, and with a feed aqueous solution with methanol mass fractions of 1%, 4%, and 10%. All the experiments were conducted with a feed flow rate of 0.30 mL/min, in order to keep the conversion below 3%. Their results showed that the reaction is weakly inhibited by hydrogen, which could be caused by the surface sites blocked by adsorbed hydrogen atoms and, in addition, by decreasing the surface concentrations of reactive intermediates formed from dehydrogenation of the methanol. They also found that the liquid-phase environment favors the water–gas shift reaction, mitigating the potential blocking of surface sites by adsorbed CO species. Since then, Pt-based catalysts for methanol APR have been improved, mostly concerning the catalyst support, where molybdenum carbide (MoC) [26], $NiAl_2O_4$ spinel [27], and zeolite [28] supports were reported. Ni-based catalysts have also been explored for the methanol APR as they favor the WGS reaction [14] and are inexpensive compared to noble metal-based catalysts. However, Ni-based catalysts suffer from lower stability (especially since they are more sensitive to oxidation under APR operating conditions) and lower hydrogen selectivity. These catalysts were studied mostly by Stekrova et al. [24] and Coronado et al. [13,22].

The present work studies the methanol APR reaction kinetics on Pt/Al_2O_3. The performance of the catalyst, with a nominal mass fraction of 5%, was investigated for the temperatures of 190 °C, 210 °C and 230 °C. The effect on the catalyst performance of the methanol mass fraction in the feed aqueous solution was studied as well, using a low value of 5% and a high value of 55%. The operating pressures used were above the bubble point of the corresponding feed stream, a minimum of 3.2 MPa for the low methanol mass fraction feed and a minimum of 5.2 MPa for the high methanol mass fraction feed. The performance of Pt/Al_2O_3 was characterized in terms of methanol conversion, hydrogen production rate, and hydrogen selectivity for a wide range of weight hourly space velocity (WHSV) values.

2. Results and Discussion

2.1. Physicochemical Characterization of the Catalyst

The H_2-TPR of a 5 wt.% Pt/Al_2O_3 sample is displayed in Figure 1; it exhibits the presence of a single sharp peak at ca. 300 °C, starting at ca. 250 °C and ending at ca. 325 °C. The presence of this single peak indicates that only one type of active metal is present in the catalyst. This TPR result is in good agreement with the XRD of the catalyst, showing only Pt and alumina—discussed in the following.

The information regarding the reducibility of Pt/Al_2O_3 was used to activate the catalyst in situ before the APR tests. Controlling the reduction temperature is important since the catalyst reduction (activation process) is an exothermic process that influences the catalyst's ability to perform the reforming reaction. Since small catalyst particle sizes are more prone to sintering, the reduction process was performed using diluted hydrogen at 300 °C.

XRD patterns of fresh and spent Pt/Al_2O_3 samples are shown in Figure 2. The spent catalyst was subjected to methanol APR for 7.5 h at 230 °C and 3.2 MPa using a 5 wt.% methanol aqueous solution supplied at 0.3 h^{-1} WHSV. Comparing fresh and spent catalysts, the XRD diffractogram of the spent sample shows that alumina support underwent phase change and Pt metal was sintered and leached.

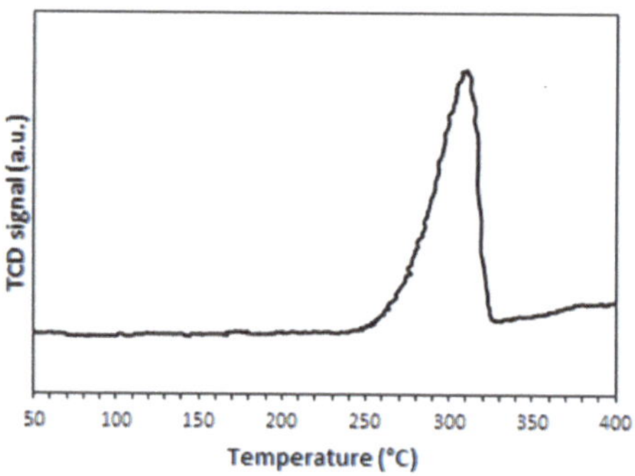

Figure 1. H_2-TPR of a 5 wt.% Pt/Al_2O_3 catalyst sample.

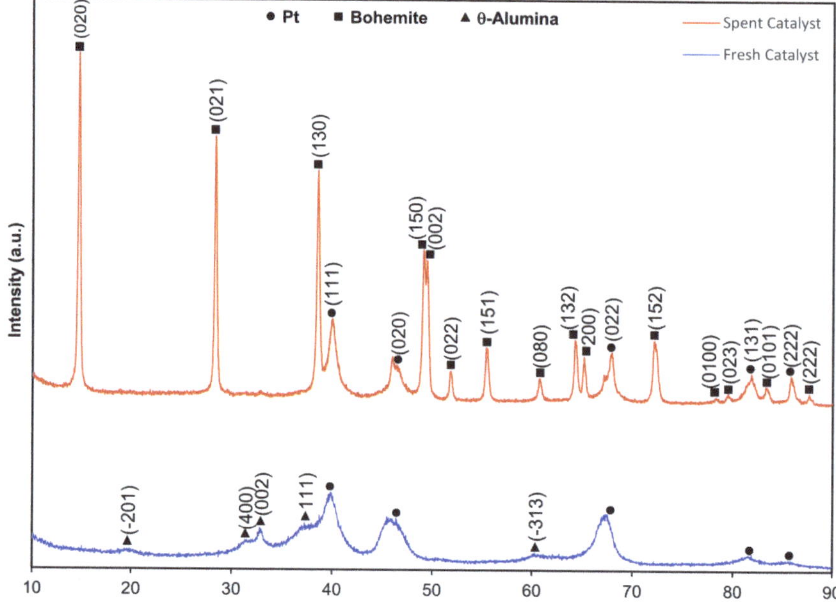

Figure 2. XRD diffractograms of fresh (blue line) and spent (red line) Pt/Al_2O_3 samples.

In both fresh and spent catalysts, characteristic Pt metallic peaks were detected at $2\theta = 39.8°$, $46.5°$, and $67.8°$ (JCPDS 01-1190) [29]. However, these peaks are broader for the fresh catalyst compared to the spent catalyst. These could be due to poor crystallinity, overlapping with some of the alumina peaks [30], or smaller crystallite size. The calculated Pt crystallite size was estimated to be ca. 5 nm in the fresh catalyst sample, as shown in Table 1.

Table 1. Pt crystallize size and mass fraction for fresh and spent catalysts.

Pt Crystallite Size (nm)	Fresh Catalyst	Spent Catalyst
	5 ($\pm$2)	12 ($\pm$1)
Pt mass Fraction (%)	Fresh Catalyst	Spent Catalyst
	4.5 ($\pm$0.5)	3.1 ($\pm$1)

The spent catalyst has a Pt crystallite size of ca. 12 nm, indicating that Pt particles were sintered under the APR reaction conditions. The calculated amount of Pt in the fresh catalyst was ca. 4.5 wt.%, which is according to the value reported by the catalyst supplier. However, the spent catalyst showed approximately 3 wt.% of Pt mass, suggesting that leaching of Pt occurred during the APR reaction. Moreover, the spent catalysts showed peaks related to boehmite, while fresh catalyst only showed peaks for θ-alumina. The presence of this crystalline structure shows that alumina was transformed in boehmite, which is a thermodynamically more stable phase under high humidity levels, such as under APR reaction conditions [31].

The N_2 adsorption–desorption isotherm for the catalyst Pt/Al_2O_3 is shown in Figure 3 and the BET surface area, average pore size, and pore volume are listed in Table 2.

According to IUPAC classification, Pt/Al_2O_3 has a type (IV) isotherm which indicates the mesoporous structure of the catalyst with the pore size wider than 4 nm. The catalyst also showed an H3-type hysteresis loop. This is usually observed for the irregular pore system with agglomerated particles that are slit-shaped pores of irregular sizes or shapes [32].

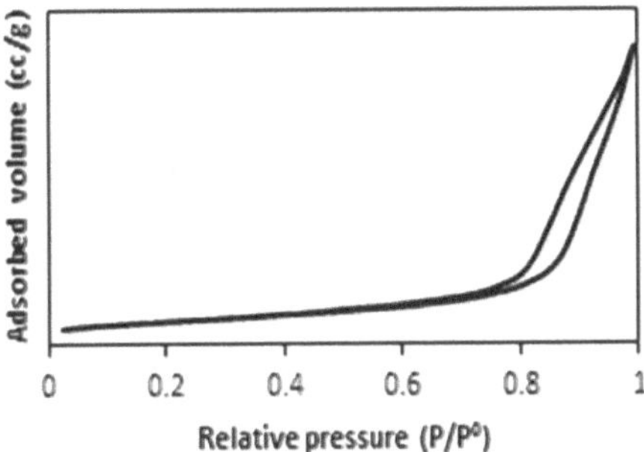

Figure 3. Nitrogen adsorption–desorption isotherm of Pt/Al_2O_3.

Table 2. Physicochemical properties of Pt/Al_2O_3.

Pt/Al_2O_3	BET Surface Area (m²/g)	Pore Volume (cc/g)	Pore Diameter (nm)
	205.9	0.87	13.1

2.2. Kinetic Characterization

The methanol–water vapor pressure was obtained for temperatures of 190 °C, 210 °C, and 230 °C using the Aspen Technology software. The plot of the bubble pressure versus methanol mole fraction at the three different temperatures is shown in Figure 4. This plot is fundamental to assure that the kinetic characterization of methanol APR using the Pt/Al_2O_3 catalyst is properly carried out at all experimental conditions, i.e., with the feed methanol water mixture in liquid phase. As the methanol concentration increases, the required pressure to keep the mixture in a liquid state increases as well. In the literature, most experiments were performed using a 5 wt.% methanol aqueous solution, yet commercial HT-PEMFC systems that use methanol reformers for in situ hydrogen production typically run using 55 wt.% methanol aqueous solutions [33,34].

For a 55 wt.% methanol aqueous solution at 230 °C, it is required to keep the system at >5.1 MPa. On the other hand, for a 5 wt.% methanol aqueous solution, >3.1 MPa is sufficient for the same temperature.

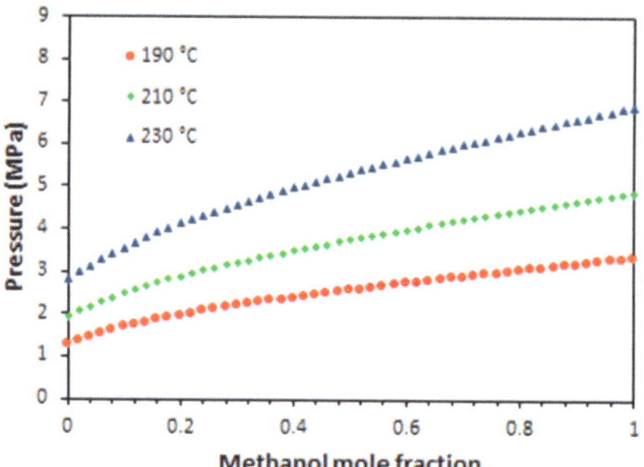

Figure 4. The bubble point pressures of binary methanol–water mixture for the different mole fractions of methanol for temperatures of 190 °C, 210 °C, and 230 °C were predicted using Aspen Technology.

2.2.1. Effect of Temperature

The effect of temperature on methanol APR conversion is depicted in Figure 5 for the temperatures 190 °C, 210 °C, and 230 °C. Thermodynamically and kinetically, either the desirable reforming or undesirable side reactions are favored by higher temperatures. On the other hand, a WGS reaction is disfavored at higher temperatures, though it is also favored kinetically with the temperature. An increase in the experimental methanol conversion was observed at higher temperatures. At the low temperature (190 °C), the methanol conversion does not go above 15%, even at a very low WHSV. Compared with the MSR, the performance of methanol APR at low temperatures is usually much lower; for example, even considering a different catalyst, at a WHSV of 1 h^{-1} and at 190 °C, the methanol conversion is ca. 75% for MSR in a Cu based catalyst [35], while for APR is less than 5% in the present work, Figure 5A. WHSV is defined as the weight of methanol in the feed stream normalized by the weight of the catalyst and time. Therefore, the methanol conversion should be similar for the same WHSV, independently of the methanol concentration, making negligible the effect of the methanol feed dilution on conversion. However, for the same WHSV, a change in the feed mass fraction of the aqueous methanol solution from 5 wt.% to 55 wt.% leads to an equivalent decrease of the feed flow rate for the same ratio, which may have some adverse influence in the conversion, for example, in the rate of adsorbed reaction products from the catalyst surface.

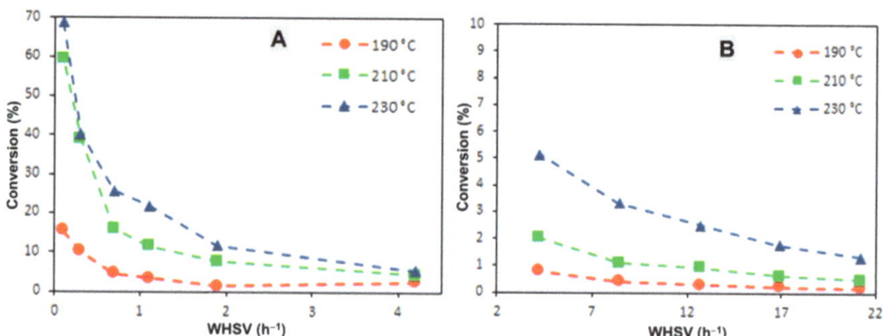

Figure 5. Methanol conversion in aqueous-phase reforming of (**A**) 5 wt.% and (**B**) 55 wt.% aqueous methanol solution over 5 wt.% Pt/Al$_2$O$_3$, N$_2$ co-feeding of 30 mL·min^{-1} and operating temperatures of 190 °C, 210 °C, and 230 °C as a function of WHSV. Operating pressure of 3.2 MPa (**A**) and 6.0 MPa (**B**). All values were obtained after achieving steady-state reaction conditions.

H$_2$ Production Rate

The APR has the potential to be integrated with an HT-PEMFC system due to its higher energy efficiency and simple recycling of unreacted reagents; this allows the APR to operate at conversions lower than 100% without significantly affecting the system efficiency. Moreover, HT-PEMFCs have a high tolerance to reforming contaminants (up to 3% [36,37]).

For 55 wt.% methanol feed aqueous solution, the maximum hydrogen production rate was ca. 500 $\mu mol \cdot min^{-1} \cdot g^{-1}_{cat}$. As depicted in Figure 6, the hydrogen production rate as a function of the WHSV peaks at WHSV $\approx$ 13 h^{-1}, especially for 210 °C and 230 °C, decreases slightly afterward. At higher methanol concentrations and for the same feed flow rate, the contact time between reactants and catalyst decreases, which adversely affects hydrogen production [17].

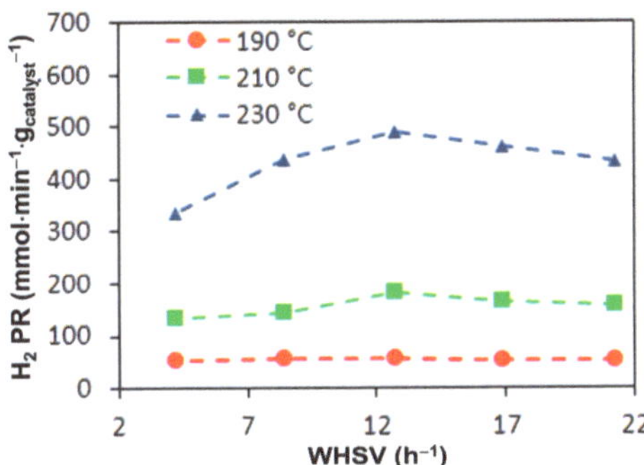

Figure 6. Hydrogen production rate in aqueous-phase reforming of 55 wt.% methanol feed aqueous solution over 5 wt.% Pt/Al$_2$O$_3$, N$_2$ co-feeding of 30 mL·min^{-1} and operating temperatures of 190 °C, 210 °C, and 230 °C at various WHSV. Operating pressure 6.0 MPa. All values were obtained after achieving steady-state reaction conditions.

By-Product Formation

Undesirable side reactions, such as methanation, are thermodynamically favored at higher temperatures and pressures. During the APR experiments, only by-products CH$_4$ and CO were detected at the GC analyzer. The low concentrations of by-products CH$_4$ and CO, inferior to 0.5%, as illustrated in the GC chromatogram of Figure 7, indicate high selectivity for the APR reaction.

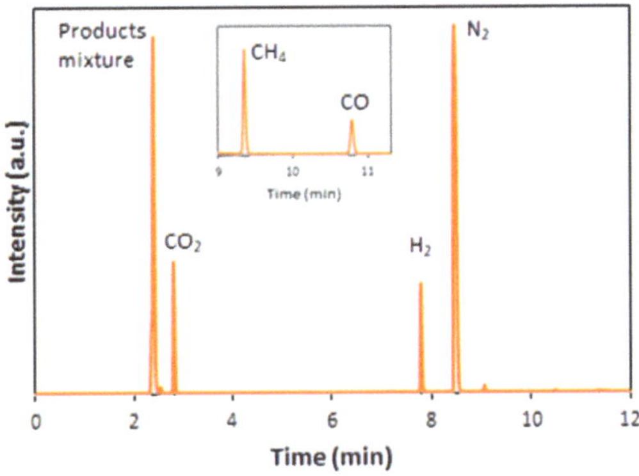

Figure 7. Gas chromatograph obtained from a 55 wt.% methanol aqueous-phase reforming on a 5 wt.% Pt/Al$_2$O$_3$ catalyst, with a N$_2$ co-feeding of 30 mL·min^{-1}, operating conditions of 230 °C, 3.2 MPa, and 21.2 h^{-1} WHSV.

Figure 8 shows the CH_4 production rate as a function of WHSV for 190 °C, 210 °C, 230 °C, and 55 wt.% methanol aqueous solution. CH_4 production can follow two paths, via cleavage of C-O bonds of methanol followed by hydrogenation (by an acid-catalyzed pathway), or via methanation, which is the hydrogenation of carbon monoxide/carbon dioxide (COx) [17]. Both paths are independent of methanol conversion since hydrogen is required to perform the hydrogenation step; therefore, CH_4 formation occurs as a parallel reaction [17]. It is noticeable that the CH_4 formation increases with temperature (Figure 8), while it shows a more complex pattern with WHSV.

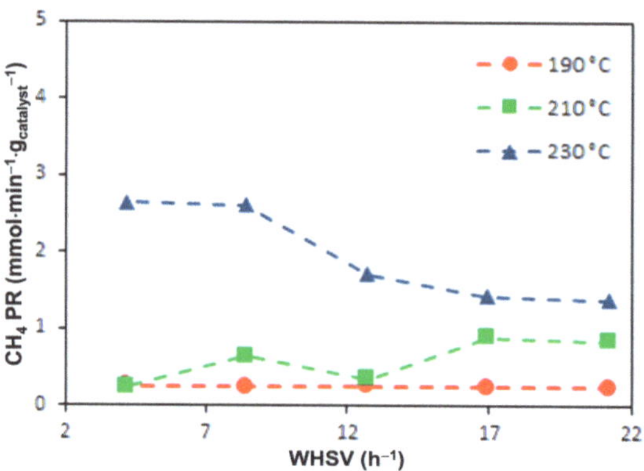

Figure 8. Methane production rate in aqueous-phase reforming fed with a 55 wt.% methanol aqueous solution, with an N_2 co-feeding of 30 mL·min^{-1}, operating at 190 °C, 210 °C, and 230 °C and at 6.0 MPa, as a function of the WHSV. All values were obtained after achieving steady-state reaction conditions.

2.2.2. Effect of Pressure

It is important to study the influence of the reaction pressure on APR performance since APR is a multiphase system with a reaction often occurring in a mass transfer-limited regime. Figure 9 shows the effects of operating pressure on methanol conversion at 230 °C. For a methanol feed mass fraction of 5 wt.%, the methanol conversion decreases from 25% at 3.2 MPa to 13% at 4.0 MPa (Figure 9A) and then it keeps more or less constant for higher pressures. For a methanol feed mass fraction of 55 wt.%, the methanol conversion is considerably lower and mostly constant with the reaction pressure, around 2% (Figure 9B). However, there is an apparent small increase in the conversion in the operating pressure region in the liquid phase. It should be emphasized, also, that a smaller conversion for a higher mass fraction of methanol in the feed stream does not necessarily lead to a lower hydrogen productivity, because of the higher amount of methanol feed to the reactor.

Performing the methanol-reforming reaction below the bubble point pressure (gaseous methanol reforming) for the methanol feed mass fraction of 55% did not yield conversions higher than those observed for methanol APR. This apparent discrepancy deserves further study in the future.

In multiphase systems, pressure affects density, bubble size distribution [38,39], and catalyst-wetting efficiency [40]. While the wetting efficiency increases with pressure [40], which is crucial for multiphase systems, increasing the operating pressure increases the hydrogen partial pressure, which negatively affects the reaction rate [17]. Hydrogen has very low solubility in water and methanol and diffuses slowly from the catalyst active site where it is formed to the bulk [18]; moreover, higher pressures reduce the bubble size, reducing the bubble disengaging rate. In this sense, the catalyst surface active sites are occupied for a longer time [17,18]. In addition, the hydrogen availability near an active site can promote the hydrogenation of intermediate species, leading to a loss in selectivity [18,41]. Since nitrogen (carrier gas) is also not soluble in the liquid reaction phase, it should leave the reactor mixed with hydrogen and CO_2, besides water and methanol vapors.

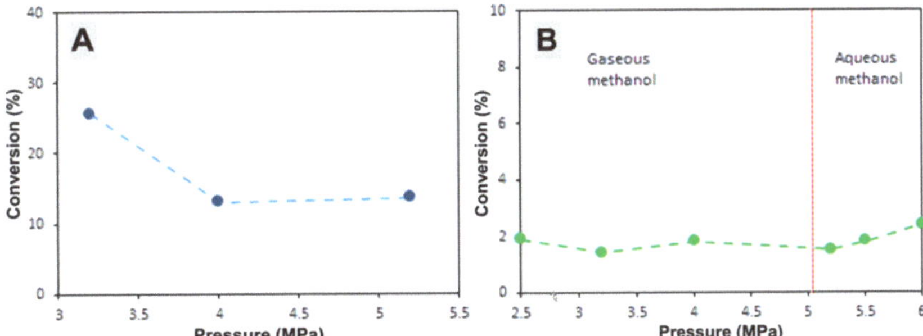

Figure 9. Methanol conversion as a function of pressure for (**A**) 5 wt.% and (**B**) 55 wt.% methanol aqueous solution over 5 wt.% Pt/Al$_2$O$_3$ catalyst. Operating temperature of 230 °C, WHSV of (**A**) 0.7 h^{-1} and (**B**) 12.7 h^{-1}. All values were obtained after achieving steady-state reaction conditions.

2.2.3. Selectivity of Methanol APR to Hydrogen

A high-methanol APR selectivity to hydrogen leads to a higher energy conversion efficiency from methanol to hydrogen and allows for directly feeding the reformate to an HT-PEMFC system. Apart from the nature of the catalyst (metal and support), the selectivity to hydrogen depends on the operating conditions [14]. Figure 10 shows the selectivity to hydrogen as a function of WHSV for different operating pressures.

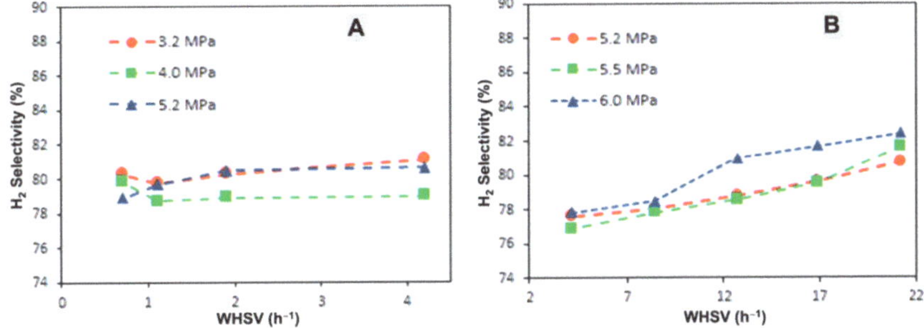

Figure 10. H$_2$ selectivity in aqueous-phase reforming of (**A**) 5 wt.% and (**B**) 55 wt.% methanol aqueous solution on a 5 wt.% Pt/Al$_2$O$_3$ catalyst, with a N$_2$ co-feeding of 30 mL·min^{-1} and at 230 °C, for various pressures and as a function of WHSV. All values were obtained after achieving steady-state reaction conditions.

For the methanol feed aqueous solution with a mass fraction of 5%, Figure 10A, the maximum change in the selectivity to hydrogen is of around 2%, either as a function of the operating pressure for a given WHSV, or as a function of the WHSV for a given operating pressure. For the methanol feed aqueous solution with a mass fraction of 50%, Figure 10B, the effect of the operating pressure for a given WHSV is also small; on the contrary, the influence of the WHSV for all three operating pressures is clearly more accentuated. For low-WHSV values (methanol mass fraction of 5%), the selectivity to hydrogen is mostly independent of the methanol conversion. Therefore, it may be concluded that hydrogen production over Pt/Al$_2$O$_3$ is not restricted by the hydrogenation of CO/CO$_2$. The selectivity to hydrogen is clearly above 75% for all experiments (Figure 10A,B), indicating that the H$_2$/CO$_2$ ratio for these experiments stayed above a factor of 3.

Apart from the nature of the catalyst (both metal and support) and operating conditions, the dependence of the selectivity to hydrogen with the reactor design has also been studied. It was reported that, by using a microchannel reactor instead of a fixed-bed reactor, the selectivity to hydrogen becomes higher by a factor of two, regardless of the feedstock conversion [42]. The selectivity to hydrogen in a microchannel reactor should increase because of a better catalyst wettability, making the mass transfer of hydrogen from the catalyst site to the gas phase faster [42]. Table 3 below summarizes the APR performance results (in terms of selectivity to CO_2, CO, CH_4) at various WHSVs and operating pressures.

Table 3. Selectivity of the CO_2, CH_4, and CO in methanol APR experiments at various operating pressures and for different WHSVs.

Product	Methanol Concentration	WHSV (h^{-1})	Selectivity (%)				
			3.2 MPa	4.0 MPa	5.2 MPa	5.5 MPa	6.0 MPa
CO_2	5 wt.%	0.7	19.7	20.8	19.1		
		1.1	20.2	21	19.4		
		1.9	19.4	21.2	20.2		
		4.2	18.5	20	20.9		
	55 wt.%	4.2			21.6	21.6	21.5
		8.4			21.3	21.0	21
		12.7			20.7	20.5	18.7
		16.9			19.9	19.6	18
		21.2			18.8	17.7	17.2
CH_4	5 wt.%	0.7	0.1	0.1	0.2		
		1.1	n/a	0.08	0.1		
		1.9	0.2	0.05	0.1		
		4.2	0.3	0.19	0.1		
	55 wt.%	4.2			0.3	0.9	0.6
		8.4			0.3	0.9	0.4
		12.7			0.3	0.7	0.3
		16.9			0.4	0.6	0.2
		21.2			0.4	0.5	0.2
CO	5 wt.%	0.7	0.05	0.02	0.02		
		1.1	0.07		0.05		
		1.9	0.04	0.01	0.05		
		4.2	0.01	0.03	0.05		
	55 wt.%	4.2			0.1	0.6	0.05
		8.4			0.1	0.2	0.04
		12.7			0.1	0.2	0.04
		16.9			0.1	0.2	0.1
		21.2			0.1	0.1	0.1

Long-Term Stability Test

The long-term stability of methanol APR was performed to study the effects on catalyst performance and methanol conversion. Figure 11 shows the methanol conversion as a function of time for a 5 wt.% methanol feed aqueous solution. Steady-state performance was achieved for >7 h.

The catalytic activity increased at the beginning and then slowly decreased until reaching a steady state. The decrease in the catalytic activity (methanol conversion) may be assigned to the phase change of the support alumina into boehmite, metal phase sintering, and Pt leaching, as discussed before.

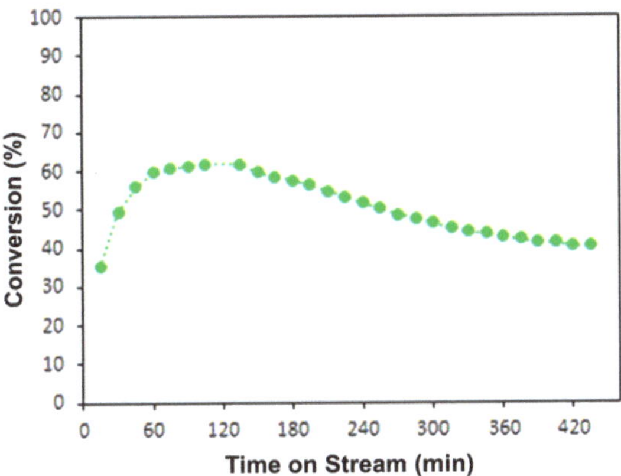

Figure 11. Methanol conversion as a function of time on stream for 5 wt.% methanol aqueous solution over 5 wt.% Pt/Al$_2$O$_3$ at 230 °C and 3.3 MPa and 0.3 h^{-1} WHSV.

2.3. Methanol Steam Reforming Versus Aqueous-Phase Reforming

Methanol has the potential to make hydrogen dispatchable, which is critical for the development of the so-called hydrogen economy. MSR is an established pathway to produce in situ hydrogen, with extensive research performed on catalysts, reactors, and purification of the produced hydrogen. Most MSR employs Cu-based catalysts due to their low cost, high hydrogen yield, and low CO production [43–45]. Methanol APR, on the other hand, is a relatively new technology where the catalyst development is still underdeveloped. Pt-based catalysts are widely used in APR due to their stability and moderate performance [25,43]. Figure 12 illustrates the interest in these two reforming processes measured by the number of publications during the last two decades.

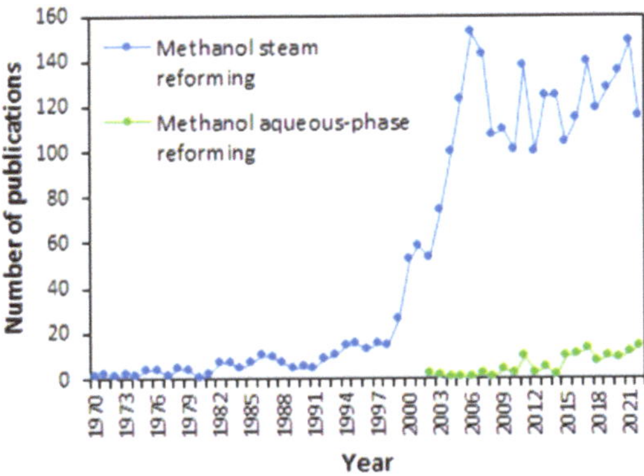

Figure 12. The number of scientific publications on methanol steam reforming and methanol aqueous-phase reforming (information retrieved from Scopus database).

Methanol APR in Pt/Al$_2$O$_3$ suffers from low stability and displays low yields, which reduces its potential interest to industrial applications.

However, the possibility of running the reforming reaction in the liquid phase and then saving the vaporization energy is still driving the research community for developing better-performing catalysts. To use HT-PEMFCs in mobile applications, which require high power output, there are certain operating condition requirements for methanol reforming that should be fulfilled: compact reforming systems, operation close to the stoichiometric concentration of methanol–water and at a temperature which allows its thermal integration with an HT-PEMFC [46–48].

At around 100 °C, the methanol steam reforming (MSR) produces just hydrogen and CO_2; the CO concentration is negligible. However, above 150 °C, there is the onset production of CO whose concentration in the product stream increases with the temperature [35,49]. Yu et al. [49] reported that the CO concentration in the product stream was 511 ppm, when the MSR reactor was run at 190 °C, for a contact time (W/F) of 180 $kg_{cat} \cdot (mol \cdot s^{-1})^{-1}$ (almost full conversion of the methanol). For 230 °C, it was 2184 ppm. Ribeirinha et al. [35] reported a CO concentration of 600 ppm at the product stream when the MSR reactor was run at 180 °C, on a $CuO/ZnO/Ga_2O_3$ catalyst, and for a W/F of 60 $kg_{cat} \cdot (mol \cdot s^{-1})^{-1}$ (ca. 80% of methanol conversion).

From the present work, we report CO production between 100 ppm (190 °C) and 1990 ppm (230 °C) [647.1 $kg_{cat} \cdot (mol \cdot s^{-1)-1}$–5.5 $kg_{cat} \cdot (mol \cdot s^{-1})^{-1}$] using a commercial Pt/Al_2O_3. Although the CO concentration is lower for the methanol APR at lower temperatures, as the temperature increases, the CO concentration in the product becomes similar for both types of methanol reforming. However, it should be emphasized that the CO concentration also depends on the reactor used for the reforming.

2.4. Perspectives

Catalyst development remains a primary focus of research for methanol APR; Pt-based catalysts are the best-performing catalysts. However, as seen in this work, support plays an equally important role in the catalyst performance. Therefore, various supports other than metal oxides should be assessed to improve the activity and stability of the catalyst. Furthermore, the long-term stability of the catalyst should be studied to better understand the catalyst deactivation under APR conditions.

In addition to the catalyst development, the mass transfer limitation of the reaction products from the active sites poses one of the greatest challenges to overcome. Nitrogen co-feeding has been employed to mitigate this issue. However, the dilution of hydrogen due to the co-feeding of nitrogen can adversely affect the performance of HT-PEMFCs during reformer/fuel cell integration. Therefore, novel reactor designs should be proposed and studied to minimize the mass transport limitation to allow higher hydrogen productivity at lower temperatures by effectively removing reaction products from the active sites. Following, the use of a membrane reactor, selective to hydrogen, may be considered as an alternative to a conventional fixed-bed reactor. Reactors such as coated microchannel reactors [50] and membrane reactors facilitate the effective removal of hydrogen from the reaction medium; for example, D´Angelo et al. [51] have reported that using a carbon-coated ceramic membrane reactor yielded 2.5 times higher hydrogen than a fixed-bed reactor for APR of sorbitol.

3. Experiment

3.1. Materials

Pt/Al_2O_3 catalyst (5 wt.% Pt loading) supplied by Sigma-Aldrich (St. Louis, MO, USA) in powder form was used without further modification.

3.2. Physicochemical Characterization

Hydrogen-temperature-programmed reduction (H_2-TPR) and X-ray diffraction (XRD) were used to determine the physicochemical properties of the Pt/Al_2O_3 catalyst. Multipoint Brunauer–Emmett–Teller (BET) provided information regarding the surface area and pore size of the catalyst. XRD patterns were recorded at room temperature, using a Panalytical MPD diffractometer equipped (Almelo, The Netherlands) with an X' Celerator detector and secondary monochromator in Bragg–Brentano geometry. $CuK\alpha_{1,2}$ radiation, λ_1 = 1.5406 Å, λ_2 = 1.5444 Å, a step size of 0.017° and 100 s/step were used. Rietveld refinements were performed using the software PowderCell 2.4 to calculate Pt crystallite size.

The mass fraction of Pt in the catalyst was estimated in a Thermo Scientific iCAP 7400 ICP-OES Duo instrument (Boston, MA, USA), coupled with an ASX-520 autosampler.

A ChemBET Pulsar TPR/TPD instrument (Anton Paar, Ashland, VA, USA) equipped with a TCD was used to perform H_2-TPR experiments. Approximately 25 mg of catalyst was placed in a U-shaped quartz tube reactor and held in place with quartz wool. The sample was heated from 50 °C to 450 °C at a heating rate of 5 °C·min^{-1} under constant H_2/Ar (1:9) flow. Multipoint BET surface measurements were performed on a Quantachrome Autosorb AS-1 instrument (Anton Paar, Ashland, VA, USA) at −196 °C. Prior to the analysis, the samples were outgassed in vacuum at 200 °C for 4 h.

3.3. Reaction Kinetic Experimental Setup

A cylindrical stainless-steel FBR (70 mm L × 7 mm ID) was used to conduct the methanol APR experiments. A 700 mg catalyst sample and an equal amount of glass beads were mixed and loaded into the midsection of the reactor. Glass wool and wire mesh were inserted at both ends of the reactor to avoid dragging the catalyst and glass beads with the product's and reactant's flow. The reactor was placed vertically inside a temperature-controlled oven. Figure 13 shows a schematic representation of the laboratory scale set-up used for this work.

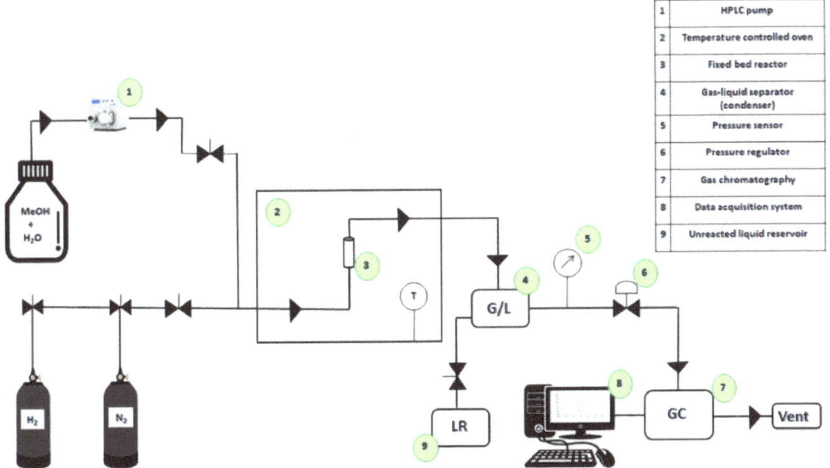

Figure 13. Schematic representation of methanol aqueous-phase reforming experimental set-up.

Before the methanol APR tests, Pt/Al$_2$O$_3$ was reduced in situ at 300 °C for 2 h in a flowing H_2/N_2 (1:1) atmosphere. Afterward, the reactor was purged with N_2 gas (50 mL·min^{-1}) for 30 min to remove the adsorbed hydrogen. For this study, methanol APR activity tests were performed at temperatures of 190 °C, 210 °C, and 230 °C, and pressures of 3.2 MPa, 4.0 MPa, 5.2 MPa, 5.5 MPa, and 6.0 MPa with N_2 co-feeding (sweep gas) at 30 mL·min^{-1}. Two different methanol aqueous solutions, 5 wt.% and 55 wt.% of methanol, were tested under the above-mentioned experimental conditions. To keep the reaction in the liquid phase, binary equilibrium plots for the methanol–water mixture were simulated using Aspen Plus (Aspen Technology Software V37.0) (Wilson-Ideal equation property package). These phase diagrams were obtained for the different methanol–water mixtures at various operating temperatures.

The methanol aqueous solution was fed upwards using an HPLC pump (KNAUER Smartline Manager 5050, Berlin, Germany) with a WHSV between 0.1 h^{-1} and 21.2 h^{-1}. The output stream was cooled down in a condenser at ca. 0 °C to separate the liquid-phase and gaseous-phase streams. The concentration of the gaseous-phase components was measured with online gas chromatography (GC) (Shimadzu GC 2010 Plus, Kyoto, Japan) equipped with two columns, Pora-Plot Q (Agilent, Santa Clara, CA, USA) and HP-Molesieve (Agilent), and a barrier ionization discharge detector (BID, Shimadzu, Kyoto, Japon). The catalytic activity was evaluated in terms of methanol conversion (%) (Equation (7)), H_2 production rate ($\mu mol \cdot min^{-1} \cdot g_{catalyst}^{-1}$) (Equation (8)), CH_4 production rate ($\mu mol \cdot min^{-1} \cdot g_{catalyst}^{-1}$) (Equation (9)), and H_2 selectivity (%) (Equation (10)).

$$\text{Methanol conversion (\%)} = \left(\frac{n\text{MeOH}_{in} - n\text{MeOH}_{out}}{n\text{MeOH}_{in}} \right) \times 100 \tag{7}$$

$$H_2 \text{ production rate } (\mu mol \cdot min^{-1} \cdot g_{catalyst}^{-1}) = \frac{n(H_2)}{m_{catalyst}} \tag{8}$$

$$CH_4 \text{ production rate } (\mu mol \cdot min^{-1} \cdot g_{catalyst}^{-1}) = \frac{n(CH_4)}{m_{catalyst}} \tag{9}$$

$$\text{Selectivity }_{Hydrogen} (\%) = \frac{n(H_2)}{\sum Gas\,Products_{out}} \times 100 \tag{10}$$

Herein, n denotes the molar flow rate, MeOH_{in} means methanol in the inlet flow, MeOH_{out} means methanol in the outlet flow, and $m_{catalyst}$ is the mass of the catalyst in grams. All values were obtained after achieving steady-state reaction conditions.

4. Conclusions

This work studies the effects of various operating conditions in methanol APR using a commercial Pt/Al_2O_3. As expected, increasing operating temperature and residence time were beneficial for APR performance. Methanol conversion was, however, below 5% while using 55 wt.% methanol feedstock at 190 °C. Despite the low methanol conversion obtained using high-methanol concentration feed (55 wt.%), the hydrogen production rate that was achieved was high. As the operating pressure was increased beyond bubble point pressure, methanol conversion dropped drastically due to the increase in mass transfer limitations. However, increasing the operating pressure did not affect the hydrogen selectivity. It was also concluded that alkane formation in APR also occurs due to parallel pathway reactions in addition to hydrogenation of CO/CO_2 but at a minute level. Slight deactivation was observed using 5 wt.% methanol feedstock after the long-term stability experiment due to the alumina phase changing into boehmite and Pt sintering and leaching.

Nonetheless, APR has come up to be an encouraging technology for hydrogen production. Several shortcomings of APR have been listed that should be researched to achieve higher hydrogen production and increased catalyst stability.

Author Contributions: The contributions of the authors for the manuscript are the following: conceptualization, A.M.; methodology, A.M., J.S., P.L. and P.R.; investigation, P.L.; resources, A.M.; writing—original draft preparation, P.L.; writing—review and editing, A.M., J.S., P.R., W.H. and J.T.; supervision, A.M., J.S. and P.R.; project administration, A.M. and P.R.; funding acquisition, W.H. and J.T. All authors have read and agreed to the published version of the manuscript.

Funding: This work was supported by: (i) Fuel Cells and Hydrogen 2 Joint Undertaking (JU) under grant agreement number 875081. The JU receives support from the European Union's Horizon 2020 research and innovation programme and Finland, Denmark, Sweden, Portugal; (ii) H2Driven Green Agenda, number C644923817-00000037, investment project number 50, financed by the Recovery and Resilience Plan (PRR) and by European Union—Next Generation EU; and (iii) LEPABE, UIDB/00511/2020 (DOI: 10.54499/UIDB/00511/2020) and UIDP/00511/2020 (DOI: 10.54499/UIDP/00511/2020) and ALiCE, LA/P/0045/2020 (DOI: 10.54499/LA/P/0045/2020).

Data Availability Statement: The data presented in this study are available on request from the corresponding author.

Conflicts of Interest: Authors Werneri Huhtinen and Johan Tallgren were employed by the company VTT Technical Research Centre of Finland Ltd. The remaining authors declare that the research was conducted in the absence of any commercial or financial relationships that could be construed as a potential conflict of interest.

Abbreviations

APR	Aqueous-phase reforming
FBR	Fixed-bed reactor
GC	Gas chromatography
HPLC	High-performance liquid chromatography
MSR	Methanol steam reforming
PR	Production rate
TPD	Temperature-programmed desorption
TPR	Temperature-programmed reduction
WGS	Water–gas shift
WHSV	Weight hourly space velocity
XRD	X-ray diffraction

References

1. Dillman, K.; Heinonen, J. A 'just' hydrogen economy: A normative energy justice assessment of the hydrogen economy. *Renew. Sustain. Energy Rev.* **2022**, *167*, 112648. [CrossRef]
2. Kar, S.K.; Harichandan, S.; Roy, B. Bibliometric analysis of the research on hydrogen economy: An analysis of current findings and roadmap ahead. *Int. J. Hydrogen Energy* **2022**, *47*, 10803–10824. [CrossRef]
3. Nastasi, B. Power to Gas and Hydrogen applications to energy systems at different scales—Building, District and National level. *Int. J. Hydrogen Energy* **2019**, *19*, 9485. [CrossRef]
4. van der Spek, M.; Banet, C.; Bauer, C.; Gabrielli, P.; Goldthorpe, W.; Mazzotti, M.; Munkejord, S.T.; Røkke, N.A.; Shah, N.; Sunny, N.; et al. Perspective on the hydrogen economy as a pathway to reach net-zero CO_2 emissions in Europe. *Energy Environ. Sci.* **2022**, *15*, 1034–1077. [CrossRef]
5. Jain, I. Hydrogen the fuel for 21st century. *Int. J. Hydrogen Energy* **2009**, *34*, 7368–7378. [CrossRef]
6. Tarhan, C.; Çil, M.A. A study on hydrogen, the clean energy of the future: Hydrogen storage methods. *J. Energy Storage* **2021**, *40*, 102676. [CrossRef]
7. Muellerlanger, F.; Tzimas, E.; Kaltschmitt, M.; Peteves, S. Techno-economic assessment of hydrogen production processes for the hydrogen economy for the short and medium term. *Int. J. Hydrogen Energy* **2007**, *32*, 3797–3810. [CrossRef]
8. Dawood, F.; Anda, M.; Shafiullah, G.M. Hydrogen Production for Energy: An Overview. *Int. J. Hydrogen Energy* **2020**, *45*, 3847–3869. [CrossRef]
9. Ajanovic, A.; Haas, R. Prospects and impediments for hydrogen and fuel cell vehicles in the transport sector. *Int. J. Hydrogen Energy* **2020**, *46*, 10049–10058. [CrossRef]
10. Griffiths, S.; Sovacool, B.K.; Kim, J.; Bazilian, M.; Uratani, J.M. Industrial decarbonization via hydrogen: A critical and systematic review of developments, socio-technical systems and policy options. *Energy Res. Soc. Sci.* **2021**, *80*, 102208. [CrossRef]
11. Goma, D.; Delgado, J.J.; Lefferts, L.; Faria, J.; Calvino, J.J.; Cauqui, M. Catalytic Performance of Ni/CeO$_2$/X-ZrO$_2$ (X = Ca, Y) Catalysts in the Aqueous-Phase Reforming of Methanol. *Nanomaterials* **2019**, *9*, 1582. [CrossRef] [PubMed]
12. Kubacka, A.; Fernández-García, M.; Martínez-Arias, A. Catalytic hydrogen production through WGS or steam reforming of alcohols over Cu, Ni and Co catalysts. *Appl. Catal. A Gen.* **2016**, *518*, 2–17. [CrossRef]
13. Coronado, I.; Stekrova, M.; Moreno, L.G.; Reinikainen, M.; Simell, P.; Karinen, R.; Lehtonen, J. Aqueous-phase reforming of methanol over nickel-based catalysts for hydrogen production. *Biomass Bioenergy* **2017**, *106*, 29–37. [CrossRef]
14. Davda, R.; Shabaker, J.; Huber, G.; Cortright, R.; Dumesic, J. A review of catalytic issues and process conditions for renewable hydrogen and alkanes by aqueous-phase reforming of oxygenated hydrocarbons over supported metal catalysts. *Appl. Catal. B Environ.* **2005**, *56*, 171–186. [CrossRef]
15. Martín, M.; Grossmann, I.E. Optimal Simultaneous Production of Hydrogen and Liquid Fuels from Glycerol: Integrating the Use of Biodiesel Byproducts. *Ind. Eng. Chem. Res.* **2014**, *53*, 7730–7745. [CrossRef]
16. Cortright, R.D.; Davda, R.R.; Dumesic, J.A. Hydrogen from catalytic reforming of biomass-derived hydrocarbons in liquid water. *Mater. Sustain. Energy* **2010**, *418*, 289–292. [CrossRef]
17. Shabaker, J.; Davda, R.; Huber, G.; Cortright, R.; Dumesic, J. Aqueous-phase reforming of methanol and ethylene glycol over alumina-supported platinum catalysts. *J. Catal.* **2003**, *215*, 344–352. [CrossRef]
18. Shabaker, J.W.; Dumesic, J.A. Kinetics of Aqueous-Phase Reforming of Oxygenated Hydrocarbons: Pt/Al$_2$O$_3$ and Sn-Modified Ni Catalysts. *Ind. Eng. Chem. Res.* **2004**, *43*, 3105–3112. [CrossRef]
19. Shabaker, J.; Huber, G.; Dumesic, J. Aqueous-phase reforming of oxygenated hydrocarbons over Sn-modified Ni catalysts. *J. Catal.* **2004**, *222*, 180–191. [CrossRef]
20. Pipitone, G.; Zoppi, G.; Pirone, R.; Bensaid, S. A critical review on catalyst design for aqueous phase reforming. *Int. J. Hydrogen Energy* **2022**, *47*, 151–180. [CrossRef]
21. Davda, R.; Shabaker, J.; Huber, G.; Cortright, R.; Dumesic, J. Aqueous-phase reforming of ethylene glycol on silica-supported metal catalysts. *Appl. Catal. B Environ.* **2003**, *43*, 13–26. [CrossRef]

22. Coronado, I.; Pitínová, M.; Karinen, R.; Reinikainen, M.; Puurunen, R.L.; Lehtonen, J. Aqueous-phase reforming of Fischer-Tropsch alcohols over nickel-based catalysts to produce hydrogen: Product distribution and reaction pathways. *Appl. Catal. A Gen.* **2018**, *567*, 112–121. [CrossRef]

23. Kotowicz, J.; Brzęczek, M. Methods to increase the efficiency of production and purification installations of renewable methanol. *Renew. Energy* **2021**, *177*, 568–583. [CrossRef]

24. Stekrova, M.; Rinta-Paavola, A.; Karinen, R. Hydrogen production via aqueous-phase reforming of methanol over nickel modified Ce, Zr and La oxide supports. *Catal. Today* **2018**, *304*, 143–152. [CrossRef]

25. Coronado, I.; Stekrova, M.; Reinikainen, M.; Simell, P.; Lefferts, L.; Lehtonen, J. A review of catalytic aqueous-phase reforming of oxygenated hydrocarbons derived from biorefinery water fractions. *Int. J. Hydrogen Energy* **2016**, *41*, 11003–11032. [CrossRef]

26. Lin, L.; Zhou, W.; Gao, R.; Yao, S.; Zhang, X.; Xu, W.; Zheng, S.; Jiang, Z.; Yu, Q.; Li, Y.-W.; et al. Low-temperature hydrogen production from water and methanol using Pt/α-MoC catalysts. *Nature* **2017**, *544*, 80–83. [CrossRef] [PubMed]

27. Li, D.; Li, Y.; Liu, X.; Guo, Y.; Pao, C.-W.; Chen, J.-L.; Hu, Y.; Wang, Y. NiAl$_2$O$_4$ Spinel Supported Pt Catalyst: High Performance and Origin in Aqueous-Phase Reforming of Methanol. *ACS Catal.* **2019**, *9*, 9671–9682. [CrossRef]

28. Tang, Z.; Monroe, J.; Dong, J.; Nenoff, T.; Weinkauf, D. Platinum-Loaded NaY Zeolite for Aqueous-Phase Reforming of Methanol and Ethanol to Hydrogen. *Ind. Eng. Chem. Res.* **2009**, *48*, 2728–2733. [CrossRef]

29. El Doukkali, M.; Iriondo, A.; Cambra, J.; Gandarias, I.; Jalowiecki-Duhamel, L.; Dumeignil, F.; Arias, P. Deactivation study of the Pt and/or Ni-based γ-Al$_2$O$_3$ catalysts used in the aqueous phase reforming of glycerol for H$_2$ production. *Appl. Catal. A Gen.* **2014**, *472*, 80–91. [CrossRef]

30. Zambare, R.S.; Vaidya, P.D. Aqueous-phase reforming of model compounds of wet biomass to hydrogen on alumina-supported metal catalysts. *Int. J. Chem. Kinet.* **2024**, *56*, 265–278. [CrossRef]

31. Pipitone, G.; Zoppi, G.; Ansaloni, S.; Bocchini, S.; Deorsola, F.A.; Pirone, R.; Bensaid, S. Towards the sustainable hydrogen production by catalytic conversion of C-laden biorefinery aqueous streams. *Chem. Eng. J.* **2019**, *377*, 120677. [CrossRef]

32. Thommes, M.; Kaneko, K.; Neimark, A.V.; Olivier, J.P.; Rodriguez-Reinoso, F.; Rouquerol, J.; Sing, K.S.W. Physisorption of gases, with special reference to the evaluation of surface area and pore size distribution (IUPAC Technical Report). *Pure Appl. Chem.* **2015**, *87*, 1051–1069. [CrossRef]

33. Justesen, K.K.; Andreasen, S.J.; Pasupathi, S.; Pollet, B.G. Modeling and control of the output current of a Reformed Methanol Fuel Cell system. *Int. J. Hydrogen Energy* **2015**, *40*, 16521–16531. [CrossRef]

34. Andreasen, S.J.; Kær, S.K.; Sahlin, S. Control and experimental characterization of a methanol reformer for a 350 W high temperature polymer electrolyte membrane fuel cell system. *Int. J. Hydrogen Energy* **2013**, *38*, 1676–1684. [CrossRef]

35. Ribeirinha, P.; Mateos-Pedrero, C.; Boaventura, M.; Sousa, J.; Mendes, A. CuO/ZnO/Ga2O3 catalyst for low temperature MSR reaction: Synthesis, characterization and kinetic model. *Appl. Catal. B Environ.* **2018**, *221*, 371–379. [CrossRef]

36. Ribeirinha, P.; Abdollahzadeh, M.; Pereira, A.; Relvas, F.; Boaventura, M.; Mendes, A. High temperature PEM fuel cell integrated with a cellular membrane methanol steam reformer: Experimental and modelling. *Appl. Energy* **2018**, *215*, 659–669. [CrossRef]

37. Li, Q.; He, R.; Gao, J.-A.; Jensen, J.O.; Bjerrum, N.J. The CO Poisoning Effect in PEMFCs Operational at Temperatures up to 200 °C. *J. Electrochem. Soc.* **2003**, *150*, A1599. [CrossRef]

38. Iliuta, I. Performance of Fixed Bed Reactors with Two-Phase Upflow and Downflow. *J. Chem. Technol. Biotechnol.* **1997**, *68*, 47–56. [CrossRef]

39. Iliuta, I.; Thyrion, F. Flow regimes, liquid holdups and two-phase pressure drop for two-phase cocurrent downflow and upflow through packed beds: Air/Newtonian and non-Newtonian liquid systems. *Chem. Eng. Sci.* **1997**, *52*, 4045–4053. [CrossRef]

40. Dudukovic, M.P.; Larachi, F.; Mills, P.L. Multiphase reactors—Revisited. *Chem. Eng. Sci.* **1999**, *54*, 1975–1995. [CrossRef]

41. García, L.; Valiente, A.; Oliva, M.; Ruiz, J.; Arauzo, J.; García, L.; Valiente, A.; Oliva, M.; Ruiz, J.; Arauzo, J. Influence of operating variables on the aqueous-phase reforming of glycerol over a Ni/Al coprecipitated catalyst. *Int. J. Hydrogen Energy* **2018**, *43*, 20392–20407. [CrossRef]

42. D'Angelo, M.F.N.; Ordomsky, V.; van der Schaaf, J.; Schouten, J.C.; Nijhuis, T.A. Aqueous phase reforming in a microchannel reactor: The effect of mass transfer on hydrogen selectivity. *Catal. Sci. Technol.* **2013**, *3*, 2834–2842. [CrossRef]

43. Lakhtaria, P.; Ribeirinha, P.; Huhtinen, W.; Viik, S.; Sousa, J.; Mendes, A. Hydrogen production via aqueous-phase reforming for high-temperature proton exchange membrane fuel cells–A review. *Open Res. Eur.* **2022**, *23*, 81. [CrossRef] [PubMed]

44. Iulianelli, A.; Ribeirinha, P.; Mendes, A.; Basile, A. Methanol steam reforming for hydrogen generation via conventional and membrane reactors: A review. *Renew. Sustain. Energy Rev.* **2014**, *29*, 355–368. [CrossRef]

45. Palo, D.R.; Dagle, R.A.; Holladay, J.D. Methanol Steam Reforming for Hydrogen Production. *Chem. Rev.* **2007**, *107*, 3992–4021. [CrossRef]

46. Brett, D.J.L.; Brandon, N.P.; Hawkes, A.D.; Staffell, I. Fuel cell systems for small and micro combined heat and power (CHP) applications. In *Small and Micro Combined Heat and Power (CHP) Systems—Advanced Design, Performance, Materials and Applications*; Beith, R., Ed.; Woodhead Publishing Limited: Cambridge, UK, 2011; pp. 233–261. [CrossRef]

47. Ribeirinha, P.; Alves, I.; Vázquez, F.V.; Schuller, G.; Boaventura, M.; Mendes, A. Heat integration of methanol steam reformer with a high-temperature polymeric electrolyte membrane fuel cell. *Energy* **2017**, *120*, 468–477. [CrossRef]

48. Avgouropoulos, G.; Paxinou, A.; Neophytides, S. In situ hydrogen utilization in an internal reforming methanol fuel cell. *Int. J. Hydrogen Energy* **2014**, *39*, 18103–18108. [CrossRef]

49. Yu, K.M.K.; Tong, W.; West, A.; Cheung, K.; Li, T.; Smith, G.; Guo, Y.; Tsang, S.C.E. Non-syngas direct steam reforming of methanol to hydrogen and carbon dioxide at low temperature. *Nat. Commun.* **2012**, *3*, 1230. [CrossRef]

50. D'Angelo, M.F.N.; Ordomsky, V.; Paunovic, V.; van der Schaaf, J.; Schouten, J.C.; Nijhuis, T.A. Hydrogen Production through Aqueous-Phase Reforming of Ethylene Glycol in a Washcoated Microchannel. *ChemSusChem* **2013**, *6*, 1708–1716. [CrossRef]

51. D'Angelo, M.F.N.; Ordomsky, V.; Schouten, J.C.; van der Schaaf, J.; Nijhuis, T.A. Carbon-Coated Ceramic Membrane Reactor for the Production of Hydrogen by Aqueous-Phase Reforming of Sorbitol. *ChemSusChem* **2014**, *7*, 2007–2015. [CrossRef]

 *catalysts*

Article

Dry Reforming of Methane (DRM) over Hydrotalcite-Based Ni-Ga/(Mg, Al)O$_x$ Catalysts: Tailoring Ga Content for Improved Stability

Ahmed Y. Elnour [1], Ahmed E. Abasaeed [1], Anis H. Fakeeha [1], Ahmed A. Ibrahim [1,*], Salwa B. Alreshaidan [2] and Ahmed S. Al-Fatesh [1,*]

[1] Chemical Engineering Department, College of Engineering, King Saud University (KSU), P.O. Box 800, Riyadh 11421, Saudi Arabia; aelnour@ksu.edu.sa (A.Y.E.); abasaeed@ksu.edu.sa (A.E.A.); anishf@ksu.edu.sa (A.H.F.)

[2] Department of Chemistry, Faculty of Science, King Saud University (KSU), P.O. Box 800, Riyadh 11451, Saudi Arabia; chem241@ksu.edu.sa

* Correspondence: aidid@ksu.edu.sa (A.A.I.); aalfatesh@ksu.edu.sa (A.S.A.-F.)

Citation: Elnour, A.Y.; Abasaeed, A.E.; Fakeeha, A.H.; Ibrahim, A.A.; Alreshaidan, S.B.; Al-Fatesh, A.S. Dry Reforming of Methane (DRM) over Hydrotalcite-Based Ni-Ga/(Mg, Al)O$_x$ Catalysts: Tailoring Ga Content for Improved Stability. *Catalysts* **2024**, *14*, 721. https://doi.org/10.3390/catal14100721

Academic Editors: Georgios Bampos, Paraskevi Panagiotopoulou and Eleni A. Kyriakidou

Received: 15 August 2024
Revised: 13 October 2024
Accepted: 14 October 2024
Published: 16 October 2024

Abstract: Dry reforming of methane (DRM) is a promising way to convert methane and carbon dioxide into syngas, which can be further utilized to synthesize value-added chemicals. One of the main challenges for the DRM process is finding catalysts that are highly active and stable. This study explores the potential use of Ni-based catalysts modified by Ga. Different Ni-Ga/(Mg, Al)O$_x$ catalysts, with various Ga/Ni molar ratios (0, 0.1, 0.3, 0.5, and 1), were synthesized by the co-precipitation method. The catalysts were tested for the DRM reaction to evaluate their activity and stability. The Ni/(Mg, Al)O$_x$ and its Ga-modified Ni-Ga/(Mg, Al)O$_x$ were characterized by N$_2$ adsorption–desorption, Fourier Transform Infrared Spectroscopy (FTIR), H$_2$-temperature-programmed reduction (TPR), X-ray diffraction (XRD), thermogravimetric analysis (TGA) and Raman techniques. The test of catalytic activity, at 700 °C, 1 atm, GHSV of 42,000 mL/h/g, and a CH$_4$: CO$_2$ ratio of 1, revealed that Ga incorporation effectively enhanced the catalyst stability. Particularly, the Ni-Ga/(Mg, Al)O$_x$ catalyst with Ga/Ni ratio of 0.3 exhibited the best catalytic performance, with CH$_4$ and CO$_2$ conversions of 66% and 74%, respectively, and an H$_2$/CO ratio of 0.92. Furthermore, the CH$_4$ and CO$_2$ conversions increased from 34% and 46%, respectively, when testing at 600 °C, to 94% and 96% when the catalytic activity was operated at 850 °C. The best catalyst's 20 h stream performance demonstrated its great stability. DFT analysis revealed an alteration in the electronic properties of nickel upon Ga incorporation, the d-band center of the Ga modified catalyst (Ga/Ni ratio of 0.3) shifted closer to the Fermi level, and a charge transfer from Ga to Ni atoms was observed. This research provides valuable insights into the development of Ga-modified catalysts and emphasizes their potential for efficient conversion of greenhouse gases into syngas.

Keywords: methane reforming; syngas; bimetallic catalyst; gallium; nickel

1. Introduction

In recent years, the quest for sustainable energy solutions has gained unprecedented momentum, prompting researchers and scientists to explore innovative approaches for mitigating the environmental impact of industrial processes. Among the various challenges faced, the transformation of greenhouse gases, such as methane (CH$_4$) and carbon dioxide (CO$_2$), into valuable products has emerged as a crucial area of investigation. In this context, the dry reforming of methane (DRM) reaction has garnered significant attention for its potential to simultaneously convert these two greenhouse gases, methane and carbon dioxide, into valuable synthesis gas (syngas), a mixture of hydrogen (H$_2$) and carbon monoxide (CO), which is a vital precursor for the production of liquid fuels, chemicals, and other value-added products [1–3]. DRM can be represented by the following endothermic reaction (Equation (1)). However, the DRM reaction is challenging due to the required high operating temperatures, as well as issues with catalyst deactivation from coke formation and sintering [3].

$$CH_4 + CO_2 \rightleftharpoons 2CO + 2H_2 \ (\triangle H°_{298} = 247 \ kJ/mol) \tag{1}$$

Nickel-based catalysts are attractive for DRM due to their high catalytic activity, economic viability, and natural abundance. Nonetheless, they are susceptible to issues related to catalyst deactivation, through coking and sintering, which severely limit their practical applications [4–6]. One approach to improving the performance of DRM catalysts is the use of bimetallic systems, where the addition of a second metal can enhance catalytic activity, stability, and selectivity.

Recently, a computational screening study, based on density functional theory (DFT) predictions, has shown that a nickel–gallium (Ni-Ga) bimetallic system could be a promising candidate for catalyzing the DRM reaction and could help mitigate these issues [7,8].

This has been confirmed through numerous experimental investigations, which showed that the incorporation of gallium (Ga) as a secondary metal into a Ni-based catalyst has a positive impact on the catalytic performance during the DRM reaction and promotes the catalysts with a strong carbon-deposition resistance [9–14].

The support material has a pivotal role in enhancing the performance of nickel catalysts, as it affects the dispersion of active sites, thermal stability, and interaction with reactants. The good textural qualities, high surface area, and mechanical and thermal stability of Al_2O_3 make it a popular choice for supporting nickel [15,16]. Generally, as an acidic support, Al_2O_3 catalyzes the unwanted side reaction of methane cracking (Equation (2)) that can produce coke [17,18]. As such, unless the deposition mechanism is effectively inhibited or the deposited carbon is successfully removed by the reversed Boudouard reaction (Equation (3)), catalyst deactivation takes place [19–21]. In this sense, a suitable basic oxide, such as MgO, is needed to neutralize the acid sites of Al_2O_3.

$$CH_4 \rightarrow C + 2H_2 \ (\triangle H°_{298} = 75.0 \ kJ/mol) \tag{2}$$

$$C + CO_2 \rightarrow 2CO \ (\triangle H°_{298} = 172.0 \ kJ/mol) \tag{3}$$

It is well documented that catalysts supported on MgO–Al_2O_3 mixed oxides show higher catalytic activity and higher carbon resistance in the CO_2 reforming of CH_4. The role of MgO is thought to provide moderate basicity that can enhance the dissociative adsorption of CO_2, and thereby the carbon deposition can be suppressed by promoting a carbon gasification reaction [22–24]. Furthermore, it has been reported that the application of MgO as a promoter for alumina could not only prevent the amount of deposited carbon but also alter the nature of carbon toward less graphitic and more destructive forms [25].

Hydrotalcites, also known as layered double hydroxides (LDHs), are natural or synthetic laminar materials, with a similar structure to brucite ($Mg(OH)_2$) and natural hydrotalcite ($Mg_6Al_2(OH)_{16}CO_3.4H_2O$) [26]. They emerge as a promising class of catalyst precursors. Their chemical composition may be presented by the general formula $[M^{2+}_{1-x} M^{3+}_{x} (OH)_2]^{x+} (A^{n-})_{x/n}.yH_2O$ where M^{2+} and M^{3+} are divalent and trivalent metal cations, respectively; x is the mole fraction of the trivalent cation; A^{n-} is the anion of compensation; and y is the degree of hydration. Generally, M^{2+} may be Ni^{2+}, Co^{2+}, Cu^{2+}, or Zn^{2+}; M^{3+} may be Al^{3+}, Ga^{3+}, Ce^{+3}, or Fe^{3+}; and A^{n-} may be CO_3^{-2} or NO^{3-} [27]. The layered structure of hydrotalcite facilitates the incorporation of metal ions and the formation of well-dispersed mixed oxides upon thermal decomposition. These mixed oxides exhibit high surface areas, tunable acid-base properties, and structural stability, which can enhance the overall catalytic performance in DRM [28,29].

To our knowledge, no one has reported on the modification of NiMgAl hydrotalcite (HT), which uses a fixed M^{2+}/M^{3+} molar ratio, with Ga as the DRM catalyst. In this work, Ni/(Mg, Al)O$_x$ and gallium-modified Ni-Ga/(Mg, Al)O$_x$ catalysts from their corresponding hydrotalcite precursors are prepared. HTlcs with different Ga/Ni molar ratios (i.e., Ga/Ni = 0, 0.1, 0.3, 0.5, and 1.0) were synthesized by the co-precipitation method. The prepared HTLcs, their calcined forms, and the spent catalysts were subjected to characterization using N_2 adsorption, FTIR, XRD, H_2-TPR, TPO, TGA, and Raman spectroscopy. The catalytic performance of these materials in terms of activity, selectivity, and stability for the dry reforming of methane (DRM) reaction was evaluated at 700 °C within a vertical fixed-bed reactor.

2. Results

2.1. Characterization of the Catalysts

FTIR is a powerful technique used to analyze the functional groups in materials and to gain knowledge of their structure and composition. The FTIR spectroscopy of the as-prepared materials, measured in the range from 4000 to 400 cm^{-1}, is shown in Figure 1. As can be seen from the figure, all the prepared samples showed similar FTIR spectra, which is typical for hydrotalcite-like materials. The broad and strong absorption band centered at around 3500 cm^{-1} is ascribed to the stretching vibration of (OH$^-$) groups, indicating that they result from hydrogen-bonded, interlayered H_2O molecules or are due to the M-OH (M = Mg, Al) vibrations [30]. The band centered at around 1650 cm^{-1} is due to the H–O–H bending vibration modes, confirming the presence of water molecules in the interlayer region [31]. The bands centered at 1520 cm^{-1} and 1380 cm^{-1} are associated with the carbonate species (CO_3^{-2}) in the interlayer spaces; the former is attributed to the v_3 antisymmetric stretching modes, while the latter is associated with the O–C–O asymmetric vibrations [32,33].

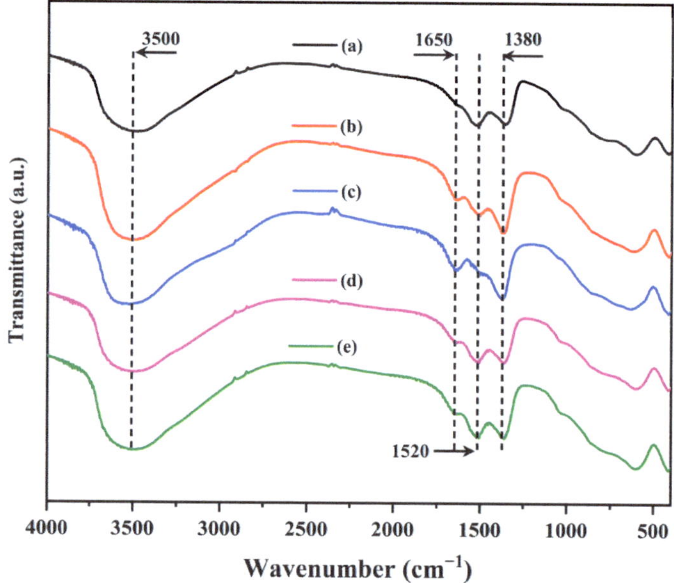

Figure 1. FTIR spectra of the as-prepared hydrotalcites: (a) Ni-Mg-Al, (b) Ni-Ga-Mg-Al (Ga/Ni = 0.1), (c) Ni-Ga-Mg-Al (Ga/Ni = 0.3), (d) Ni-Ga-Mg-Al (Ga/Ni = 0.5), and (e) Ni-Ga-Mg-Al (Ga/Ni = 1.0).

These results demonstrate the successful preparation of layered double hydroxides (LDHs) as precursor catalysts.

The XRD patterns of the as-prepared hydrotalcites are demonstrated in Figure 2. As can be seen from the diffractograms, all the patterns are very similar and show reflection peaks typical of the 3R rhombohedral layered double hydroxide structure. The diffraction peaks match well with those reported in the reference data (JCDPS: 96-900-9273).

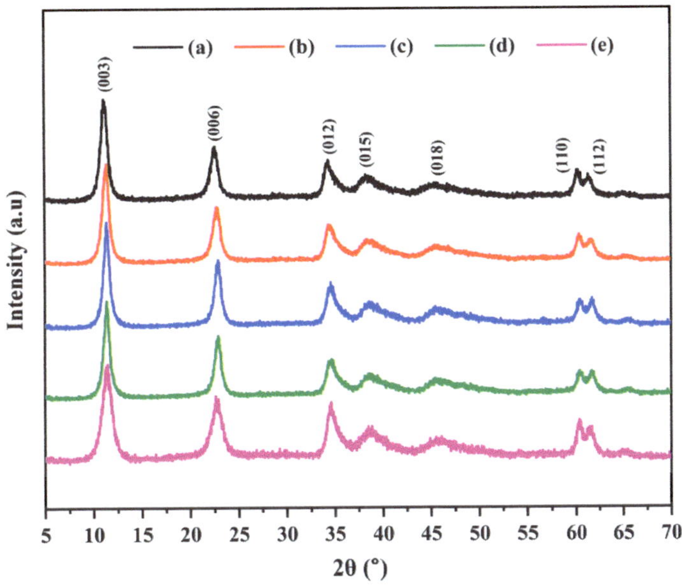

Figure 2. XRD of the as-prepared hydrotalcites: (a) Ni-Mg-Al, (b) Ni-Ga-Mg-Al (Ga/Ni = 0.1), (c) Ni-Ga-Mg-Al (Ga/Ni = 0.3), (d) Ni-Ga-Mg-Al (Ga/Ni = 0.5), and (e) Ni-Ga-Mg-Al (Ga/Ni = 1.0).

The reflection peaks at 11.6°, 23.2°, 34.5°, 39.1°, 46.5°, 59.9°, and 60.9° are associated with the plane families (003), (006), (009), (015), (018), (110), and (0015), respectively. No peaks of other phases are observed, irrespective of the Ga content, indicating the high purity of the as-prepared LDHs and pointing to the successful incorporation of both Ni^{2+} and Ga^{3+} cations into the brucite layers [31,32].

The BET-specific surface area, pore volume, and pore size distribution for the calcined samples were determined by nitrogen physisorption. The BET adsorption–desorption isotherms and pore size distribution for the calcined samples are presented in Figure 3A,B, and the corresponding results are summarized in Table 1. While the BET adsorption–desorption isotherms for the as-prepared hydrotalcites are provided in Figure S2. As can be seen from Figure 3A, all the calcined catalysts show a type IV adsorption–desorption isotherm with an H1-type hysteresis loop [34], indicative of the presence of mesoporous structures in these catalysts, except for Ni-Ga/(Mg, Al)O_x (Ga/Ni = 1.0), which displays a hysteresis loop of type H3, often associated with materials with platelet-like porous structures [35].

Table 1. Physicochemical properties of dried and calcined catalysts: BET surface area, total pore volume, and pore size.

Sample	Surface Area (m²/g)		Pore Volume (cm³)		Pore Size (nm)	
	As-Prepared	Calcined	As-Prepared	Calcined	As-Prepared	Calcined
Ni/(Mg, Al)O_x	131.2	215.4	0.51	0.68	14.2	12.4
Ni-Ga/(Mg, Al)O_x (Ga/Ni = 0.1)	124.2	205.7	0.55	0.73	16.8	15.4
Ni-Ga/(Mg, Al)O_x (Ga/Ni = 0.3)	150.4	189.9	0.66	0.70	16.6	15.4
Ni-Ga/(Mg, Al)O_x (Ga/Ni = 0.5)	122.2	197.7	0.59	0.80	17.8	17.0
Ni-Ga/(Mg, Al)O_x (Ga/Ni = 1.0)	150.4	253.1	0.64	0.96	15.9	15.6

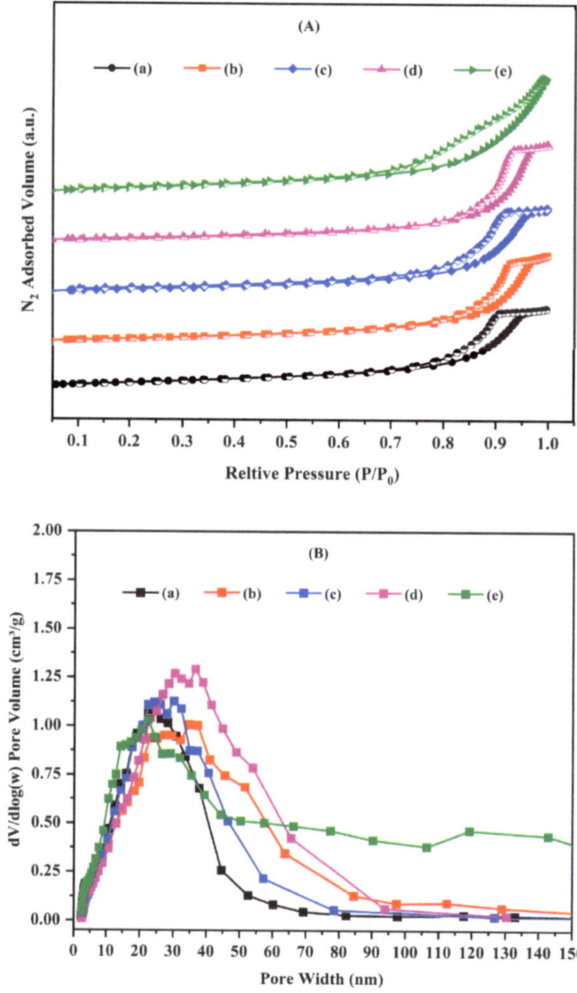

Figure 3. (**A**) N_2 adsorption-desorption isotherms: adsorption (filled); desorption (half filled) symbols) and (**B**) pore size distribution profiles of (a) Ni/(Mg, Al)O_x, (b) Ni-Ga/(Mg, Al)O_x (Ga/Ni = 0.1), (c) Ni Ga/(Mg, Al)O_x (Ga/Ni = 0.3), (d) Ni-Ga/(Mg, Al)O_x (Ga/Ni = 0.5), and (e) Ni-Ga/(Mg, Al)O_x (Ga/Ni = 1.0).

Table 1 also compares the textural characteristics of the as-prepared and calcined catalysts in terms of specific surface areas and pore volumes. All calcined catalyst samples exhibit a larger specific area and pore volume than the as-prepared samples, which is attributed to the dehydroxylation and decarbonization process that takes place during the thermal decomposition upon calcination [36]. The specific surface areas for the as-synthesized samples are in the range of 124–150 m^2/g, while those for calcined samples are in the range of 190–253 m^2/g. Furthermore, the pore volumes for the as-synthesized samples are in the range of 0.51–0.66 cm^3/g and increased to a range of 0.68–0.97 cm^3/g for the calcined samples. The pore size distribution, shown in Figure 3B, illustrated that all samples exhibited a unimodal pore size in the range of 12.4–17 nm, and the incorporation of gallium broadened the pore size distribution. The relatively high specific surface areas and pore volumes for these catalysts would facilitate the mass transport process of the reactants and products during the DRM reaction.

The H_2-temperature-programmed reduction profile for the hydrotalcite samples is shown in Figure 4A, while that for the calcined ones is depicted in Figure 4B. In the TPR of HT samples, Figure 4A, two significant negative peaks, at temperature ranges of ~165 and ~340 °C, are observed. These two peaks might be attributed to the dehydroxylation (removal of interlayer water) and decarbonization (removal of CO_2) processes during the H_2-TPR analysis [37]. The presence of interlayer water and carbonate anions in the HT structures leads to the evolution of H_2O and CO_2 upon thermal decomposition during the H_2-TPR analysis, which contributes to the appearance of negative responses in the TPR profile, unlike with the case for the calcined samples, shown in Figure 4B, where H_2 is consumed, due to the reduction process, which appears as a positive response. Generally, the strongly interacted NiO species are harder to reduce than the moderately interacted NiO species. For the unmodified catalyst sample, a significant reduction peak appears at a temperature of ~850 °C. This peak corresponds to the hydrogen uptake upon reduction of nickel oxide to metallic nickel (NiO to Ni^0), which is in strong interaction with the mixed oxide support [38]. Furthermore, a small reduction peak at ~550 °C is detected, ascribed to the reduction of surface NiO with weak interaction with the support [39]. The absence of a reduction peak below 400 °C indicates that no free NiO species exist over the catalyst surface. Upon Ga modification, the high-temperature peak, the peak at ~850 °C, is shifted toward higher reduction temperatures, and this shifting gradually increases with increased Ga loading. It has been reported that the reduction process of Ga_2O_3 requires higher temperatures [40]. Therefore, the shifting in reduction temperatures for the Ga-modified samples toward higher temperatures can be attributed to the simultaneous reduction of Ni^{2+} and Ga^{3+} to metallic Ni and Ga in the form of a Ni-Ga alloy. This also is in good accordance with the XRD of reduced catalysts, as will be discussed later. Similarly, as shown in Table 2, hydrogen consumption also increases upon Ga modification. Notably, the small reduction peak at the low-temperature range (at ~550 °C) diminishes for the Ga-modified catalyst. These results indicate the enhancement of the metal-support interaction and a growing number of active sites upon Ga incorporation into the catalyst. All the observed reduction peaks are assigned to Ni and Ga oxide species reduction into metallic Ni and Ga since the Mg-Al oxides are not reducible in the H_2-TPR conditions. Similar TPR profiles are also observed for the calcined samples, considering the high-temperature reduction peak at ~850 °C, as shown in Figure 4B, except for the disappearance of the two negative peaks at low temperatures, related to dehydroxylation and decarbonization. The quantitative summary of H_2-TPR analysis for the calcined samples is also provided in Table S2 of the Supplementary Materials.

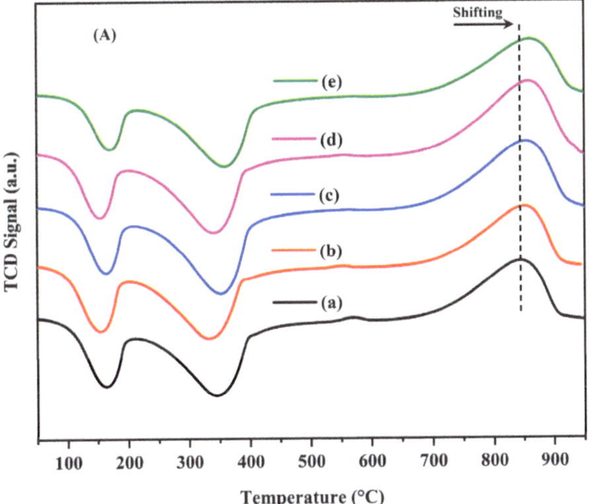

Figure 4. *Cont.*

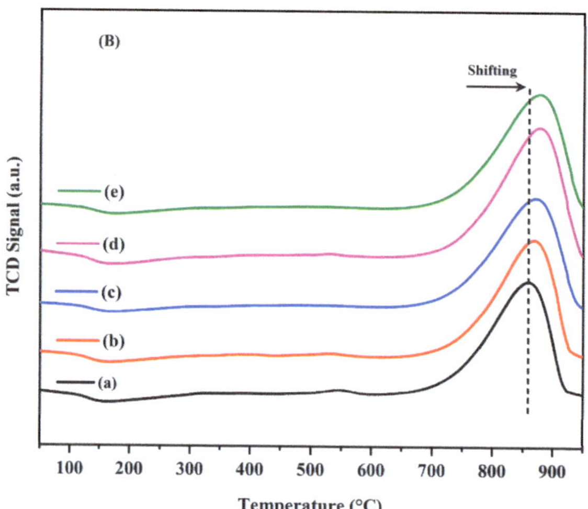

Figure 4. H_2-TPR profiles of (**A**) hydrotalcite and (**B**) calcined samples: (a) Ni-Mg-Al, (b) Ni-Ga-Mg-Al (Ga/Ni = 0.1), (c) Ni-Ga-Mg-Al (Ga/Ni = 0.3), (d) Ni-Ga-Mg-Al (Ga/Ni = 0.5), and (e) Ni-Ga-Mg-Al (Ga/Ni = 1.0).

Table 2. Quantitative summary of H_2-TPR analysis for the as-prepared hydrotalcite samples.

HT Sample	Peak Temperature (°C)	Total H_2 Consumed (mmol/g_{cat}) [a]
Ni/(Mg, Al)	845	0.92
Ni-Ga/(Mg, Al) (Ga/Ni = 0.1)	851	0.95
Ni-Ga/(Mg, Al) (Ga/Ni = 0.3)	855	1.35
Ni-Ga/(Mg, Al) (Ga/Ni = 0.5)	858	1.10
Ni-Ga/(Mg, Al) (Ga/Ni = 1.0)	860	1.21

[a] Calculated from the H_2-TPR profiles over the whole temperature range.

The XRD diffractograms of the HT-like precursors after calcination at 550 °C for 5 h are demonstrated in Figure 5A. All the samples showed similar XRD patterns, except that the peak intensity decreases when Ga content increases. The diffraction peaks at 2θ angles of 34.7, 43.3, 62.6, and 79.4° are ascribed to the (111), (020), (022), and (222) planes of MgO periclase (JCDPS: 96-900-7060) [31,32]. These peaks can also indicate the formation of periclase-like mixed oxides, in the form of a (Mg(Ni,Ga,Al)-O)-type structure. The diffraction peaks at 2θ angles of ca. 11.4 and 23.1° are associated to the (003) and (006) plane families of hydrotalcites, which can either be attributed to the incomplete decomposition of the HT-precursors or to the partial reconstruction of the calcined mixed oxides into HTs, upon exposure to the air atmosphere, due to the so-called memory effect [41].

To reveal the crystalline structure of reduced Ni/(Mg, Al)O_x and Ni-Ga/(Mg, Al)O_x (with different Ni:Ga molar ratios) catalysts, firstly a reductive treatment was performed for the as-prepared LDHs at 700 °C for 2 h in 30 mL of H_2 flow. The XRD patterns of the resulting catalysts are shown in Figure 5B. The distinct diffraction peaks for Ni/(Mg, Al)O_x catalysts observed at 2θ angles of 43.5° and 76.5° correspond to the (111) and (220) atomic planes of the metallic Ni with fcc structure (JCDPS: 00-04-0850), respectively [42]. Particularly, these two peaks are slightly shifted toward lower 2θ angles for the Ni-Ga/(Mg, Al)O_x (with different Ni: Ga molar ratios) catalyst samples, suggesting the formation of Ni-Ga intermetallic compounds due to the larger atomic radius of Ga (0.135 nm) than that of Ni (0.125 nm) [9,43,44]. The sharp reflection peaks at 2θ angles of 36.9, 43.5, 62.9, and 76.5° could also be ascribed to the periclase-like phase of mixed oxides of the (Mg(Ni,Ga,Al)-O)-type structure [45]. The reflection peak located at ca. 63.3° could also be attributed to the (220) facets of the face-centered cubic NiO phase (JCDPS: 01-089-5881), due to the incomplete reduction process [46].

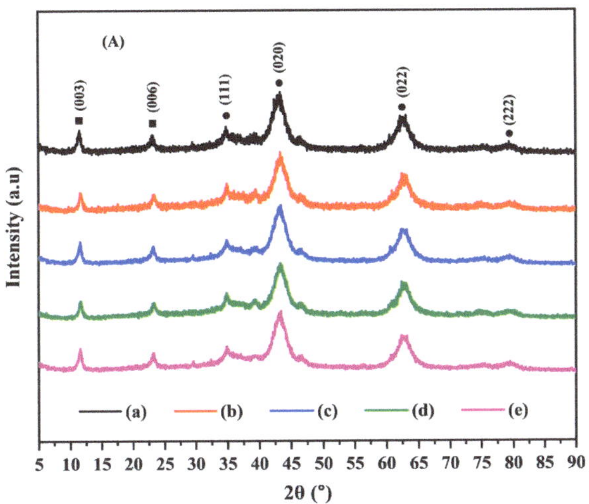

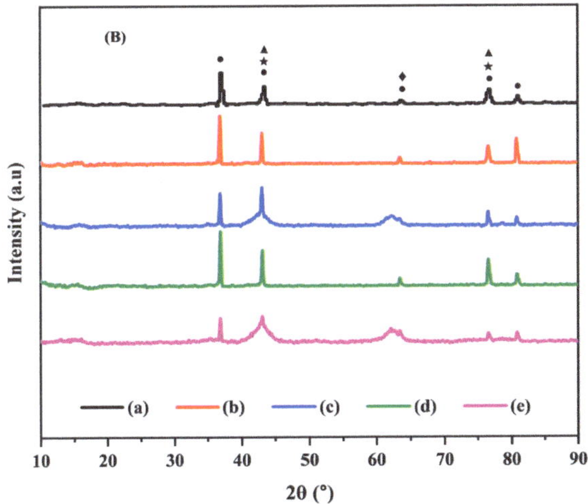

Figure 5. XRD patterns of calcined (**A**) and reduced catalyst samples (**B**): (a) Ni/(Mg, Al)Ox, (b) Ni-Ga/(Mg, Al)O$_x$ (Ga/Ni = 0.1), (c) Ni-Ga/(Mg, Al)O$_x$ (Ga/Ni = 0.3), (d) Ni-Ga/(Mg, Al)O$_x$ (Ga/Ni = 0.5), and (e) Ni-Ga/(Mg, Al)O$_x$ (Ga/Ni = 1.0). ▲ Ni, ★ Ni-Ga alloy, ● Mg (GaAl)O periclase-like, ◆ NiO, ■ hydrotalcite.

2.2. Catalytic Activity Tests

The activity of Ni/(Mg, Al)O$_x$ and Ni-Ga/(Mg, Al)O$_x$ (Ga/Ni = 0.1, 0.3, 0.5, and 1.0) catalysts in terms of CH$_4$ conversion and CO$_2$ conversion are displayed in Figure 6. For all studied catalysts, the CH$_4$ conversion is always lower than the CO$_2$ conversion, which is due to the reverse water gas shift (RWGS) (CO$_2$ + H$_2$ → CO + H$_2$O) reaction accompanying the DRM. The unpromoted catalyst, Ni/(Mg, Al)O$_x$, showed initial CH$_4$ and CO$_2$ conversions of ∼65% and 74%, respectively; however, these conversions tend to gradually decrease over time on stream.

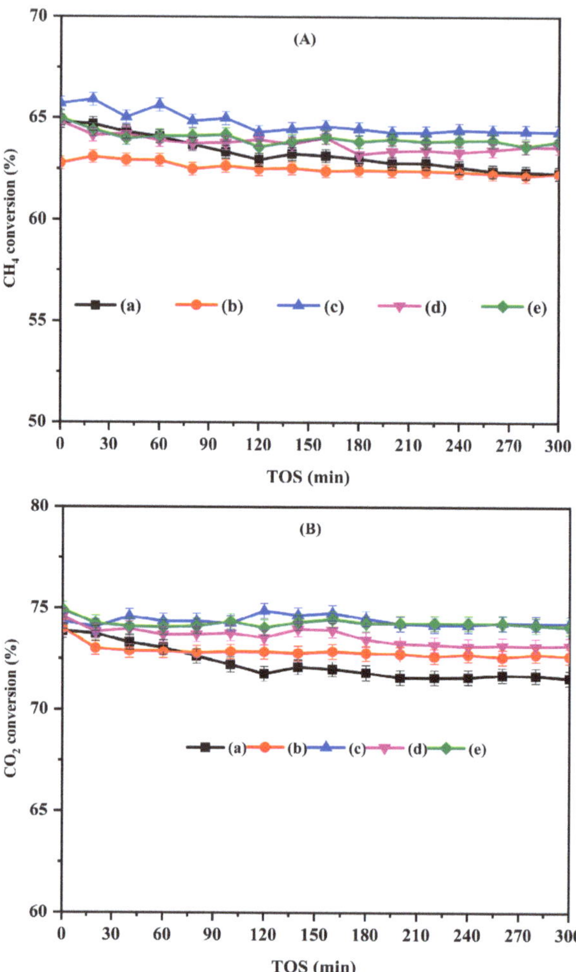

Figure 6. Catalytic activity tests for (**A**) CH_4 conversion, (**B**) CO_2 conversion at 700 °C, 1 atm, and GHSV of 42,000 mL/h/g for (a) Ni/(Mg, Al)Ox, (b) Ni-Ga/(Mg, Al)O_x (Ga/Ni = 0.1), (c) Ni-Ga/(Mg, Al)O_x (Ga/Ni = 0.3), (d) Ni-Ga/(Mg, Al)O_x (Ga/Ni = 0.5), and (e) Ni-Ga/(Mg, Al)O_x (Ga/Ni = 1.0).

For Ga-promoted catalysts, Ni-Ga/(Mg, Al)O_x (Ga/Ni = 0.1, 0.3, 0.5, and 1), initial conversions were comparable to the unpromoted catalyst. However, after 5 h on stream, significant differences in catalytic stability emerged. Of all the catalysts tested, the Ni-Ga/(Mg, Al)O_x catalysts with (Ga/Ni = 0.3) exhibited the best performance in terms of catalytic stability without noticeable deactivation.

The improved catalytic stability on Ga incorporation might be attributed to the enhanced process of CO_2 activation. As pointed out by many researchers, the hydrotalcite-derived mixed oxides substituted with Ga have good CO_2 sorption (capture) characteristics at elevated temperatures [47,48]. This in turn will facilitate carbon removal through the gasification reaction (C + CO_2 $\rightleftharpoons$ 2CO). A proposed mechanism for the enhanced stability of this system involves the formation of $Ni_3GaC_{0.25}$ through the interaction of carbon from methane decomposition with the Ni_3Ga alloy. This carbide is subsequently oxidized to CO by CO_2 in a cyclic process, preventing carbon accumulation and polymerization [8].

The catalytic performance of the Ni-Ga/(Mg, Al)O_x catalyst with Ga/Ni = 0.3 was further tested at 700 °C and GHSV of 42,000 mL/h/g for a 20 h-duration TOS. The results provided in Figure S4 of the Supplementary Materials show stable CH_4 and CO_2 conversion profiles over the entire reaction period.

The influence of reaction temperature on the catalytic performance of Ni-Ga/(Mg, Al)O_x catalyst with Ga/Ni = 0.3 is displayed in Figure 7. As can be seen from the figure, on increasing the reaction temperature from 600 to 850 °C, CH_4 conversion rises from 35% to 94%, whereas CO_2 conversion increases from 46% to 96%. This is due to the endothermic nature of the DRM reaction. Comparing CH_4 and CO_2 conversions displays that CH_4 conversion is always lower than CO_2 conversion, for the studied temperature range, which indicates the coexistence of the reverse water gas shift (RWGS) reaction under these reaction conditions. Furthermore, the ratio H_2/CO increases with reaction temperature from 0.77 at 600 °C to 1.00 at 850 °C. A comparison of the catalytic activity of the Ni-Ga/(Mg, Al)O_x (Ga/Ni = 0.3) catalyst with other Ga-modified catalysts, which was reported in previous studies, is provided in Table S3 of the Supplementary Materials. As shown in the table, the catalytic performance of the Ni-Ga/(Mg, Al)O_x catalyst with Ga/Ni = 0.3 outperforms the analogs presented in the literature, which is most likely due to the promotional effect and improved structural properties upon Ga modification.

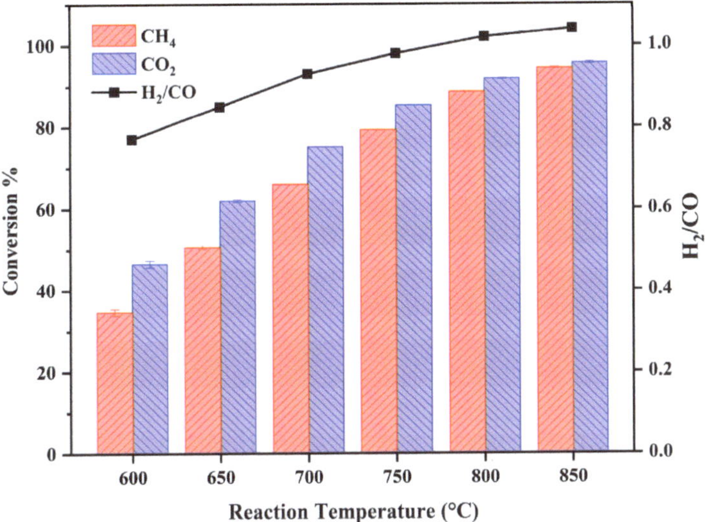

Figure 7. Influence of reaction temperature on the catalytic performance for Ni-Ga/(Mg, Al)O_x (Ga/Ni = 0.3) (GHSV of 42,000 mL/h/g_{cat}, CH_4/CO_2/N_2 molar ratio of 1, and 1 atm).

2.3. Characterization of Spent Catalysts

Ni-based catalysts typically suffer from severe coking during the DRM reaction, due to the accumulation of carbonaceous species on the catalyst's surface, which lead to the deactivation of the catalytic active sites [3]. Therefore, for a specific catalyst, the catalytic activity and stability are closely related to the amount of carbon species formed on its surface during the DRM reaction.

Thermogravimetric analysis (TGA) could provide useful information on quantifying the deposited carbon on spent catalysts. The TGA and the differential thermal analysis (DTA) of spent catalysts are shown in Figure 8A,B. Figure 8A shows that the addition of Ga enhances the catalytic stability compared to the unmodified catalyst. The unmodified catalyst showed a relatively high weight loss of ~24 wt.%, indicating its poor resistance to coke deposition. The incorporation of Ga resulted in good coke resistance, indicated by the reduced amount of deposited carbon. The Ni-Ga/(Mg, Al)O_x catalyst with Ga/Ni ratio = 0.3 showed the best catalytic stability, with only ~13 wt.% weight loss. From the DTA profiles, Figure 8B, a peak at a temperature of ~300 °C, with comparable intensity, appears for all the catalyst samples, which is associated with combustion of the less-ordered (amorphous) carbon. The high-temperature peak is related to the highly ordered form of carbon (>650 °C) (graphitic carbon) that requires higher temperatures to decompose [49]. The intensity of the later peak is the highest for the unmodified $Ni/(Mg, Al)O_x$ catalyst and is significantly reduced upon Ga incorporation. Interestingly, a graphitic carbon peak is not detected for the $Ni-Ga/(Mg, Al)O_x$ catalyst with Ga/Ni ratio = 0.3, indicating that the carbon deposited on such a catalyst is mainly amorphous and can easily be gasified.

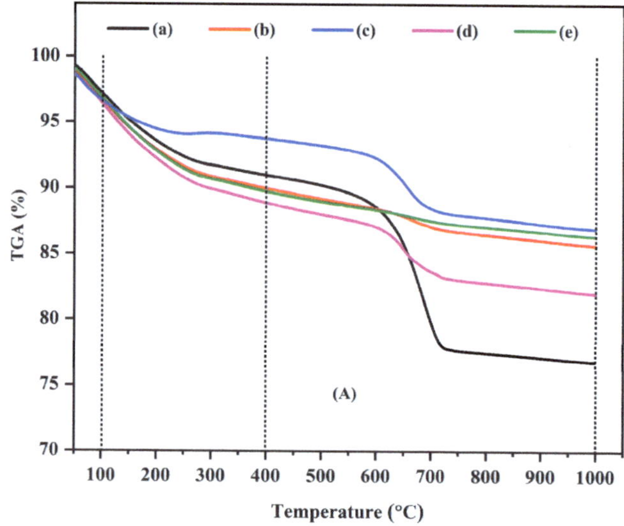

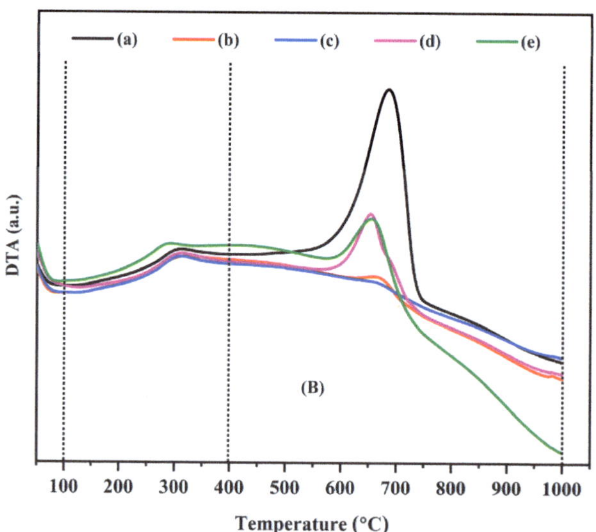

Figure 8. (**A**) TGA analysis and (**B**) DTA of spent catalysts: (a) $Ni/(Mg, Al)O_x$, (b) $Ni-Ga/(Mg, Al)O_x$ (Ga/Ni = 0.1), (c) $Ni-Ga/(Mg, Al)O_x$ (Ga/Ni = 0.3), (d) $Ni-Ga/(Mg, Al)O_x$ (Ga/Ni = 0.5), and (e) $Ni-Ga/(Mg, Al)O_x$ (Ga/Ni = 1.0).

TPO analysis could provide additional information on the type and amount of deposited carbon on the spent catalysts. The TPO profiles of spent catalysts are depicted in Figure 9. The TPO profiles for all the catalysts show two significant weight loss regions, the low-temperature region between 100–400 °C, corresponding to the combustion of the amorphous carbon or metal carbide species, and the high-temperature above 500 °C, related to the combustion of the graphitic carbon.

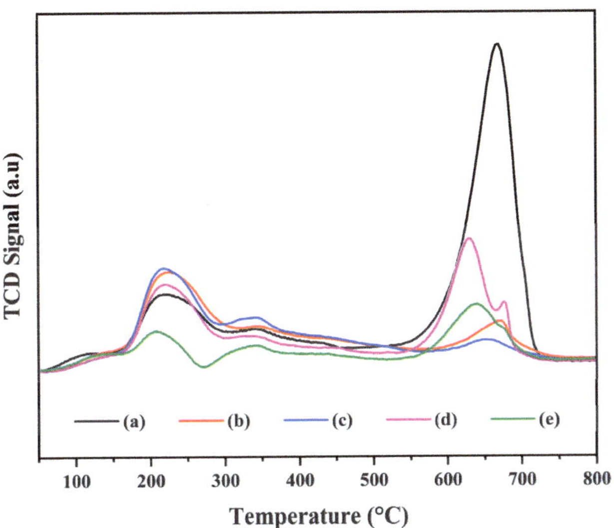

Figure 9. TPO profiles of used catalysts: (a) Ni/(Mg, Al)O$_x$, (b) Ni-Ga/(Mg, Al)O$_x$ (Ga/Ni = 0.1), (c) Ni-Ga/(Mg, Al)O$_x$ (Ga/Ni = 0.3), (d) Ni-Ga/(Mg, Al)O$_x$ (Ga/Ni = 0.5), and (e) Ni-Ga/(Mg, Al)O$_x$ (Ga/Ni = 1.0).

The latter type of carbon (graphitic carbon) is less reactive and responsible for catalyst deactivation [49]. From the TPO profiles, it can be seen that the type and amount of each carbon species are significantly altered upon gallium addition. For the unmodified catalyst, Ni/(Mg, Al)O$_x$, the main coke species deposited on its surface is graphitic. Also, the Ni-Ga/(Mg, Al)O$_x$ catalyst with Ga/Ni ratio = 0.3 had both the least graphic carbon type and the highest amount of amorphous carbon. This result is in good consistency with the TGA analysis and indicates that catalyst deactivation is significantly reduced upon Ga incorporation, due to the formation of reactive-type carbon instead of a graphitic one, which in turn secures in situ gasification of deposited carbon and hence better catalytic stability. These results are in good accordance with the catalytic activity tests discussed previously.

Raman spectroscopy is a useful tool that provides information about the structural disorders and defects in carbon-based materials. Raman spectra for the spent catalysts are shown in Figure 10. The peak at the Raman shift of ~1380 (D-band) is due to disordered carbon species (e.g., amorphous or defective filamentous carbon), whereas the peak at the Raman shift of ~1589 (G-band) arises from the first-order scattering E2g mode of ordered graphite [49,50]. Generally, the relative intensity between the D and G bands is an indicator of the graphitic degree of the deposited carbon; the lower the value, the more graphitic the structure [51]. As shown in the Raman spectra, the unmodified catalyst showed a more ordered graphite structure with a high graphitization nature, indicated by the lowest value of I$_D$/I$_G$ ratio of 0.93. On Ga modification, this ratio increased to greater values (>1). The Ni-Ga/(Mg, Al)O$_x$ (with Ga/Ni = 0.3), among all the studied catalysts, has the highest value of I$_D$/I$_G$ of 1.80. This indicates that for this specific catalyst the degree of crystallinity for the deposited carbon is the least during the DRM reaction. This result is also in good agreement with the trends observed in the DTA and TPO analyses discussed previously, shown in Figures 8 and 9, respectively.

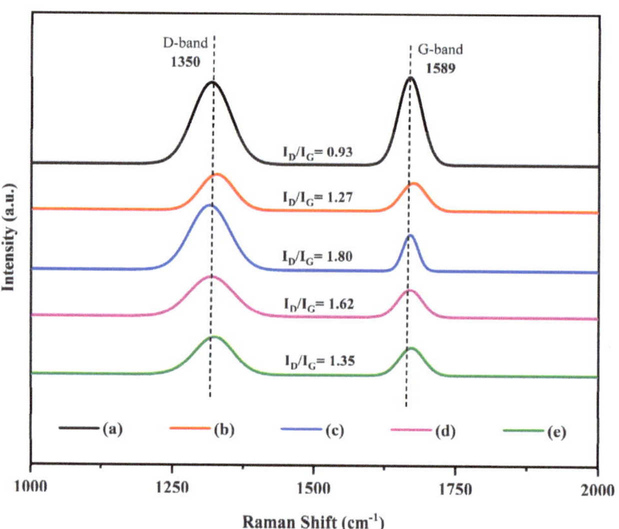

Figure 10. Raman spectra of used catalysts: (a) Ni/(Mg, Al)O$_x$, (b) Ni-Ga/(Mg, Al)O$_x$ (Ga/Ni = 0.1), (c) Ni-Ga/(Mg, Al)O$_x$ (Ga/Ni = 0.3), (d) Ni-Ga/(Mg, Al)O$_x$ (Ga/Ni = 0.5), and (e) Ni-Ga/(Mg, Al)O$_x$ (Ga/Ni = 1.0).

2.4. DFT Mechanistic Insights on the Effect of Ga Modification

DFT calculations were performed to obtain mechanistic insights into the enhanced CO_2 activation due to Ga incorporation into the Ni_3Ga catalyst. The density of states (DOS), d-band center, charge density difference, and Bader charge were calculated, and the results are presented in Figure 11. As can be seen from the projected density of states (PDOS) profiles shown in Figure 11A, there is a clear overlapping between the d-band of Ni atoms and the p-bands of Ga atoms in the $Ni_3Ga(111)$ system; this orbital overlapping indicates the hybridization of occupied Ga 4p orbitals with the unoccupied Ni 3d. This orbital hybridization results in electron transfer from Ga atoms to Ni ones. Furthermore, the d-band center of Ni on the $Ni_3Ga(111)$ surface shifts closer to the Fermi level with a value of ~−1.11 eV compared with ~−1.37 eV for that of the Ni(111) surface. From the Bader charge analysis, Figure S3, the estimated specific charge amount of the Ni atoms increased from ~10.04 e on the Ni (111) surface to ~10.12 e on the Ni_3Ga surface.

The electron transfer from Ga to Ni atoms is further confirmed through the analysis of the charge density, shown in Figure 11. The charge density difference shows a clear charge localization, indicated by the electronic clouds around Ga atoms in the figure, which is due to electron redistribution between Ni and Ga atoms, where a partial electronic charge is transferred from Ga atoms to Ni ones.

Overall, it is believed that this charge localization will render Ni atoms to act as Lewis basic centers "pushing" electrons to the electrophilic carbon atom of CO_2, while the Lewis acidic Ga centers facilitate stabilization of the bound CO_2 by "pulling" electron density from the electron-rich CO_2 molecule. Therefore, The Lewis acid and Lewis base synergistically interact with CO_2 boosting its chemisorption and activation. This will eventually result in an in-situ carbon removal process, preventing carbon deposits from self-polymerizing, and consequently leading to enhanced catalytic stability. This result is in good agreement with prior experimental X-ray absorption near-edge structure (XANES) investigations [52,53] and DFT theoretical calculations [54,55].

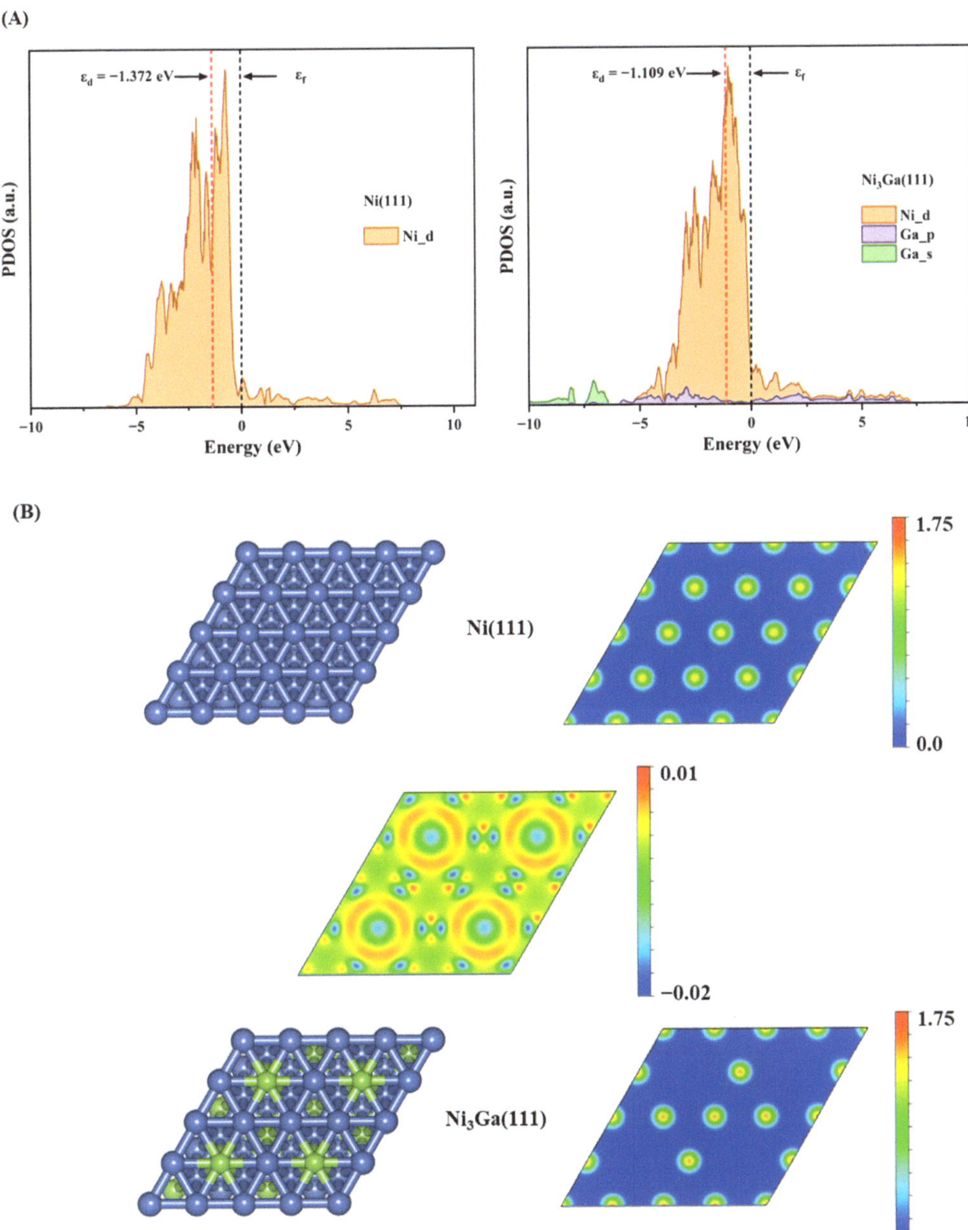

Figure 11. (**A**) Projected electronic densities of states of Ga p and s orbitals and those of the Ni d orbitals on Ni(111) and Ni_3Ga(111) surfaces and (**B**) charge density distribution on Ni(111) (top) and Ni_3Ga(111) (bottom) surfaces (in e bohr^{-3}) and their charge density difference (center); (from blue to red indicates the transition from electron depletion to accumulation).

3. Materials and Methods

3.1. Catalyst Preparation

The monometallic Ni and bimetallic Ni-Ga catalysts were synthesized from their respective layered double hydroxide (LDH) precursors via the following procedure. The LDH samples, with a targeted M^{2+}/M^{3+} molar ratio of 3 and a total active metal loading of 10 wt.%, were prepared through the co-precipitation method at the high supersaturation method. In this method, five LDH precursors were synthesized based on the targeted molar ratio of gallium to nickel on the final catalyst (i.e., Ga/Ni = 0, 0.10, 0.20, 0.50, and 1.0). According to the calculations of stoichiometric ratios (Table S1 in the Supplementary Materials), the required weight of metal nitrates, $Ni(NO_3)_2 \cdot 6H_2O$ (Sigma Aldrich Chemical Company Inc., Milwaukee, WI, USA), $Ga(NO_3)_3 \cdot xH_2O$ (Sigma Aldrich Chemical Company Inc., Milwaukee, WI, USA), $Al(NO_3)_3 \cdot 9H_2O$ (LOBA Chemie PVT. Ltd., Mumbai, India), and $Mg(NO_3)_2 \cdot 6H_2O$ (Sigma Aldrich Chemical Company Inc., Milwaukee, WI, USA), were mixed in 100 mL of deionized water. An anionic precursor solution of Na_2CO_3 (VWR Chemicals BDH, Radnor, PA, USA) and NaOH (LOBA Chemie PVT. Ltd., Mumbai, India) was also prepared, in which Na_2CO_3 was used in excess, twice that of the calculated stoichiometric ratio, to ensure that the charge compensation anion was carbonate CO_3^{-2} instead of nitrate NO_3^- Both solutions were kept on a magnetic stirrer for 15 min before final mixing. A nitrate solution containing the divalent and trivalent cations was added drop-wisely to the anionic precursor solution under vigorous stirring using a graduated burette. During the addition, the solution medium was kept at a temperature of 65 °C and a pH of 9.5–10.5 by adding a 0.2 M NaOH solution when required. The final mixture was then aged for approximately 18 h at room temperature with continuous stirring, which resulted in a suspension of the aforementioned hydrotalcite. The resulting suspension was then centrifuged, and the solid mixture was washed several times with distilled water, to ensure the entire removal of nitrate and sodium ions. After this, the samples were dried overnight at 110 °C. After drying, the samples were divided into two parts: one was calcined at 550 °C for a duration of 5 h for further characterizations, while for the other part, the dried samples were crushed, sieved to the desired fraction (~0.15 μm size), and stored in a desiccator for the subsequent activity tests. The unmodified catalyst samples are denoted as $Ni/(Mg, Al)O_x$, and the gallium-modified ones are denoted as $Ni-Ga/(Mg, Al)O_x$ (Ga/Ni = y), where y refers to the above-stated Ga/Ni molar ratio (Ga/Ni ratio = 0.1, 0.3, 0.5 and 1). More details on the prepared catalysts are provided in Table S1 in the Supplementary Materials.

3.2. Catalyst Characterization

The BET surface area, pore volume, and pore diameter of the catalyst samples were measured from N_2 adsorption–desorption data obtained using a Micromeritics Tristar II 3020 instrument (Micrometrics, Norcross, GA, USA). For each analysis, ~70 mg of catalyst was used. The analysis was performed at −196 °C, and before the measurement, the samples were degassed at 250 °C for 180 min to remove the moisture and other gases adsorbed on the surface of the catalysts.

The FTIR measurements were performed in a Shimadzu IRPrestige-21 (Shimadzu Corporation, Kyoto, Japan) infrared spectrophotometer. The pellets were prepared by mixing the sample with spectroscopic grade KBr. The collected spectra of all the studied samples were measured under ambient conditions in a wavelength range between 400 and 4000 cm^{-1} with a scan resolution of 4 cm^{-1} and a total of 32 scans.

Hydrogen-temperature-programmed reduction (H_2-TPR) was performed in an Autochem II 2920 chemisorption analyzer (Micromeritics, Norcross, GA, USA). Typically, 70 mg of the calcined catalyst was purged with a pure argon stream at 150 °C for 60 min, followed by cooling to ambient temperature. Then, a mixture of 10 vol.% H_2/Ar flowing at 40 mL/min was introduced to the sample while the temperature was increased from room temperature to 1000 °C, with the heating rate being 10 °C/min. The signal of H_2 consumption was monitored by a thermal conductivity detector (TCD). A similar approach was adopted for performing the oxygen-temperature-programmed-oxidation (O_2-TPO) analysis, except that degassing was performed with He, and a mixture of 10% O_2/He was used.

The crystalline phases of the as-prepared HT, calcined, and reduced samples were determined using a Miniflex Rigaku X-ray powder diffraction (XRD) instrument (Rigaku, Tokyo, Japan) using a Cu Kα radiation source operated at 40 kV and 40 mA. The scans were collected in the 2θ range 5–90° (step size: 0.05° and time per step 0.8 s).

Thermogravimetric analysis (TGA) was performed for the used catalyst samples on a Shimadzu TGA-51 (Shimadzu Corporation, Japan). Typically, 10–15 mg of the spent catalyst sample was placed in an alumina macro crucible and heated from room temperature up to 1000 °C (at 20 °C/min temperature ramp) in a flowing air atmosphere. The amount of coke deposited was calculated based on the mass change of the spent catalyst sample during the experiment.

The graphitization degree and the type of carbon deposited over the surface of spent catalysts were characterized using a Laser Raman (NMR-4500) spectrometer (JASCO, Tokyo, Japan). The analysis was carried out using a diode laser beam with a wavelength of 532 nm. The spectra were collected under ambient conditions using a Raman shift in the range of 500–3500 cm^{-1}. The collected spectra were an accumulation of 20 with 5 s acquiring time per each.

3.3. Catalytic Activity Performance

The DRM catalytic tests were performed in a fixed-bed reactor, with an inner diameter of 0.92 cm, at a temperature of 700 °C and under atmospheric pressure. Initially, 100 mg of catalyst sample was reduced in situ for 2 h under H2 flow (20 mL/min) at 700 °C. Subsequently, nitrogen gas (50 mL/min) was introduced for 15 min to purge the residual H_2. Then, a mixture of CH_4 (30 mL/min), CO_2 (30 mL/min), and N_2 (10 mL/min) was introduced simultaneously into the reactor with a total GHSV of 42,000 mL/h/g. Nitrogen gas was used as the internal standard for volume-change estimation in the reaction. All volumetric flow rates given in this study are related to 25 °C and atmospheric pressure. The best-performing catalysts were also tested as a function of temperature in a stepwise mode from 600 to 850 °C. The outlet gases were analyzed by an online gas chromatograph (Shimadzu Technologies, Santa Clara, CA, USA). The reactant conversions (X) and ratio of hydrogen to carbon monoxide were calculated based on the formulas given below:

$$X_{CH_4}(\%) = \frac{F_{CH_4,in} - F_{CH_4,out}}{F_{CH_4,in}} \times 100 \tag{4}$$

$$X_{CO_2}(\%) = \frac{F_{CO_2,in} - F_{CO_2,out}}{F_{CO_2,in}} \times 100 \tag{5}$$

$$\frac{H_2}{CO} = \frac{F_{H_2,out}}{F_{CO,out}} \tag{6}$$

where $F_{i,in}$ and $F_{i,out}$ are the inlet and outlet gas flow rates of i gas (CH_4, CO_2, H_2, and CO).

3.4. DFT Calculations

DFT calculations were performed using the Vienna Ab initio Simulation Package (VASP.6.1.0) [56–58]. The generalized gradient approximation with the Perdew–Burke–Ernzerhof functional (GGA-PBE) was used to account for the electron exchange and correlation [59]. The core electrons were described using the projector augmented wave (PAW) method [60,61], with a plane-wave basis set that was expanded up to a energy cutoff of 400 eV. The reciprocal first Brillouin zones were approximated by a sum over special k points chosen using the Monkhorst–Pack scheme [62]. The bulk Ni and Ni_3Ga systems were sampled using $6 \times 6 \times 6$ k-point meshes, while for Ni (111) and $Ni_3Ga(111)$ surfaces, a mesh of $5 \times 5 \times 1$ k-points was used. More details on the DFT calculation method and the modeled slab surfaces are provided in the Supplementary Materials, paragraph S1 and Figure S1.

4. Conclusions

This study investigated the impact of Ga modification on the catalytic performance of Ni/(Mg, Al) O_x catalysts for dry reforming of methane. Catalysts were prepared from hydrotalcite precursors with Ga/Ni ratios, i.e., 0, 0.1, 0.3, 0.5, and 1.0, through the co-precipitation method. The prepared catalysts were extensively characterized both before and after the DRM reaction. The characterization results of the as-prepared catalysts indicated the successful preparation of hydrotalcite precursors. The addition of Ga did not alter the mesoporous structure of the catalysts, as evidenced by the BET results. Moreover, the H_2-TPR analysis revealed an improvement in the metal-support interactions due to Ga incorporation. Among the various Ga-modified catalysts, the catalyst with a Ga/Ni ratio of 0.3 showed the best catalytic performance with superior carbon-deposition resistance compared to the unmodified one, which resulted in the highest CH_4 and CO_2 conversions at the reaction temperature of 700 °C. Characterizations on spent catalysts from TGA/DTA, TPO, and Raman spectroscopy showed enhanced carbon-deposition resistance and reduced degree graphitization of the deposited carbon on Ga-modified catalysts. The firm stability of 20 h time on stream showed stable CH_4 and CO_2 conversion profiles. DFT calculations revealed a charge transfer from Ga to Ni atoms, which is thought to enhance CO_2 chemisorption and activation, ultimately resulting in better catalytic stability due to the in situ process of carbon removal. These findings provide valuable insights into the role of Ga promotion in Ni-based catalysts and highlight the potential of Ga-modified catalysts for enhancing DRM efficiency. Future research should focus on exploring different supports, understanding deactivation mechanisms, and conducting longer-term stability tests. In situ characterization and advanced computational modeling can further refine our understanding and guide catalyst development.

Supplementary Materials: The following supporting information can be downloaded at: https://www.mdpi.com/article/10.3390/catal14100721/s1. Paragraph S1: DFT calculation method. Figure S1: Schematic illustration for the modeled surface slab of (a) Ni(111) and (b) Ni$_3$Ga1(111). The surface slab models are 2 × 2 slabs with 4 layers; the top two layers are free to move while the bottom two are kept fixed; Ni atoms in blue and Ga atoms in green. Figure S2: The adsorption isotherm for the as-prepared hydrotalcites: (a) Ni-Mg-Al, (b) Ni-Ga-Mg-Al (Ga/Ni = 0.1), (c) Ni-Ga-Mg-Al (Ga/Ni = 0.3), (d) Ni-Ga-Mg-Al (Ga/Ni = 0.5), and (e) Ni-Ga-Mg-Al. Figure S3: Bader charges analysis of (a) Ni(111) and (b) Ni$_3$Ga(111) surfaces; the numbers labeled indicate the valence electrons on the atoms (in e); the arrows point out the electron transfer direction. Figure S4: CH_4 and CO_2 conversion of Ni-Ga/(Mg, Al)O_x (Ga/Ni = 0.3) catalyst for long-time stream. Table S1: List of prepared catalysts and their nominal composition. Table S2: Quantitative summary of H_2-TPR analysis for the calcined catalysts. Table S3: Comparison of the activity of the Ni-Ga/(Mg, Al)O_x (Ga/Ni = 0.3) catalyst with other Ga-modified catalysts reported in previous studies on DRM. References [9–11,63–66] are cited in the Supplementary Materials.

Author Contributions: Conceptualization, A.Y.E., S.B.A. and A.S.A.-F.; methodology, A.Y.E. and A.S.A.-F.; software, A.Y.E.; validation, A.S.A.-F., A.E.A. and A.A.I.; formal analysis, A.Y.E. and S.B.A.; investigation, A.Y.E.; resources, A.S.A.-F. and S.B.A.; data curation, A.Y.E. and A.A.I.; writing—original draft preparation, A.Y.E.; writing—review and editing, A.E.A., A.H.F. and A.S.A.-F.; visualization, A.E.A., A.A.I. and A.S.A.-F.; supervision, A.S.A.-F. and A.E.A.; project administration, A.S.A.-F., A.E.A. and A.H.F.; funding acquisition, A.H.F. and S.B.A. All authors have read and agreed to the published version of the manuscript.

Funding: Researchers Supporting Project number (RSP2024R368), King Saud University.

Data Availability Statement: Data are contained within the article and Supplementary Materials.

Acknowledgments: The authors would like to extend their sincere appreciation to the Researchers Supporting Project number (RSP2024R368), King Saud University, Riyadh, Saudi Arabia.

Conflicts of Interest: The authors affirm that their work in this study was not influenced by any known conflicting financial interests or personal relationships. The authors declare that they have no conflicts of interest to disclose.

References

1. Yentekakis, I.V.; Panagiotopoulou, P.; Artemakis, G. A review of recent efforts to promote dry reforming of methane (DRM) to syngas production via bimetallic catalyst formulations. *Appl. Catal. B Environ.* **2021**, *296*, 120210. [CrossRef]
2. Hussien, A.G.S.; Polychronopoulou, K. A Review on the Different Aspects and Challenges of the Dry Reforming of Methane (DRM) Reaction. *Nanomaterials* **2022**, *12*, 3400. [CrossRef] [PubMed]
3. Xu, Z.; Park, E.D. Recent Advances in Coke Management for Dry Reforming of Methane over Ni-Based Catalysts. *Catalysts* **2024**, *14*, 176. [CrossRef]
4. Huang, L.; Li, D.; Tian, D.; Jiang, L.; Li, Z.; Wang, H.; Li, K. Optimization of Ni-Based Catalysts for Dry Reforming of Methane via Alloy Design: A Review. *Energy Fuels* **2022**, *36*, 5102–5151. [CrossRef]
5. Torrez-Herrera, J.J.; Korili, S.A.; Gil, A. Recent progress in the application of Ni-based catalysts for the dry reforming of methane. *Catal. Rev.* **2023**, *65*, 1300–1357. [CrossRef]
6. Nguyen, T.; Luu, C.L.; Phan, H.P.; Nguyen, P.A.; Van Nguyen, T.T. Methane dry reforming over nickel-based catalysts: Insight into the support effect and reaction kinetics. *React. Kinet. Mech. Catal.* **2020**, *131*, 707–735. [CrossRef]
7. Yu, Y.-X.; Yang, J.; Zhu, K.-K.; Sui, Z.-J.; Chen, D.; Zhu, Y.-A.; Zhou, X.-G. High-Throughput Screening of Alloy Catalysts for Dry Methane Reforming. *ACS Catal.* **2021**, *11*, 8881–8894. [CrossRef]
8. Wang, W.-Y.; Wang, G.-C. Ni$_3$X-type alloy catalysts for dry reforming of methane with high catalytic activity and stability: A density functional theory study. *Appl. Surf. Sci.* **2024**, *648*, 158958. [CrossRef]
9. Liu, X.; Jiang, F.; Liu, K.; Zhao, G.; Liu, J.; Xu, H. NiGa nano-alloy for the dry reforming of methane: Influences of Ga alloying with Ni on the catalytic activity and stability. *Chem. Eng. J.* **2024**, *487*, 150351. [CrossRef]
10. Bai, Y.; Sun, K.; Wu, J.; Zhang, M.; Zhao, S.; Kim, Y.D.; Liu, Y.; Gao, J.; Liu, Z.; Peng, Z. The Ga-promoted Ni/CeO$_2$ catalysts for dry reforming of methane with high stability induced by the enhanced CO$_2$ activation. *Mol. Catal.* **2022**, *530*, 112577. [CrossRef]
11. Kim, K.Y.; Lee, J.H.; Lee, H.; Noh, W.Y.; Kim, E.H.; Ra, E.C.; Kim, S.K.; An, K.; Lee, J.S. Layered Double Hydroxide-Derived Intermetallic Ni$_3$GaC$_{0.25}$ Catalysts for Dry Reforming of Methane. *ACS Catal.* **2021**, *11*, 11091–11102. [CrossRef]
12. Zeng, F.; Wei, B.; Lan, D.; Ge, J. Highly Dispersed NixGay Catalyst and La$_2$O$_3$ Promoter Supported by LDO Nanosheets for Dry Reforming of Methane: Synergetic Catalysis by Ni, Ga, and La$_2$O$_3$. *Langmuir* **2021**, *37*, 9744–9754. [CrossRef] [PubMed]
13. Al-Fatesh, A.S.; Rajput, Y.B.; Bayazed, M.O.; Alrashed, M.M.; Abu-Dahrieh, J.K.; Elnour, A.Y.; Ibrahim, A.A.; Fakeeha, A.H.; Abasaeed, A.E.; Kumar, R. Impact of gallium loading and process conditions on H$_2$ production from dry reforming of methane over Ni/ZrO$_2$-Al$_2$O$_3$ catalysts. *Appl. Catal. A Gen.* **2024**, *681*, 119794. [CrossRef]
14. Al-Fatesh, A.S.; Chava, R.; Alwan, S.M.; Ibrahim, A.A.; Fakeeha, A.H.; Abu-Dahrieh, J.K.; Yagoub Elnour, A.; Abasaeed, A.E.; Al-Othman, O.; Appari, S. Effect of Ga-Promoted on Ni/Zr + Al$_2$O$_3$ Catalysts for Enhanced CO$_2$ Reforming and Process Optimization. *Catal. Lett.* **2024**, 1–19. [CrossRef]
15. Shabani, S.; Mirkazemi, S.M.; Rezaie, H.; Vahidshad, Y.; Trasatti, S. A comparative study on the thermal stability, textural, and structural properties of mesostructured γ-Al$_2$O$_3$ granules in the presence of La, Sn, and B additives. *Ceram. Int.* **2022**, *48*, 6638–6648. [CrossRef]
16. Alreshaidan, S.B.; Ibrahim, A.A.; Fakeeha, A.H.; Almutlaq, A.M.; Ali, F.A.A.; Al-Fatesh, A.S. Effect of Modified Alumina Support on the Performance of Ni-Based Catalysts for CO$_2$ Reforming of Methane. *Catalysts* **2022**, *12*, 1066. [CrossRef]
17. Gao, X.; Ge, Z.; Zhu, G.; Wang, Z.; Ashok, J.; Kawi, S. Anti-Coking and Anti-Sintering Ni/Al$_2$O$_3$ Catalysts in the Dry Reforming of Methane: Recent Progress and Prospects. *Catalysts* **2021**, *11*, 1003. [CrossRef]
18. Ahmed, W.; Noor El-Din, M.R.; Aboul-Enein, A.A.; Awadallah, A.E. Effect of textural properties of alumina support on the catalytic performance of Ni/Al$_2$O$_3$ catalysts for hydrogen production via methane decomposition. *J. Nat. Gas Sci. Eng.* **2015**, *25*, 359–366. [CrossRef]
19. Bang, S.; Hong, E.; Baek, S.W.; Shin, C.-H. Effect of acidity on Ni catalysts supported on P-modified Al$_2$O$_3$ for dry reforming of methane. *Catal. Today* **2018**, *303*, 100–105. [CrossRef]
20. Pereira, S.C.; Ribeiro, M.F.; Batalha, N.; Pereira, M.M. Catalyst regeneration using CO$_2$ as reactant through reverse-Boudouard reaction with coke. *Greenh. Gases Sci. Technol.* **2017**, *7*, 843–851. [CrossRef]
21. Osaki, T.; Mori, T. Kinetics of the reverse-Boudouard reaction over supported nickel catalysts. *React. Kinet. Catal. Lett.* **2006**, *89*, 333–339. [CrossRef]
22. Park, H.-R.; Kim, B.-J.; Lee, Y.-L.; Ahn, S.-Y.; Kim, K.-J.; Hong, G.-R.; Yun, S.-J.; Jeon, B.-H.; Bae, J.W.; Roh, H.-S. CO$_2$ Reforming of CH$_4$ Using Coke Oven Gas over Ni/MgO-Al$_2$O$_3$ Catalysts: Effect of the MgO:Al$_2$O$_3$ Ratio. *Catalysts* **2021**, *11*, 1468. [CrossRef]
23. Kim, B.-J.; Park, H.-R.; Lee, Y.-L.; Ahn, S.-Y.; Kim, K.-J.; Hong, G.-R.; Roh, H.-S. Effect of surface properties of Ni-MgO-Al$_2$O$_3$ catalyst for simultaneous H$_2$ production and CO$_2$ utilization using dry reforming of coke oven gas. *Catal. Today* **2023**, *411–412*, 113855. [CrossRef]
24. Jin, B.; Li, S.; Liang, X. Enhanced activity and stability of MgO-promoted Ni/Al$_2$O$_3$ catalyst for dry reforming of methane: Role of MgO. *Fuel* **2021**, *284*, 119082. [CrossRef]

25. Charisiou, N.D.; Papageridis, K.N.; Tzounis, L.; Sebastian, V.; Hinder, S.J.; Baker, M.A.; AlKetbi, M.; Polychronopoulou, K. Goula, M.A. Ni supported on CaO-MgO-Al$_2$O$_3$ as a highly selective and stable catalyst for H$_2$ production via the glycerol steam reforming reaction. *Int. J. Hydrogen Energy* **2019**, *44*, 256–273. [CrossRef]
26. Basile, F.; Benito, P.; Fornasari, G.; Vaccari, A. Hydrotalcite-type precursors of active catalysts for hydrogen production. *Appl. Clay Sci.* **2010**, *48*, 250–259. [CrossRef]
27. Long, H.; Xu, Y.; Zhang, X.; Hu, S.; Shang, S.; Yin, Y.; Dai, X. Ni-Co/Mg-Al catalyst derived from hydrotalcite-like compound prepared by plasma for dry reforming of methane. *J. Energy Chem.* **2013**, *22*, 733–739. [CrossRef]
28. Abbas, M.; Sikander, U.; Mehran, M.T.; Kim, S.H. Exceptional stability of hydrotalcite derived spinel Mg(Ni)Al$_2$O$_4$ catalyst for dry reforming of methane. *Catal. Today* **2022**, *403*, 74–85. [CrossRef]
29. Dębek, R.; Motak, M.; Duraczyska, D.; Launay, F.; Galvez, M.E.; Grzybek, T.; Da Costa, P. Methane dry reforming over hydrotalcite-derived Ni–Mg–Al mixed oxides: The influence of Ni content on catalytic activity, selectivity and stability. *Catal. Sci. Technol.* **2016**, *6*, 6705–6715. [CrossRef]
30. Pérez-Ramírez, J.; Mul, G.; Moulijn, J.A. In situ Fourier transform infrared and laser Raman spectroscopic study of the thermal decomposition of Co–Al and Ni–Al hydrotalcites. *Vib. Spectrosc.* **2001**, *27*, 75–88. [CrossRef]
31. Wiyantoko, B.; Kurniawati, P.; Purbaningtias, T.E.; Fatimah, I. Synthesis and Characterization of Hydrotalcite at Different Mg/Al Molar Ratios. *Procedia Chem.* **2015**, *17*, 21–26. [CrossRef]
32. Cavani, F.; Trifirò, F.; Vaccari, A. Hydrotalcite-type anionic clays: Preparation, properties and applications. *Catal. Today* **1991**, *11*, 173–301. [CrossRef]
33. Dębek, R.; Zubek, K.; Motak, M.; Da Costa, P.; Grzybek, T. Effect of nickel incorporation into hydrotalcite-based catalyst systems for dry reforming of methane. *Res. Chem. Intermed.* **2015**, *41*, 9485–9495. [CrossRef]
34. Thommes, M.; Kaneko, K.; Neimark, A.V.; Olivier, J.P.; Rodriguez-Reinoso, F.; Rouquerol, J.; Sing, K.S.W. Physisorption of gases, with special reference to the evaluation of surface area and pore size distribution (IUPAC Technical Report). *Pure Appl. Chem.* **2015**, *87*, 1051–1069. [CrossRef]
35. Jayaprakash, S.; Dewangan, N.; Jangam, A.; Das, S.; Kawi, S. LDH-derived Ni–MgO–Al$_2$O$_3$ catalysts for hydrogen-rich syngas production via steam reforming of biomass tar model: Effect of catalyst synthesis methods. *Int. J. Hydrogen Energy* **2021**, *46*, 18338–18352. [CrossRef]
36. Abdelsadek, Z.; Holgado, J.P.; Halliche, D.; Caballero, A.; Cherifi, O.; Gonzalez-Cortes, S.; Masset, P.J. Examination of the Deactivation Cycle of NiAl- and NiMgAl-Hydrotalcite Derived Catalysts in the Dry Reforming of Methane. *Catal. Lett.* **2021**, *151*, 2696–2715. [CrossRef]
37. Zotin, F.M.Z.; Tournayan, L.; Varloud, J.; Perrichon, V.; Fréty, R. Temperature-programmed reduction: Limitation of the technique for determining the extent of reduction of either pure ceria or ceria modified by additives. *Appl. Catal. A Gen.* **1993**, *98*, 99–114. [CrossRef]
38. Kim, S.M.; Abdala, P.M.; Margossian, T.; Hosseini, D.; Foppa, L.; Armutlulu, A.; van Beek, W.; Comas-Vives, A.; Copéret, C.; Müller, C. Cooperativity and Dynamics Increase the Performance of NiFe Dry Reforming Catalysts. *J. Am. Chem. Soc.* **2017**, *139*, 1937–1949. [CrossRef]
39. Guo, J.; Lou, H.; Zhao, H.; Chai, D.; Zheng, X. Dry reforming of methane over nickel catalysts supported on magnesium aluminate spinels. *Appl. Catal. A Gen.* **2004**, *273*, 75–82. [CrossRef]
40. Kang, K.; Kakihara, S.; Higo, T.; Sampei, H.; Saegusa, K.; Sekine, Y. Equilibrium unconstrained low-temperature CO$_2$ conversion on doped gallium oxides by chemical looping. *Chem. Commun.* **2023**, *59*, 11061–11064. [CrossRef]
41. Kwon, D.; Kang, J.Y.; An, S.; Yang, I.; Jung, J.C. Tuning the base properties of Mg–Al hydrotalcite catalysts using their memory effect. *J. Energy Chem.* **2020**, *46*, 229–236. [CrossRef]
42. Tsyganok, A.I.; Tsunoda, T.; Hamakawa, S.; Suzuki, K.; Takehira, K.; Hayakawa, T. Dry reforming of methane over catalysts derived from nickel-containing Mg–Al layered double hydroxides. *J. Catal.* **2003**, *213*, 191–203. [CrossRef]
43. Wang, L.; Li, F.; Chen, Y.; Chen, J. Selective hydrogenation of acetylene on SiO$_2$-supported Ni-Ga alloy and intermetallic compound. *J. Energy Chem.* **2019**, *29*, 40–49. [CrossRef]
44. Sharafutdinov, I.; Elkjær, C.F.; Pereira de Carvalho, H.W.; Gardini, D.; Chiarello, G.L.; Damsgaard, C.D.; Wagner, J.B.; Grunwaldt, J.-D.; Dahl, S.; Chorkendorff, I. Intermetallic compounds of Ni and Ga as catalysts for the synthesis of methanol. *J. Catal.* **2014**, *320*, 77–88. [CrossRef]
45. López-Asensio, R.; Cecilia-Buenestado, J.A.; Herrera-Delgado, C.; Larrubia-Vargas, M.Á.; García-Sancho, C.; Maireles-Torres, P.J.; Moreno-Tost, R. Mixed Oxides Derived from Hydrotalcites Mg/Al Active in the Catalytic Transfer Hydrogenation of Furfural to Furfuryl Alcohol. *Catalysts* **2023**, *13*, 45. [CrossRef]
46. Stepanova, L.N.; Kobzar, E.O.; Trenikhin, M.V.; Leont'eva, N.N.; Serkova, A.N.; Salanov, A.N.; Lavrenov, A.V. Catalysts Based on Ni(Mg)Al-Layered Hydroxides Prepared by Mechanical Activation for Furfural Hydrogenation. *Catalysts* **2023**, *13*, 497. [CrossRef]
47. Miguel, C.V.; Trujillano, R.; Rives, V.; Vicente, M.A.; Ferreira, A.F.P.; Rodrigues, A.E.; Mendes, A.; Madeira, L.M. High temperature CO$_2$ sorption with gallium-substituted and promoted hydrotalcites. *Sep. Purif. Technol.* **2014**, *127*, 202–211. [CrossRef]
48. Yavuz, C.T.; Shinall, B.D.; Iretskii, A.V.; White, M.G.; Golden, T.; Atilhan, M.; Ford, P.C.; Stucky, G.D. Markedly Improved CO$_2$ Capture Efficiency and Stability of Gallium Substituted Hydrotalcites at Elevated Temperatures. *Chem. Mater.* **2009**, *21*, 3473–3475. [CrossRef]

49. Guo, J.; Lou, H.; Zheng, X. The deposition of coke from methane on a Ni/MgAl$_2$O$_4$ catalyst. *Carbon* **2007**, *45*, 1314–1321. [CrossRef]
50. Wang, P.; Tanabe, E.; Ito, K.; Jia, J.; Morioka, H.; Shishido, T.; Takehira, K. Filamentous carbon prepared by the catalytic pyrolysis of CH$_4$ on Ni/SiO$_2$. *Appl. Catal. A Gen.* **2002**, *231*, 35–44. [CrossRef]
51. Akri, M.; El Kasmi, A.; Batiot-Dupeyrat, C.; Qiao, B. Highly Active and Carbon-Resistant Nickel Single-Atom Catalysts for Methane Dry Reforming. *Catalysts* **2020**, *10*, 630. [CrossRef]
52. Zimmerli, N.K.; Rochlitz, L.; Checchia, S.; Müller, C.R.; Copéret, C.; Abdala, P.M. Structure and Role of a Ga-Promoter in Ni-Based Catalysts for the Selective Hydrogenation of CO$_2$ to Methanol. *JACS Au* **2024**, *4*, 237–252. [CrossRef] [PubMed]
53. Li, C.; Chen, Y.; Zhang, S.; Zhou, J.; Wang, F.; He, S.; Wei, M.; Evans, D.G.; Duan, X. Nickel–Gallium Intermetallic Nanocrystal Catalysts in the Semihydrogenation of Phenylacetylene. *ChemCatChem* **2014**, *6*, 824–831. [CrossRef]
54. Rao, D.-M.; Zhang, S.-T.; Li, C.-M.; Chen, Y.-D.; Pu, M.; Yan, H.; Wei, M. The reaction mechanism and selectivity of acetylene hydrogenation over Ni-Ga intermetallic compound catalysts: A density functional theory study. *Dalton Trans.* **2018**, *47*, 4198–4208. [CrossRef]
55. Chang, Y.K.; Lin, K.P.; Pong, W.F.; Tsai, M.-H.; Hseih, H.H.; Pieh, J.Y.; Tseng, P.K.; Lee, J.F.; Hsu, L.S. Charge transfer and hybridization effects in Ni$_3$Al and Ni$_3$Ga studies by x-ray-absorption spectroscopy and theoretical calculations. *J. Appl. Phys.* **2000**, *87*, 1312–1317. [CrossRef]
56. Kresse, G.; Hafner, J. Ab initio molecular dynamics for open-shell transition metals. *Phys. Rev. B* **1993**, *48*, 13115–13118. [CrossRef]
57. Kresse, G.; Furthmüller, J. Efficiency of ab-initio total energy calculations for metals and semiconductors using a plane-wave basis set. *Comput. Mater. Sci.* **1996**, *6*, 15–50. [CrossRef]
58. Kohn, W.; Sham, L.J. Self-Consistent Equations Including Exchange and Correlation Effects. *Phys. Rev.* **1965**, *140*, A1133–A1138. [CrossRef]
59. Perdew, J.P.; Burke, K.; Ernzerhof, M. Generalized Gradient Approximation Made Simple. *Phys. Rev. Lett.* **1996**, *77*, 3865–3868. [CrossRef]
60. Kresse, G.; Joubert, D. From ultrasoft pseudopotentials to the projector augmented-wave method. *Phys. Rev. B* **1999**, *59*, 1758–1775. [CrossRef]
61. Blöchl, P.E. Projector augmented-wave method. *Phys. Rev. B* **1994**, *50*, 17953–17979. [CrossRef] [PubMed]
62. Monkhorst, H.J.; Pack, J.D. Special points for Brillouin-zone integrations. *Phys. Rev. B* **1976**, *13*, 5188–5192. [CrossRef]
63. Al-Fatesh, A.S.; Ibrahim, A.A.; Abu-Dahrieh, J.K.; Al-Awadi, A.S.; El-Toni, A.M.; Fakeeha, A.H.; Abasaeed, A.E. Gallium-Promoted Ni Catalyst Supported on MCM-41 for Dry Reforming of Methane. *Catalysts* **2018**, *8*, 229. [CrossRef]
64. Fakeeha, A.H.; Kurdi, A.; Al-Baqmaa, Y.A.; Ibrahim, A.A.; Abasaeed, A.E.; Al-Fatesh, A.S. Performance Study of Methane Dry Reforming on Ni/ZrO$_2$ Catalyst. *Energies* **2022**, *15*, 3841. [CrossRef]
65. Al-Fatesh, A.S.; Fakeeha, A.H.; Ibrahim, A.A.; Abasaeed, A.E. Ni supported on La$_2$O$_3$+ZrO$_2$ for dry reforming of methane: The impact of surface adsorbed oxygen species. *Int. J. Hydrogen Energy* **2021**, *46*, 3780–3788. [CrossRef]
66. Pan, Y.-x.; Kuai, P.; Liu, Y.; Ge, Q.; Liu, C.-j. Promotion effects of Ga$_2$O$_3$ on CO$_2$ adsorption and conversion over a SiO$_2$-supported Ni catalyst. *Energy Environ. Sci.* **2010**, *3*, 1322–1325. [CrossRef]

 *catalysts*

Article

Are Rh Catalysts a Suitable Choice for Bio-Oil Reforming? The Case of a Commercial Rh Catalyst in the Combined H_2O and CO_2 Reforming of Bio-Oil

José Valecillos *, Leire Landa, Gorka Elordi , Aingeru Remiro , Javier Bilbao and Ana Guadalupe Gayubo *

Department of Chemical Engineering, University of the Basque Country (UPV/EHU), P.O. Box 644, 48080 Bilbao, Spain; leire.landa@ehu.eus (L.L.); gorka.elordi@ehu.eus (G.E.); aingeru.remiro@ehu.eus (A.R.); javier.bilbaoe@ehu.eus (J.B.)
* Correspondence: jose.valecillos@ehu.eus (J.V.); anaguadalupe.gayubo@ehu.eus (A.G.G.)

Abstract: Bio-oil combined steam/dry reforming (CSDR) with H_2O and CO_2 as reactants is an attractive route for the joint valorization of CO_2 and biomass towards the sustainable production of syngas (H_2 + CO). The technological development of the process requires the use of an active and stable catalyst, but also special attention should be paid to its regeneration capacity due to the unavoidable and quite rapid catalyst deactivation in the reforming of bio-oil. In this work, a commercial Rh/ZDC (zirconium-doped ceria) catalyst was tested for reaction–regeneration cycles in the bio-oil CSDR in a fluidized bed reactor, which is beneficial for attaining an isothermal operation and, moreover, minimizes catalyst deactivation by coke deposition compared to a fixed-bed reactor. The fresh, spent, and regenerated catalysts were characterized using either N_2 physisorption, H_2-TPR, TPO, SEM, TEM, or XRD. The Rh/ZDC catalyst is initially highly active for the syngas production (yield of 77% and H_2/CO ratio of 1.2) and for valorizing CO_2 (conversion of 22%) at 700 °C, with space time of 0.125 $g_{catalyst}$ h $(g_{oxygenates})^{-1}$ and $CO_2/H_2O/C$ ratio of 0.6/0.5/1. The catalyst activity evolves in different periods that evidence a selective deactivation of the catalyst for the reforming reactions of the different compounds, with the CH_4 reforming reactions (with both steam and CO_2) being more rapidly affected by catalyst deactivation than the reforming of hydrocarbons or oxygenates. After regeneration, the catalyst's textural properties are not completely restored and there is a change in the Rh–support interaction that irreversibly deactivates the catalyst for the CH_4 reforming reactions (both SR and DR). As a result, the coke formed over the regenerated catalyst is different from that over the fresh catalyst, being an amorphous mass (of probably turbostractic nature) that encapsulates the catalyst and causes rapid deactivation.

Keywords: bio-oil; steam reforming; dry reforming; syngas; coke deactivation; regeneration; irreversible deactivation

Citation: Valecillos, J.; Landa, L.; Elordi, G.; Remiro, A.; Bilbao, J.; Gayubo, A.G. Are Rh Catalysts a Suitable Choice for Bio-Oil Reforming? The Case of a Commercial Rh Catalyst in the Combined H_2O and CO_2 Reforming of Bio-Oil. *Catalysts* 2024, 14, 571. https://doi.org/10.3390/catal14090571

Academic Editors: Georgios Bampos, Paraskevi Panagiotopoulou and Eleni A. Kyriakidou

Received: 30 July 2024
Revised: 22 August 2024
Accepted: 26 August 2024
Published: 29 August 2024

1. Introduction

Syngas, a blend of H_2 and CO, is a basic (petro)chemical platform for the syntheses of alcohols (mainly methanol), ethers, carboxylic acids, other various carbonyl compounds, synthetic fuels, ammonia, and urea, and is also a fuel employed in gas engines for energy generation. Its production is still highly dependent on fossil resources, mostly by reforming of natural gas and petroleum derivatives, with a significant contribution to the global CO_2 emissions. Hence, sustainable production options are urged for the transition towards chemical and energy industries that are completely clean and renewable. Among many alternatives, the reforming of biomass and its derivatives replacing the traditional fossil-based feedstock is an attractive option as it meets the goal of net zero CO_2 emissions [1]. Its feasibility for a prompt implementation takes advantage of starting from an existing reforming technology with the challenge of making improvements to minimize costs and environmental impacts.

One promising route is the use of lignocellulose biomass wastes, which does not interfere with food chains and can be processed by fast pyrolysis in simple and decentralized facilities yielding large quantities of bio-oil (a complex mixture of oxygenates comprising carboxylic acids, aldehydes, alcohols, ketones, esters, furfurals, phenols, and saccharides [2,3]). The subsequent bio-oil reforming may be targeted at the production of syngas with suitable H_2/CO ratios for the syntheses of fuels or chemicals, depending on the reforming strategy [4,5]. The CO_2 reforming, commonly known as dry reforming (DR), has been proposed as an attractive alternative reforming strategy that allows the production of useful syngas from bio-oil oxygenates ($C_nH_mO_k$) while valorizing CO_2 simultaneously (Equation (1)), avoiding the excessive side CO_2 production when using the steam reforming (SR) strategy [4]. The reverse water gas shift (r-WGS) reaction also contributes to the CO_2 conversion at high temperatures (reverse Equation (2)). Nevertheless, the inherent H_2O content in the bio-oil, which varies depending on its origin, also promotes the SR of oxygenates (Equations (3) and (4)), and thus a combined steam/dry reforming (CSDR) takes place. Moreover, in the conversion of bio-oil, the decomposition/cracking of oxygenates into H_2, CO, CO_2, CH_4, hydrocarbons (C_aH_b), other oxygenates ($C_xH_yO_z$), and carbon (coke) should be considered, as represented by Equation (5). Therefore, the conversion of CH_4 and hydrocarbons by DR (Equations (6) and (7), respectively) and SR (Equations (8) and (9), respectively) also contributes to the overall kinetic scheme. Likewise, coke formation may be favored by CH_4 decomposition (Equation (10)), hydrocarbon decomposition (Equation (11)), and the CO disproportionation (Boudouard) reaction (Equation (12)), whereas its gasification may occur with steam (Equation (13)) or CO_2 (reverse Equation (12)). Consequently, the co-feeding of CO_2 is expected to favor coke removal.

$$C_nH_mO_k + xCO_2 \rightarrow (n + x)CO + (m/2 - (x + k - n))H_2 + (x + k - n)H_2O \tag{1}$$

$$CO + H_2O \leftrightarrow CO_2 + H_2 \tag{2}$$

$$C_nH_mO_k + (n - k)H_2O \rightarrow nCO + (n + m/2 - k)H_2 \tag{3}$$

$$C_nH_mO_k + (2n - k)H_2O \rightarrow nCO_2 + (2n + m/2 - k)H_2 \tag{4}$$

$$C_nH_mO_k \rightarrow C_xH_yO_z + (CO, CO_2, CH_4, C_aH_b, H_2) + C(coke) \tag{5}$$

$$CH_4 + CO_2 \leftrightarrow 2CO + 2H_2 \tag{6}$$

$$C_aH_b + aCO_2 \leftrightarrow (2a)CO + (b/2)H_2 \tag{7}$$

$$CH_4 + H_2O \leftrightarrow CO + H_2 \tag{8}$$

$$C_aH_b + aH_2O \leftrightarrow aCO + (a + b/2)H_2 \tag{9}$$

$$CH_4 \rightarrow 2H_2 + C \tag{10}$$

$$C_aH_b \rightarrow (b/2)H_2 + aC \tag{11}$$

$$2CO \leftrightarrow C + CO_2 \tag{12}$$

$$C + H_2O \rightarrow CO + H_2 \tag{13}$$

The catalysts for bio-oil reforming are commonly based on non-noble or noble metals, or bimetallic compositions [6–11]. Nevertheless, the CSDR of real bio-oil has been scarcely studied experimentally on Ni catalysts (non-noble metal catalysts) [12–14], showing promising results, and no studies with a noble metal catalyst have been reported so far. However, it could be hypothesized that the use of noble metal catalysts, such as Rh catalysts, may improve the performance of Ni catalysts grounded in previous studies reporting a remarkable activity for the H_2/syngas production from real bio-oil by conventional SR [15,16], oxidative SR (OSR) [17–19], and sequential cracking [20,21]. Likewise, Rh catalysts have been successfully used in the DR of CH_4 [22–24] and ethanol [25], whose results may be extrapolated to the case of bio-oil, in particular due to the role of CH_4 as a reaction intermediate in bio-oil reforming. The advantage of noble metal catalysts relies on their high dehydrogenation and oxidation capacities, which are translated into high yields of H_2 and CO/CO_2 without CH_4 production (CH_4 is converted) and low coke formation, preventing catalyst deactivation. Additionally, Rh sites may also adsorb and dissociate CO_2, which is a paramount step in DR reactions [26]. Likewise, catalyst support may play a role in the catalysis by providing acidic or basic sites, hydrophilicity, or oxygen mobility [22]. For example, the use of CeO_2, TiO_2, and ZrO_2 provides the latter features enhancing H_2O or CO_2 adsorption and dissociation generating OH or CO species facilitating their conversions, which would globally enhance the performance of the bio-oil CSDR. Kartavova et al. [27] highlight the importance of the support on the surface distribution and crystal size of Rh in the catalysts used in the cyclohexane ring opening reaction. CeO_2 is an interesting support because of its oxygen storage and mobility capacity due to the fast Ce^{4+}/Ce^{3+} redox cycling. Its limited thermal stability is improved by the formation of the $Ce_xZr_{1-x}O_2$ solid solution, with excellent performance as a support for chromium oxide in the dehydrogenation of propane with CO_2 [28].

In spite of the high activity of Rh catalysts, reversible and irreversible deactivation has been observed in the OSR and SR of bio-oil [19,29,30]. The reversible deactivation is associated with the formation and deposition of coke encapsulating the active sites, whereas the irreversible deactivation has been related to irreversible structural changes in the catalyst due to the harsh reaction conditions (high temperature and presence of steam). These changes mainly involve the aging of the support partially occluding Rh sites and possible changes in the Rh structure. Likewise, Rh sintering is a cause of deactivation at 750 °C in the OSR bio-oil [19]. In other catalytic systems for different applications, it has been found that Rh catalysts are irreversibly deactivated by the partial encapsulation of Rh nanoparticles by CeO_2 [31,32], which has also been reported for various noble metals [33]. In the case of reforming strategies, it is believed that high steam concentrations may favor this phenomenon of irreversible deactivation of Rh catalysts, which represents a drawback of using these catalysts.

In this work, we have studied the performance of a commercial catalyst made of Rh supported on zirconium-doped ceria (Rh/ZDC) for the CSDR of a real bio-oil targeted at syngas production, with focus on its activity, deactivation, and regeneration capacity. It is expected that the low steam concentration in the CSDR strategy lessens the phenomenon of irreversible deactivation. As previously mentioned, the novelty of this work is stressed by the use of a noble metal catalyst for this process with a feed of real bio-oil, with a formulation that is expected to enhance the CSDR performance. The bio-oil CSDR tests were carried out in a fluidized bed reactor at 700 °C, comprising a reaction/regeneration cycle. To investigate the causes of deactivation, the fresh, deactivated, and regenerated catalyst samples have been analyzed by means of temperature-programmed reduction (H_2-TPR), N_2 physisorption, temperature-programmed oxidation (TPO), X-ray diffraction (XRD), scanning electron microscopy (SEM), and transmission electron microscopy (TEM). The discussion of the results is aimed at understanding the reversibility and irreversibility of catalyst deactivation.

2. Results

The results are divided in three main subsections. The first one summarizes the findings for the performance of the Rh/ZDC catalyst in the CSDR of bio-oil at 700 °C considering the operation in reaction–regeneration cycles and a comparison with the conventional SR of bio-oil at the same conditions. The second one collects the data obtained from the characterization of the fresh and spent catalysts by using different techniques in order to shed light on the carbon formation that causes a reversible catalyst deactivation. The last subsection explores the reasons for the irreversible catalyst deactivation observed for the regenerated catalyst, based on the analysis of the Rh sites by H_2-TPR measurements.

2.1. Performance of the Rh Catalyst in Bio-Oil Reforming

Figure 1 shows the performance of the Rh/ZDC catalyst in the CSDR of bio-oil at 700 °C in two successive reactions with a catalyst regeneration in between. The performance is analyzed in terms of the evolution over time on stream of the product yields (Figure 1a for the reaction with fresh catalyst (first reaction) and Figure 1c for the reaction with the regenerated catalyst (second reaction)) and syngas yield or H_2/CO ratio (Figure 1b and Figure 1d for the first and second reactions, respectively). In these data, the oxygenate conversion is estimated as the total yield of the carbon gaseous products (sum of all carbonaceous products after subtracting the molar flow rate of CO_2 in the feed), since the estimated coke content and yield are very low according to the TPO analysis shown in the next subsection.

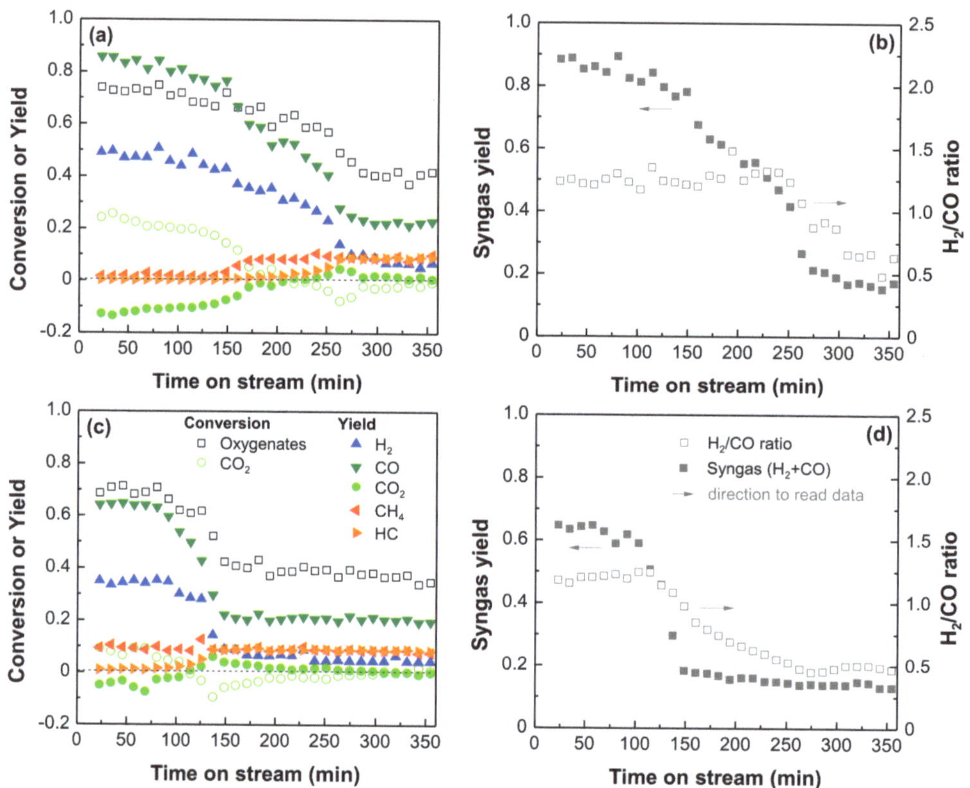

Figure 1. Performance of the Rh/ZDC catalyst in the CSDR of bio-oil at 700 °C, $CO_2/H_2O/C$ ratio of 0.6/0.5/1, space time (referred to the mass flow of oxygenates in the bio-oil), and 0.125 $g_{catalyst}$ h $(g_{oxygenates})^{-1}$: Evolution of the conversion and product yields over time on stream for (**a**) the fresh catalyst and (**c**) the regenerated catalyst; evolution of the syngas yield and H_2/CO ratio over time on stream for (**b**) the fresh catalyst and (**d**) the regenerated catalyst.

The initial oxygenate conversion is around 0.77 with the fresh catalyst (Figure 1a). This low value evidences that the space time used is low to reach an equilibrium state at these reaction conditions (particularly because of the low S/C ratio), and therefore the maximum product yields may not have been achieved, although this situation is adequate to study the catalyst deactivation. In spite of this, the initial yield values of H_2 and CO provide evidence for the high activity of Rh for the bio-oil CSDR, which has also been demonstrated for the OSR of bio-oil [17–19]. The comparison of the results in Figure 1a,b with those obtained on a Ni catalyst (Ni/Al_2O_3 catalyst derived from a $NiAl_2O_4$ spinel) under the same conditions [12] reveals that Rh exhibits notably high activity considering that less active sites are actually present in the Rh catalyst (2 wt% Rh vs. 36 wt% Ni). Thus, the initial yield and H_2/CO ratio of the syngas are almost 90% and 1.2, respectively, with this Rh catalyst compared to 92% and 0.9 with the Ni catalyst, and the CO_2 conversion is 22% against 28%. As a comparison, Figure S1 in the Supplementary Materials shows the results obtained with the Rh/ZDC catalyst at the same conditions as in Figure 1 but without CO_2 co-feeding (that is, under SR conditions with a low S/C ratio of 0.5). When comparing the SR and CSDR strategies, a slightly higher oxygenate conversion is observed in the SR reaction (around 80%, Figure S1a) than in the CSDR reaction (Figure 1a). The main difference between both reforming strategies is the increase in the CO yield in the CSDR reaction (from 62% in Figure S1a to 85% in Figure 1a), hence yielding slightly more syngas (near 90% in CSDR reaction, compared to 80% in SR reaction) with a H_2/CO ratio of about 1.2 (Figure 1b) in comparison to a H_2/CO ratio of 2 for the SR strategy (Figure S1b).

The product distribution (H_2, CO, CO_2, CH_4, and hydrocarbons) observed may be explained considering the DR (Equation (1)) and SR (Equations (3) and (4)) of oxygenates, as well as the catalytic cracking of oxygenates (Equation (5)), as the dominant routes at this high temperature, generating H_2 and various gaseous intermediates (CO, CO_2, CH_4 and hydrocarbons). Then, the SR or DR of CH_4 and hydrocarbons (Equations (6)–(9)) generates CO and H_2, and the CO_2 may be transformed into CO and H_2O by the r-WGS reaction (reverse Equation (2)), which is thermodynamically favored at this high temperature and low S/C ratio [4,5]. In spite of its high catalytic activity, the Rh catalyst undergoes deactivation during both the CSDR and SR reactions, which is clearly evidenced by the drop of the H_2 and CO yields over time on stream (Figure 1a,b). The activity decay must be mainly attributed to coke formation, which may proceed through the decomposition of oxygenates (Equation (5)), CH_4 (Equation (10)), and hydrocarbons (Equation (11)), considering the high temperature (700 °C) and low S/C ratio employed [2]. Simultaneously, coke gasification may take place with CO_2 (reverse Equation (12)) or steam (Equation (13)).

Curiously, the evolution of the product yields draws five well-distinguished periods, as has also been observed in the OSR and SR of bio-oil with the same catalyst [19,29] and in Figure S1. Thus, for the CSDR strategy, the first period (around 125 min) shows a quite stable behavior of the catalyst, with high oxygenate conversion and complete conversion of intermediates, CH_4, and hydrocarbons, by SR or DR reactions, generating the highest H_2 and CO yields with the maximum CO_2 conversion boosted by the r-WGS reaction. The second period (~125–175 min) shows a decrease in the H_2 and CO yields and in the CO_2 conversion and an increase in the CH_4 yield, indicating the selective deactivation of the catalyst for the SR or DR of CH_4 and the r-WGS reaction. The third period (~175–225 min) describes a pseudo steady state where the conversion of hydrocarbons by SR and DR reactions remains complete. Subsequently, in the fourth period (~225–275 min), the complete catalyst deactivation for the SR and DR reactions of hydrocarbons and oxygenates occurs, evidenced by the sharp decrease in the oxygenate conversion and the H_2 and CO yields and increase in the hydrocarbon yield. In this period, the CO_2 conversion falls to negative values, reaching a minimum and then increasing up to zero, which is translated into a maximum for the CO_2 yield, which suggests a selective deactivation of the DR of hydrocarbons over their SR reaction. The fifth period (above 275 min) presumably corresponds to the yields obtained in the thermal reaction of the bio-oil oxygenates in the presence of both steam and CO_2. Consequently, the syngas yield decreases over time following the same trend of these five periods (Figure 1b), whereas the H_2/CO ratio is almost constant during the first two periods, slightly increases in the third period, and decays noticeably in the fourth period when the catalyst undergoes complete deactivation.

These periods are also observed in the SR of bio-oil (Figure S1), but in this case CO_2 is only a product. Particularly, the CO_2 yield increases during the second period (partial deactivation for CH_4 conversion) and keeps a higher stable value in the third period in comparison with the first period. This result suggests that the r-WGS reaction (which is thermodynamically favored in these reaction conditions) is also affected by deactivation in the second period. This behavior can also be observed in the CSDR reaction (Figure 1a), though the CO_2 yield values are negative. Strikingly, the H_2/CO ratio evolution follows a trend similar to that of the CO_2 yield, which confirms the partial catalyst deactivation for the r-WGS reaction in the second period.

After the CSDR reaction with the fresh catalyst (Figure 1a), the spent catalyst was regenerated in situ (in the reactor) by coke combustion with air at 600 °C and subsequent reduction in a H_2/N_2 flow at 700 °C. Upon regeneration (Figure 1c), the catalyst does not fully recover its initial activity, which is particularly evidenced by the absence of two different deactivation periods described above. In turn, the reaction with the regenerated catalyst initiates at a stable period in which hydrocarbons are fully converted by SR or DR reactions, whereas CH_4 is not converted (third period in Figure 1a), followed by the complete catalyst deactivation (fourth period in Figure 1a). The syngas yield (Figure 1d) is initially stable and then decreases over time on stream as described by the third and fourth periods in Figure 1b, and the H_2/CO ratio also slightly increases and then decreases following the trend described by the CO_2 yield. This evidences that the Rh/ZDC catalyst undergoes both reversible and irreversible deactivation. The former is due to the deposition of coke so that the catalyst partially recovers its reforming activity after regeneration. The latter is presumably associated with irreversible changes in the catalyst structure or metal sites. In the following sections, we further investigate the causes of both deactivation phenomena by analyzing the spent catalyst from each reaction test.

2.2. Chacaractarization of Deactivated Catalyst Samples

The coke formed and deposited on the catalyst during each reaction was analyzed by TPO, SEM, and TEM in order to determine its amount and/or nature. Additionally, N_2 physisorption was also employed to analyze the effect of coke deposition on the catalyst textural properties. The possible changes in the crystalline structure of the deactivated catalyst were analyzed by XRD analysis. Finally, the H_2-TPR profiles of the fresh and regenerated catalyst were compared to analyze the Rh sites. In general, the spent catalyst samples are identified with reference to their reaction test: CSDR-1 for the first CSDR reaction (with the fresh catalyst) and CSDR-2 for the second CSDR reaction (with the regenerated catalyst).

2.2.1. Coke Formation

Figure 2 shows the TPO profiles for the spent catalyst samples, in which the peaks are associated with the combustion of coke. The corresponding coke content (determined by integration of each curve) is indicated next to each profile. The TPO profiles are noticeably different for the two samples, showing combustion peaks centered at different temperatures. The CSDR-1 sample shows two separate peaks centered at 300 and 400 °C, and the CSDR-2 sample shows two overlapping peaks that are approximately centered at 415 and 430 °C. Likewise, the coke content is significantly lower in the CSDR-1 sample (93.6 mg g^{-1}) than in the CSDR-2 sample (302 mg g^{-1}). The temperature position of the combustion peaks may be related to the nature of carbon and to the possible catalytic effect of the catalyst components on the carbon combustion. Thus, the peaks at lower temperatures can be associated with the combustion of poorly developed coke (amorphous carbon, most probably encapsulating Rh sites) and the peaks at higher temperatures are associated with structured coke (carbon nanostructures or turbostratic/graphite carbon) [34,35]. Additionally, both Ce (in the ZDC support) and Rh species can catalyze the carbon combustion [36], which results in combustion peaks at lower temperature than those obtained for Ni catalysts in different bio-oil reforming strategies [12,17,37].

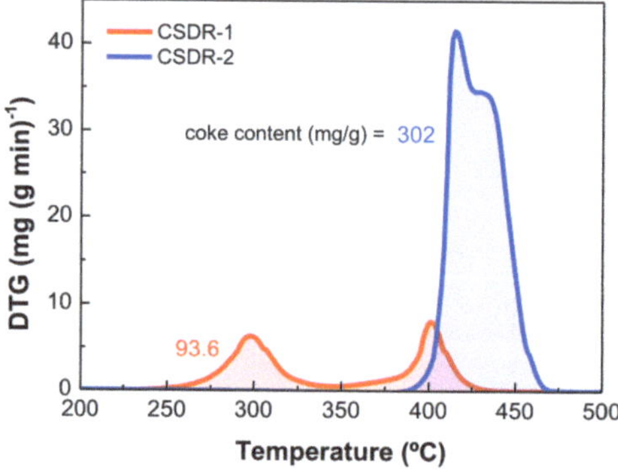

Figure 2. TPO profiles of the spent catalyst samples. Coke content is indicated close to each profile.

To discern the nature of the coke, the two spent catalyst samples have been analyzed by SEM (with a backscattered electron (BSE) or secondary electron (SE) detector) and TEM. Figure 3 shows the BSE-SEM images for the fresh and spent catalyst samples, which provide qualitative information on the degree of coke deposition on the external surface. Accordingly, the brightness intensity is indicative of the presence of heavy (bright) or light (dark) elements [2,12,35], and for this case, a high or dense presence of coke on the catalyst surface is indicated by darker particles since coke contains the lightest element (C) in these samples in comparison with the catalyst components (Rh, Ce, and Zr). In general, all the particles exhibit relatively bright intensities similar to those of the fresh catalyst (Figure 3a), indicating a poor coke deposition on the catalyst surface or the deposition of carbon with low density, although the CSDR-1 sample also exhibits some dark particles.

Figure 3. BSE-SEM images of the (**a**) fresh catalyst and spent catalyst samples from the (**b**) 1st CSDR reaction (with fresh catalyst) and (**c**) 2nd CSDR reaction (with regenerated catalyst).

Further, to explore the coke deposition, Figure 4 shows the SE-SEM images at higher zooms for all the samples. The Rh/ZDC catalyst shows a granular surface typical of porous materials (in this case the ZDC support), showing some compacted areas. The majority of the particles in the CSDR-1 sample are bright, even though there are some carbon filaments on the surface and some granular and compacted areas that may be the surface of the ZDC support (Figure 4b). The dark particles in the CSDR-1 sample show an abundant presence of carbon filaments (Figure 4c). On the other hand, the CSDR-2 sample shows a smooth granular-like surface (apparently different from that of the catalyst support) with some carbon filaments (Figure 4d), and when this surface is zoomed in upon (Figure 4e), it seems to be a blend of an amorphous mass of carbon with some nanotubes trapped in it.

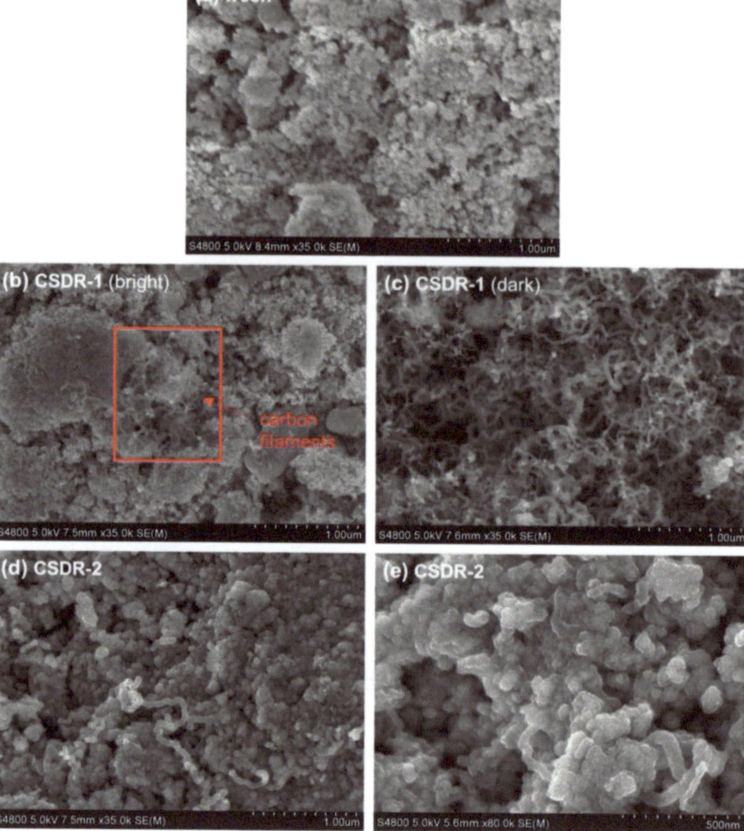

Figure 4. SE-SEM images of the (**a**) fresh catalyst and spent catalyst samples from the (**b,c**) 1st CSDR reaction (with fresh catalyst) and (**d,e**) 2nd CSDR reaction (with regenerated catalyst).

TEM has been useful to confirm the presence of these carbon types as shown in Figure 5, from which it should be considered that only tiny fragments of the samples can be analyzed without distinguishing between the external/internal surface of the particle. The Rh/ZDC catalyst shows grains of different grey tones, with the darker ones presumably being Rh crystals as this is the densest catalyst component. The CSDR-1 sample shows the presence of both carbon filaments (Figure 5b) and amorphous/turbostratic carbon encapsulating the catalyst components (Figure 5c). On the other hand, the CSDR-2 sample barely shows a mass of carbon covering the catalyst components in spite of having the highest coke content. The difficulty to observe the carbon on this latter sample is perhaps due to its nature, apparently a thin layer of carbon mass covering the entire surface evenly, as well as the limitation of the technique to analyze this type of solid.

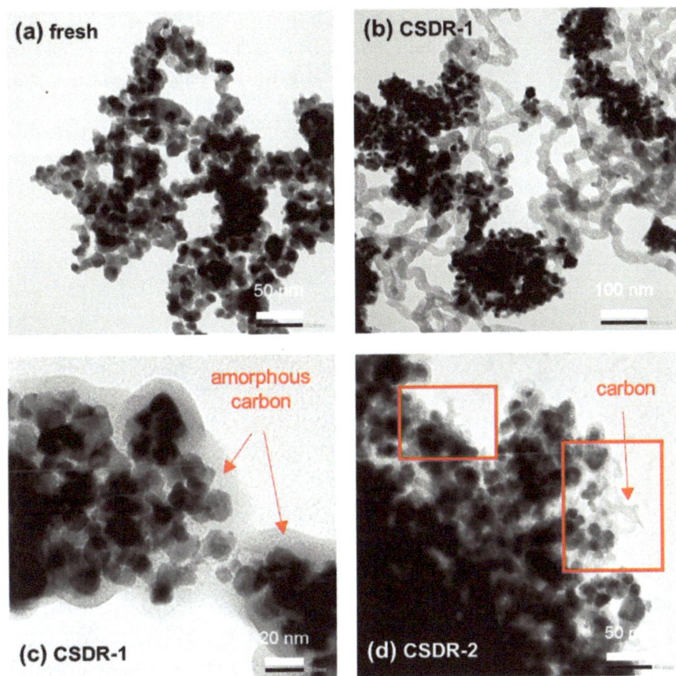

Figure 5. TEM images of the (**a**) fresh catalyst and spent catalyst samples from the (**b,c**) 1st CSDR reaction (with fresh catalyst) and (**d**) 2nd CSDR reaction (with regenerated catalyst).

The textural properties of the fresh and spent catalyst samples (Table 1) evidence the impact of coke deposition. The formation of both carbon filaments and amorphous carbon in the CSDR reaction with the fresh catalyst (CSDR-1 sample) causes the BET surface area to decrease by 33%, with a strong impact on the microporous surface area and on the volume, which both decrease by about 62%. On the other hand, the formation of a blend of amorphous/turbostratic carbon and small carbon filaments in the CSDR reaction with the regenerated catalyst (CSDR-2 sample) causes a strong decrease in the BET surface area (about 74%) and in the total volume of pores (by 83%) without affecting the microporous surface area. This suggests that in the reaction with the fresh catalyst, the amorphous carbon may be deposited on the microporous surface, whereas all the carbon is deposited on the external surface of the catalyst particles in the reaction with the regenerated catalyst. It should be noted that the reduced specific surface is a notable limitation of the CeO_2 support, as pointed out in the literature [28], and it is more pronounced in this reaction due to coke deposition.

Table 1. Textural properties of fresh, spent, and regenerated Rh/ZDC catalyst samples.

Sample	S_{BET} $(m^2\,g^{-1})$	S_{micro} $(m^2\,g^{-1})$	V_{pore} $(cm^3\,g^{-1})$	V_{micro} $(cm^3\,g^{-1})$	d_{pore} (nm)
Fresh	57.5	2.17	0.24	1.31×10^{-3}	17.0
CSDR-1	38.4	0.82	0.15	0.49×10^{-3}	15.9
CSDR-2	14.8	2.16	0.04	0.85×10^{-3}	10.9
1st regeneration (CSDR-1)	39.6	2.16	0.20	1.00×10^{-3}	20.2

S_{BET} = BET specific surface area; S_{micro} = microporous specific surface area; V_{pore} = specific volume of pores; V_{micro} = specific volume of micropores; d_{pore} = average pore diameter.

2.2.2. Characterization of Catalyst Structure and Rh Sites

The XRD analysis of the fresh and spent catalyst samples (Figure 6) evidences no changes in the crystalline properties of the ZDC support, as all the diffraction peaks observed are similar among all the samples and correspond to the ZDC structure [38]. No diffraction peaks are observed for Rh species, since the Rh content is low (2 wt%) and it is most probably well dispersed on the support. Likewise, the formation of crystalline carbon is not evidenced either, as no diffraction peaks are detected for carbon (15–30° range), which may also be due to the low coke content.

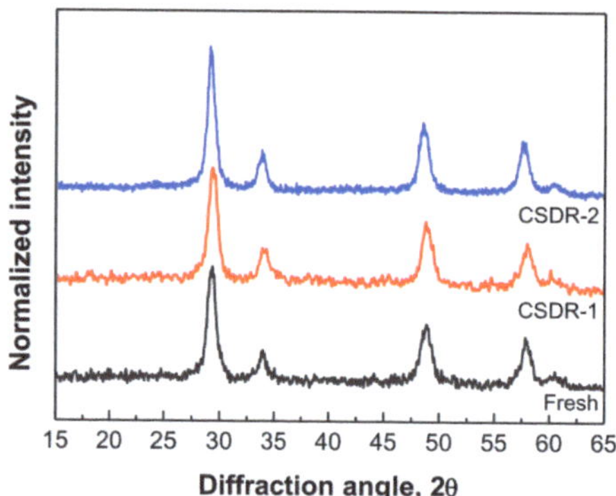

Figure 6. XRD patterns of the fresh and spent catalyst samples.

To understand the irreversible catalyst deactivation in the CSDR reaction, the H_2-TPR profiles of the fresh and regenerated catalyst are analyzed (Figure 7), providing insights on the Rh sites. The reducibility of these sites is related to the interaction between the active phase and the support, which in the case of Rh is strongly dependent on the properties of the latter [27]. For the regenerated catalyst, the CSDR-1 sample was subjected to combustion (TPO) to remove the coke and then to a H_2-TPR measurement to observe the reducibility of Rh species. In general, both H_2-TPR profiles are similar, showing a broad reduction peak between 50 and 125 °C and a shoulder between 125 and 200 °C, which are similar to the H_2-TPR profiles of Rh catalysts reported in the literature [19,29,39]. To better distinguish the reduction of different Rh species, the H_2-TPR profiles were deconvoluted into three Gauss bands that correspond to the reduction of three Rh species (identified as Rh-1, Rh-2, and Rh-3) possibly indicating different interactions with the ZDC support. The Rh-1 and Rh-2 bands may be associated with the reduction of Rh^{3+} species on the support [39], likely having different interactions with the ZDC support (weak/strong interactions). Those Rh^{3+} species with a weak support interaction would be reduced easily, and therefore they would be associated with the Rh-1 band. The Rh-3 band is associated with the reduction of Rh species strongly interacting with the ZDC support, such as Rh^{3+} species in the CeO_2 lattice [39]. The deconvolution also shows how the relative portions of the different Rh species change after catalyst regeneration (Figure 7b) in comparison with the fresh catalyst (Figure 7a). Thus, the fraction of Rh-2 species notably increases after the regeneration, which may suggest that the Rh^{3+} species have stronger interactions with the support. This would support the possible encapsulation of Rh sites, causing a decrease in their activity (particularly for the CH_4 reforming and CO_2 conversion).

The textural properties of the regenerated catalyst (Table 1) evidence that the BET specific surface area is not fully recovered, though the microporous specific surface area is fully restored after regeneration.

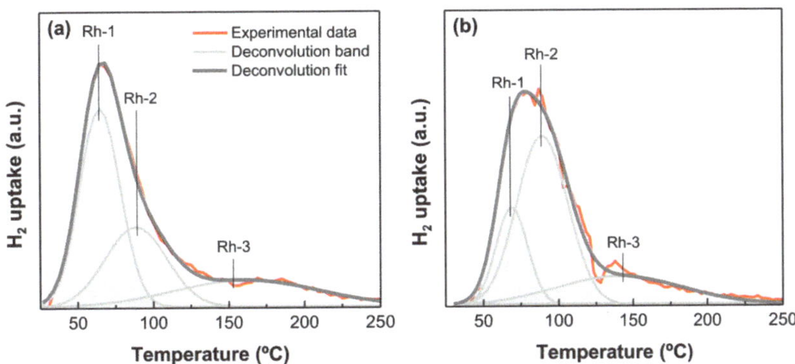

Figure 7. H_2-TPR profiles for the (**a**) fresh catalyst and (**b**) regenerated after the 1st CSDR reaction.

3. Discussion

The results in Figure 1 show that the commercial Rh/ZDC catalyst is highly active for the bio-oil CSDR, though it undergoes reversible and irreversible deactivation. The reversible deactivation is associated with the coke deposition as demonstrated in other reforming strategies of a real bio-oil feed using Rh or Ni catalysts [29,40], whereas the irreversible deactivation is similar to that observed for the OSR and SR of a real bio-oil with a Rh catalyst, which is associated with irreversible changes in the catalyst structure [19,29]. Thus, the low steam concentration in the bio-oil CSDR (steam is formed from the intrinsic H_2O content in the bio-oil) did not lessen the irreversible deactivation as it was hypothesized. In the present work, we have further observed that the coke formation in the first reaction (with the fresh catalyst) and second reaction (with the regenerated catalyst) is quite different, which is a consequence of the irreversible deactivation. The lower activity of the regenerated catalyst (unable to convert CH_4) leads to a faster coke deactivation (Figure 1c) with a higher coke deposition (three times more than in the reaction with the fresh catalyst) that promptly blocks all the catalyst surface (as revealed by the low BET surface area of CSDR-2 sample in Table 1). The analysis of Figures 2–5 and Table 1 evidences that an apparent mass of amorphous carbon combusting at 300 °C and carbon filaments combusting at 400 °C are formed in the reaction with the fresh catalyst (CSDR-1 sample), whereas an apparent blend of amorphous/turbostratic carbon and carbon filaments, covering the entire catalyst surface and combusting at 415–430 °C, is formed in the reaction with the regenerated catalyst (CSDR-2 sample). The lower temperatures required to combust the coke in comparison to the expected combustion temperatures for the types of carbons observed in the bio-oil reforming processes over other catalysts [12,17,37] can be attributed to the catalytic effect of the Ce and Rh species on the carbon combustion. This effect is more evident in the coke combustion from the reaction with the fresh catalyst because the coke deposition causes low blockage of the porous surface (Table 1), enabling good contact between O_2 and the catalyst surface during the combustion. Conversely, the higher combustion temperature of the coke formed in the reaction with the regenerated catalyst may be explained by the coke refractoriness and the poor contact between O_2 and the catalytic species during the combustion due to the higher pore blockage (Table 1), lessening the catalytic effect of Ce and Rh on the carbon combustion.

The irreversible deactivation selectively affects the CH_4 reforming reactions (DR and SR) and also partially the r-WGS reaction, which suggests that the sites that activate CH_4 and CO_2 are irreversibly modified in a given moment during the reaction with the fresh catalyst and/or the regeneration process. In resemblance with other works [19,29], the irreversible deactivation of a similar Rh/ZDC catalyst was detected to occur during the reaction of the OSR of bio-oil when the catalyst loses its activity for the CH_4 conversion. This deactivation was attributed to the aging of the ZDC support at high temperatures (700 °C, the same temperature used in this work) and presence of steam, which was evidenced by the loss of textural properties and interpreted as a collapse of the narrower pores occluding some Rh sites. In this present work, we have also evidenced the loss of textural properties after the catalyst regeneration (Table 1), directly observing an irreversible loss of surface area and an increase in the average pore size, which may be congruent with the collapse of some pores. This can also be supported by the observations of the SEM analysis (Figure 5b), in which an apparent increase in the compacted areas after the reaction with the fresh catalyst (CSDR-1 sample) can be interpreted as a reduction in the porosity. We have further analyzed the changes in the Rh sites by H_2-TPR (Figure 7), possibly suggesting partial encapsulation of Rh^{3+} species with the ZDC support, which apparently reduces the activity for the bio-oil CSDR process (unable to activate CH_4 and CO_2). It is worth mentioning that Rh sintering was not evidenced by the TEM images, since the Rh nanoparticles have similar sizes for the fresh and spent catalysts. Likewise, Rh sintering is not significant for this catalyst in the SR of bio-oil at 700 °C [29], whereas it starts to be slightly significant in the OSR of bio-oil at 750 °C [19], and therefore we may disregard the occurrence of Rh sintering in the bio-oil CSDR at 700 °C.

It is not straightforward to attribute the irreversible deactivation of this catalyst to the irreversible modifications of a sole catalyst component (either the metal (Rh) or the support (ZDC)). Focusing on Rh, the existence of this irreversible deactivation selectively affecting the CH_4 reforming was also observed in the SR of bio-oil over two different Rh catalysts, one supported on La_2O_3–αAl_2O_3 (Rh/LaAl) and another on ZDC (the one used in this work, Rh/ZDC), as shown in Figures S2 and S3 (Supplementary Materials) [30]. Both catalysts showed the deactivation periods observed for the bio-oil CSDR in this present work, with the Rh/LaAl catalyst being more stable than the Rh/ZDC catalyst in the SR reaction with each fresh catalyst. Nevertheless, the Rh/LaAl catalyst suffers a more severe irreversible deactivation than the commercial Rh/ZDC catalyst, evidenced by the shorter duration of the initial stable period in the SR reactions with the regenerated catalysts. This comparison suggests that the irreversible deactivation observed in the reforming of bio-oil may also be an intrinsic characteristic of Rh regardless of the support.

Regarding the support, the role of CeO_2 in the catalyst deactivation can be explained by various phenomena. Primarily, it is important to consider that the CeO_2 structure may collapse when subject to high temperature treatments causing the loss of surface area [41,42], which is one of the observations of our present work, and it has been interpreted to cause an occlusion of Rh sites. However, another close effect would be the phenomenon of encapsulation of noble metals by CeO_2 at high temperatures, which is recognized as a cause of irreversible deactivation [31–33]. The possible encapsulation of Rh sites may be evidenced by the increase in the interaction between Rh sites and the ZDC support as demonstrated by the H_2-TPR analysis in this present work. Curiously, this phenomenon is a method for catalyst preparation [43], yet it requires the metal does not lose its catalytic properties. Nevertheless, our results suggest that the Rh sites may be irreversibly modified and partially lose their catalytic properties when they are encapsulated by ZDC support. Considering that Rh sites have specific surfaces that are active for the CH_4 or CO_2 activation [44], this partial irreversible activity loss may occur through a simple preferential encapsulation of CeO_2 on the Rh surfaces that are active for the CH_4 and CO_2 activations, leading to the irreversible blockage of those specific active sites. Another possibility is a modification of the Rh crystallographic properties, thus changing the Rh active surfaces. For example, the loss of activity for the CO_2 methanation with a Ni/CeO_2 catalyst was attributed to the modification of Ni nanocrystals from hexagons to quasi-spheres upon partial encapsulation by CeO_2 [45]. As previously commented, this encapsulation and modification of Rh sites would occur in the first period of the reaction with the fresh catalyst and therefore is favored under reaction conditions (presence of various reactant species) as the previous reduction treatment (with H_2) seems not to have a significant effect on the Rh sites. Moreover, the modification of the temperature of catalyst regeneration by coke combustion will require detailed studies, as it cannot be ruled out that the catalyst sintering will be affected by the generation of hot spots (favored by the oxidizing activity of the catalyst).

4. Materials and Methods

The Rh supported on zirconium-doped ceria (Rh/ZDC) catalyst, with 2 wt% Rh, was supplied by Fuel Cell Materials. The catalyst samples (fresh, spent, and regenerated) were analyzed using N_2 physisorption, temperature-programmed reduction (H_2-TPR), temperature-programmed oxidation (TPO), scanning electron microscopy (SEM) using backscattered electron (BSE) or secondary electron (SE) detectors, transmission electron microscopy (TEM), or X-ray diffraction (XRD).

The N_2 physisorption equilibria measurements were carried out in Micromeritics ASAP 2010 analyzer at 77 K, and the data were treated with the BET, t-plot, and BJH methods to determine the specific surface area, volume, and average pore diameter. The H_2-TPR was performed in a TA Instruments SDT 650 thermogravimetric analyzer, and the procedure consisted of reducing in a $H_2/Ar/N_2$ flow (100 mL min^{-1}) while increasing the temperature at 5 °C min^{-1} from 25 °C to 700 °C. Prior to the reduction, the deposited coke was removed by combustion in an O_2/N_2 flow at 700 °C. The TPO measurements were carried out in a thermogravimetric analyzer (TA Instruments, Q5000, New Castle, DE, USA) coupled with a mass spectrometer (Balzers Instruments, ThermoStar GSD 300, Balzers, Liechtenstein) for monitoring the CO_2 signal, and the procedure consisted of heating up the sample at 5 °C min^{-1} up to 700 °C in a O_2/N_2 flow. The SEM images were collected with a field emission gun scanning electron microscope (Hitachi, S-4800 N, Santa Clara, CA, USA) with an accelerating voltage of 5 kV and a SE detector, and a microscope with an accelerating voltage of 15 kV using a BSE detector (Hitachi, S-3400N, Santa Clara, CA, USA). For the TEM analysis, the samples were crushed and dispersed in ethanol and a drop of the dispersion was placed on a grid covered with a carbon film, and, after allowing to dry, the TEM images were obtained with a transmission electron microscope (JEOL, 1400 Plus, Tokyo, Japan) using an accelerating voltage of 100 kV. The XRD patterns were carried out in a Bruker D8 Advance diffractometer (Tokyo, Japan) with a Cu Kα1 radiation (wavelength of 1.5406 Å) corresponding to an X-ray tube with a Cu anticathode.

The bio-oil was supplied by BTG Bioliquids BV (Enschede, The Netherlands) and was obtained by flash pyrolysis of pine sawdust in a plant provided with a conical rotatory reactor. The chemical composition was determined by gas chromatography/mass spectrometry analysis (Shimadzu, GC/MS-QP2010S, Carlsbad, CA, USA) provided with a BPX-5 column (length of 50 m, diameter of 0.22 m and thickness of 0.25 µm) and a mass selective detector. The identification of the compounds was carried out by comparison with the pattern spectra available in the NIST 147 and NIST 27 libraries. The detailed composition is described elsewhere [46], mainly containing acetic acid (16.6 wt%), levoglucosane (11.1 wt%), guaiacol (11.1 wt%), and acetol (9.4 wt%). The water content (23 wt%) was determined by Karl Fischer volumetric valorization (Metrohm, KF Titrino Plus 870, Herisau, Switzerland). The empirical formula ($C_{3.9}H_{6.9}O_{2.9}$) was obtained by CHO analysis in a Leco CHN-932 analyzer (water-free basis).

The SR and CSDR reaction tests were carried out in a continuous two-unit reaction equipment (MicroActivity Reference from PID Eng&Tech, Madrid, Spain) as described elsewhere [12]. The first unit is a U-shaped steel tube for the vaporization of bio-oil and the controlled deposition of pyrolytic lignin (PL) formed by repolymerization of some oxygenates (mainly phenolic compounds) during the vaporization of bio-oil at 500 °C [47]. The vaporized oxygenates enter the second unit, consisting of a stainless steel fluidized bed reactor. The reaction components (H_2, CO, CO_2, H_2O, CH_4, light hydrocarbons C_2–C_4) were analyzed by gas chromatography (Varian CP-490 Micro GC with three chromatographic columns, described elsewhere [48]) connected to the reactor through an insulated line (130 °C) to avoid condensation. The bio-oil was fed at 0.06 mL min^{-1} by using a Harvard Apparatus 22 injection pump and was mixed with CO_2 and/or N_2 at the entrance of the first unit. The catalyst was mixed with SiC to provide a minimum height/diameter ratio of 2 in the reactor and to improve the heat dissipation. The space time was defined as the ratio between the catalyst mass and mass flow rate of oxygenates in the bio-oil, and the value was set at 0.125 $g_{catalyst}$ h $(g_{oxygenates})^{-1}$. This low value is adequate for the study of catalyst deactivation in a short reaction time, which is required to avoid the plugging of the first unit (for bio-oil vaporization) by accumulation of the PL. All the reaction tests were carried out at 700 °C. Prior to each reaction, the catalyst was reduced in the reactor itself at 700 °C under H_2/N_2 flow (10% H_2). For the SR reaction, bio-oil was only mixed with N_2, and the S/C ratio is that provided by the H_2O content in the bio-oil (S/C = 0.5). For the CSDR reaction, the bio-oil was mixed with both CO_2 and N_2, with a CO_2/C ratio of 0.6 and S/C ratio of 0.5 (provided by the H_2O content in the bio-oil). N_2 was used as an internal pattern and the flow rates of the reaction components were calculated using the chromatographic data. The deactivated catalyst was regenerated in the reactor itself by coke combustion in air stream (100 mL min^{-1}) at 600 °C for 4 h, which assured total coke removal. A second reaction was performed with the regenerated catalyst under the same conditions as the first reaction to analyze the activity recovery.

The yields of products (Equations (14)–(17)) were calculated referring to the moles of C contained in the bio-oil, rather than the total carbon supplied (from both the bio-oil and CO_2) [5,12], which helps to relate the yields to the conversion of oxygenates in the bio-oil. Consequently, in the calculation of the CO_2 yield, it is necessary to substract the moles of CO_2 fed with the bio-oil to avoid overestimating its yield relative to the oxygenates supplied.

$$Y_{H2} = \frac{F_{H2,out}}{F^0_{H2}} \tag{14}$$

$$Y_i = \frac{F_{i,out}}{F_{C,in}} \text{ for } i = \text{CO, CH}_4, \text{ hydrocarbons} \tag{15}$$

$$Y_{CO2} = \frac{F_{CO2,out} - F_{CO2,in}}{F_{C,in}} \tag{16}$$

$$Y_{syngas} = \frac{F_{H2,out} + F_{CO,out}}{F^0_{syngas}} \tag{17}$$

In the above equations, $F_{H2,out}$ is the molar flow rate of H_2 in the outlet stream, $F_{i,out}$ is the molar flow rate in C basis of the carbonaceous products (except CO_2) in the outlet stream, $F_{C,in}$ is the molar flow rate of oxygenates in C basis in the inlet stream (subtracting the C deposited as PL), $F_{CO2,out}$ and $F_{CO2,in}$ are the molar flow rate of CO_2 in the reactor outlet and inlet, respectively, and $F_{CO,out}$ is the molar flow rate of CO in the outlet stream. Additionally, F^0_{H2} and F^0_{syngas} are the stoichiometric amounts of H_2 and syngas, respectively, that would be formed per mole of C in the oxygenate feed in the SR reaction, which, according to the stoichiometry of Equations (4) (for H_2) and (3) (for syngas), are:

$$F^0_{H2} = F^0_{syngas} = \frac{2n + \frac{m}{2} - k}{n} F_{C,in}, \tag{18}$$

where n, m, and k are the atoms in the generic empirical formula for oxygenates ($C_nH_mO_k$).

The CO_2 conversion was calculated as:

$$X_{CO2} = \frac{F_{CO2,in} - F_{CO2,out}}{F_{CO2,in}}, \tag{19}$$

The conversion of oxygenates entering the reforming reactor is estimated by the total yield of carbonaceous gas products

$$X_{oxygenates} = \frac{F_{C,out} - F_{CO2,in}}{F_{in}} = \sum Y_i \tag{20}$$

where $F_{C,out}$ is the total molar flow rate in C basis of the gas carbon products in the outlet stream. The above definition can be used when the coke deposition is low and therefore the coke yield can be considered negligible (as occurs in the conditions of this study).

5. Conclusions

The Rh/ZDC catalyst is highly active for the combined steam/dry reforming of bio-oil, producing a syngas yield close to 90% with a H_2/CO ratio of 1.2 (suitable for synthesis of useful chemicals) at 700 °C with CO_2/CO ratio of 0.6, S/C ratio of 0.5 (as a result of the water content in the bio-oil), and low space time of 0.125 $g_{catalyst}$ h $(g_{oxygenates})^{-1}$. However, it undergoes deactivation, evidenced by the decrease in syngas yields over time on stream in two well-distinguished deactivation periods. The first deactivation period involves the complete loss of activity for CH_4 reforming reactions (both SR or DR) and partially for the r-WGS reaction. The second period leads to the complete catalyst deactivation for the reforming of oxygenates and hydrocarbons. The first period corresponds to an irreversible deactivation, and, therefore, it is not observed after catalyst regeneration, whereas the second period is a reversible deactivation, as the activity of the catalyst at the beginning of this period is recovered after the regeneration of the catalyst.

The reversible deactivation is associated with the formation and deposition of coke blocking access to the active sites, and its removal by combustion allows for the catalyst to partially recover its activity. The irreversible catalyst deactivation is generated by changes to the Rh sites and to the textural properties of the ZDC support. Thus, the regenerated catalyst does not recover the specific surface area, which is probably a consequence of the collapse of some pores due to the high temperatures during the reaction and regeneration. Likewise, the H_2-TPR of the regenerated catalyst evidences a change in the Rh species, increasing those with high interactions with the support, which possibly suggests the partial encapsulation of Rh sites by CeO_2.

This work corroborates the irreversible deactivation of Rh catalysts for different reforming strategies of complex feeds, such as bio-oil, which has also been observed for other Rh supported catalysts. Thus, although Rh catalysts are very active, their use for bio-oil reforming may not be practical because they might not meet the need for a good regeneration capacity due to irreversible deactivation.

Supplementary Materials: The following supporting information can be downloaded at: https://www.mdpi.com/article/10.3390/catal14090571/s1, Figure S1: Performance of the Rh/ZDC catalyst in the SR of bio-oil at 700 °C, H_2O/C ratio of 0.5/1, and space time of 0.125 $g_{catalyst}$ h $(g_{oxygenates})^{-1}$: (a) Evolution of the product yields over time on stream; (b) evolution of the syngas yield and H_2/CO ratio over time on stream; Figure S2: Evolution with time on stream (TOS) of bio-oil conversion and H_2 yield in two reaction steps of raw bio-oil SR with intermediate regeneration (coke combustion with air) over Rh catalysts supported on zirconia-doped ceria (ZDC) or La_2O_3-αAl_2O_3 (LaAl). Reaction conditions: 700°C; S/C, 6; 0.15 $g_{catalyst}$ h $(g_{oxygenates})^{-1}$; Figure S3: Evolution with time on stream (TOS) of CH_4 and hydrocarbons (HCs) yields in two reaction steps of raw bio-oil SR with intermediate regeneration (coke combustion with air) over Rh catalyst supported on La_2O_3-αAl_2O_3. Reaction conditions: 700 °C; S/C, 6; space time, 0.15 $g_{catalyst}$ h $(g_{oxygenates})^{-1}$.

Author Contributions: Conceptualization, J.V., J.B. and A.G.G.; methodology, L.L.; formal analysis, L.L. and J.V.; investigation, L.L. and J.V.; resources, A.G.G.; data curation, L.L. and A.G.G.; writing—original draft preparation, J.V. and G.E.; writing—review and editing, A.R., J.B. and A.G.G.; visualization, L.L., J.V. and G.E.; supervision, A.R. and A.G.G.; project administration, A.G.G.; funding acquisition, A.G.G. All authors have read and agreed to the published version of the manuscript.

Funding: This research was funded by the Ministry of Science and Innovation of the Spanish Government, grant number PID2021-127005OB-I00, funded by MCIN/AEI/10.13039/501100011033 and by "ERDF A way of making Europe"; the European Commission (HORIZON H2020-MSCA RISE 2018), contract number 823745; the Department of Education, Universities and Investigation of the Basque Government, grant number IT1645-22 and PhD grant number PRE_2021_2_0147 for L.L.

Data Availability Statement: Available upon request.

Acknowledgments: The authors are thankful for the technical and human support provided by SGIker (UPV/EHU/ERDF, EU).

References

1. Bachmann, M.; Völker, S.; Kleinekorte, J.; Bardow, A. Syngas from What? Comparative Life-Cycle Assessment for Syngas Production from Biomass, CO_2, and Steel Mill Off-Gases. *ACS Sustain. Chem. Eng.* **2023**, *11*, 5356–5366. [CrossRef]
2. García-Gómez, N.; Valecillos, J.; Valle, B.; Remiro, A.; Bilbao, J.; Gayubo, A.G. Combined effect of bio-oil composition and temperature on the stability of Ni spinel derived catalyst for hydrogen production by steam reforming. *Fuel* **2022**, *326*, 124966. [CrossRef]
3. Baloch, H.A.; Nizamuddin, S.; Siddiqui, M.T.H.; Riaz, S.; Jatoi, A.S.; Dumbre, D.K.; Mubarak, N.M.; Srinivasan, M.P.; Griffin, G.J. Recent advances in production and upgrading of bio-oil from biomass: A critical overview. *J. Environ. Chem. Eng.* **2018**, *6*, 5101–5118. [CrossRef]
4. Landa, L.; Remiro, A.; De Torre, R.; Aguado, R.; Bilbao, J.; Gayubo, G. Global vision from the thermodynamics of the effect of the bio-oil composition and the reforming strategies in the H 2 production and the energy requirement. *Energy Convers. Manag.* **2021**, *239*, 114181. [CrossRef]
5. Landa, L.; Remiro, A.; Valecillos, J.; Bilbao, J.; Gayubo, A.G. Thermodynamic study of the CO_2 valorization in the combined steam-dry reforming of bio-oil into syngas. *J. CO2 Util.* **2023**, *72*, 102503. [CrossRef]
6. Zhao, Z.; Situmorang, Y.A.; An, P.; Chaihad, N.; Wang, J.; Hao, X.; Xu, G.; Abudula, A.; Guan, G. Hydrogen production from catalytic steam reforming of bio-oils: A critical review. *Chem. Eng. Technol.* **2020**, *43*, 625–640. [CrossRef]
7. Silva, J.; Rocha, C.; Soria, M.A.; Madeira, L.M. Catalytic Steam Reforming of Biomass-Derived Oxygenates for H2 Production: A Review on Ni-Based Catalysts. *ChemEngineering* **2022**, *6*, 39. [CrossRef]
8. Pafili, A.; Charisiou, N.D.; Douvartzides, S.L.; Siakavelas, G.I.; Wang, W.; Liu, G.; Papadakis, V.G.; Goula, M.A. Recent progress in the steam reforming of bio-oil for hydrogen production: A review of operating parameters, catalytic systems and technological innovations. *Catalysts* **2021**, *11*, 1526. [CrossRef]
9. Setiabudi, H.D.; Aziz, M.A.A.; Abdullah, S.; Teh, L.P.; Jusoh, R. Hydrogen production from catalytic steam reforming of biomass pyrolysis oil or bio-oil derivatives: A review. *Int. J. Hydrogen Energy* **2020**, *45*, 18376–18397. [CrossRef]
10. Zhang, C. Review of catalytic reforming of biomass pyrolysis oil for hydrogen production. *Front. Chem.* **2022**, *10*, 1–6. [CrossRef]
11. Kumar, R.; Strezov, V. Thermochemical production of bio-oil: A review of downstream processing technologies for bio-oil upgrading, production of hydrogen and high value-added products Oil in Water. *Renew. Sustain. Energy Rev.* **2021**, *135*, 110152. [CrossRef]
12. Landa, L.; Remiro, A.; Valecillos, J.; Bilbao, J.; Gayubo, A.G. Syngas production through combined steam-dry reforming of raw bio-oil over a $NiAl_2O_4$ spinel derived catalyst. *J. CO2 Util.* **2023**, *78*, 102637. [CrossRef]
13. Hu, X.; Lu, G. Bio-oil steam reforming, partial oxidation or oxidative steam reforming coupled with bio-oil dry reforming to eliminate CO_2 emission. *Int. J. Hydrogen Energy* **2010**, *35*, 7169–7176. [CrossRef]
14. Qingli, X.; Peng, F.; Wei, Q.; Kai, H.; Shanzhi, X.; Yongjie, Y. Catalyst deactivation and regeneration during CO_2 reforming of bio-oil. *Int. J. Hydrogen Energy* **2019**, *44*, 10277–10285. [CrossRef]
15. Rioche, C.; Kulkarni, S.; Meunier, F.C.; Breen, J.P.; Burch, R. Steam reforming of model compounds and fast pyrolysis bio-oil on supported noble metal catalysts. *Appl. Catal. B Environ.* **2005**, *61*, 130–139. [CrossRef]
16. Rennard, D.; French, R.; Czernik, S.; Josephson, T.; Schmidt, L. Production of synthesis gas by partial oxidation and steam reforming of biomass pyrolysis oils. *Int. J. Hydrogen Energy* **2010**, *35*, 4048–4059. [CrossRef]
17. Remiro, A.; Arandia, A.; Bilbao, J.; Gayubo, A.G. Comparison of Ni based and Rh based catalyst performance in the oxidative steam reforming of raw bio-oil. *Energy Fuels* **2017**, *31*, 7147–7156. [CrossRef]
18. Arandia, A.; Remiro, A.; Oar-Arteta, L.; Bilbao, J.; Gayubo, A.G. Reaction conditions effect and pathways in the oxidative steam reforming of raw bio-oil on a Rh/CeO_2-ZrO_2 catalyst in a fluidized bed reactor. *Int. J. Hydrogen Energy* **2017**, *42*, 29175–29185. [CrossRef]
19. Remiro, A.; Arandia, A.; Oar-Arteta, L.; Bilbao, J.; Gayubo, A.G. Stability of a Rh/CeO_2-ZrO_2 Catalyst in the Oxidative Steam Reforming of Raw Bio-oil. *Energy Fuels* **2018**, *32*, 3588–3598. [CrossRef]
20. Iojoiu, E.E.; Domine, M.E.; Davidian, T.; Guilhaume, N.; Mirodatos, C. Hydrogen production by sequential cracking of biomass-derived pyrolysis oil over noble metal catalysts supported on ceria-zirconia. *Appl. Catal. Gen.* **2007**, *323*, 147–161. [CrossRef]
21. Domine, M.E.; Iojoiu, E.E.; Davidian, T.; Guilhaume, N.; Mirodatos, C. Hydrogen production from biomass-derived oil over monolithic Pt- and Rh-based catalysts using steam reforming and sequential cracking processes. *Catal. Today* **2008**, *133–135*, 565–573. [CrossRef]
22. Ranjekar, A.M.; Yadav, G.D. Dry reforming of methane for syngas production: A review and assessment of catalyst development and efficacy. *J. Indian Chem. Soc.* **2021**, *98*, 100002. [CrossRef]

23. Wang, C.; Wang, Y.; Chen, M.; Liang, D.; Yang, Z.; Cheng, W.; Tang, Z.; Wang, J.; Zhang, H. Recent advances during CH_4 dry reforming for syngas production: A mini review. *Int. J. Hydrogen Energy* **2021**, *46*, 5852–5874. [CrossRef]

24. Fan, M.S.; Abdullah, A.Z.; Bhatia, S. Catalytic technology for carbon dioxide reforming of methane to synthesis gas. *ChemCatChem* **2009**, *1*, 192–208. [CrossRef]

25. da Silva, A.M.; de Souza, K.R.; Jacobs, G.; Graham, U.M.; Davis, B.H.; Mattos, L.V.; Noronha, F.B. Steam and CO_2 reforming of ethanol over Rh/CeO_2 catalyst. *Appl. Catal. B Environ.* **2011**, *102*, 94–109. [CrossRef]

26. Díaz López, E.; Comas-Vives, A. CO_2 activation dominating the dry reforming of methane catalyzed by Rh(111) based on multiscale modelling. *Catal. Sci. Technol.* **2023**, *13*, 7162–7171. [CrossRef]

27. Kartavova, K.E.; Mashkin, M.Y.; Kostin, M.Y.; Finashina, E.D.; Kalmykov, K.B.; Kapustin, G.I.; Pribytkov, P.V.; Tkachenko, O.P.; Mishin, I.V.; Kustov, L.M.; et al. Rhodium-Based Catalysts: An Impact of the Support Nature on the Catalytic Cyclohexane Ring Opening. *Nanomaterials* **2023**, *13*, 936. [CrossRef]

28. Mashkin, M.Y.; Tedeeva, M.A.; Fedorova, A.A.; Fatula, E.R.; Egorov, A.V.; Dvoryak, S.V.; Maslakov, K.I.; Knotko, A.V.; Baranchikov, A.E.; Kapustin, G.I.; et al. Synthesis of $Ce_xZr_{1-x}O_2/SiO_2$ supports for chromium oxide catalysts of oxidative dehydrogenation of propane with carbon dioxide. *J. Chem. Technol. Biotechnol.* **2023**, *98*, 1247–1259. [CrossRef]

29. Remiro, A.; Ochoa, A.; Arandia, A.; Castaño, P.; Bilbao, J.; Gayubo, A.G. On the dynamics and reversibility of the deactivation of a Rh/CeO_2–ZrO_2 catalyst in raw bio-oil steam reforming. *Int. J. Hydrogen Energy* **2019**, *44*, 2620–2632. [CrossRef]

30. García-Gómez, N.; Valle, B.; Remiro, A.; Bilbao, J.; Gayubo, A.G. The effect of the support in the deactivation and regeneration of Rh catalysts used in the steam reforming of bio-oil. In Proceedings of the 3rd International Congress Chemistry & Environment ANQUE-ICCE3, Santander, Spain, 19–21 June 2019.

31. Machida, M.; Yoshida, H.; Kamiuchi, N.; Fujino, Y.; Miki, T.; Haneda, M.; Tsurunari, Y.; Iwashita, S.; Ohta, R.; Yoshida, H.; et al. Thermal Aging of Rh/ZrO_2-CeO_2 Three-Way Catalysts under Dynamic Lean/Rich Perturbation Accelerates Deactivation via an Encapsulation Mechanism. *ACS Catal.* **2023**, *13*, 3806–3814. [CrossRef]

32. Machida, M.; Yoshida, H.; Kamiuchi, N.; Fujino, Y.; Miki, T.; Haneda, M.; Tsurunari, Y.; Iwashita, S.; Ohta, R.; Yoshida, H.; et al. Rh Nanoparticles Dispersed on ZrO_2-CeO_2 Migrate to Al_2O_3 Supports to Mitigate Thermal Deactivation via Encapsulation. *ACS Appl. Nano Mater.* **2023**, *6*, 9805–9815. [CrossRef]

33. Craciun, R.; Daniell, W.; Knözinger, H. The effect of CeO_2 structure on the activity of supported Pd catalysts used for methane steam reforming. *Appl. Catal. A Gen.* **2002**, *230*, 153–168. [CrossRef]

34. He, L.; Liao, G.; Li, H.; Ren, Q.; Hu, S.; Han, H.; Xu, J.; Jiang, L.; Su, S.; Wang, Y.; et al. Evolution characteristics of different types of coke deposition during catalytic removal of biomass tar. *J. Energy Inst.* **2020**, *93*, 2497–2504. [CrossRef]

35. García-Gómez, N.; Valecillos, J.; Remiro, A.; Valle, B.; Bilbao, J.; Gayubo, A.G. Effect of reaction conditions on the deactivation by coke of a $NiAl_2O_4$ spinel derived catalyst in the steam reforming of bio-oil. *Appl. Catal. B Environ.* **2021**, *297*, 120445. [CrossRef]

36. Gong, X.; Guo, Z.; Wang, Z. Reactivity of pulverized coals during combustion catalyzed by CeO_2 and Fe_2O_3. *Combust. Flame* **2010**, *157*, 351–356. [CrossRef]

37. Arandia, A.; Remiro, A.; García, V.; Castaño, P.; Bilbao, J.; Gayubo, A. Oxidative Steam Reforming of Raw Bio-Oil over Supported and Bulk Ni Catalysts for Hydrogen Production. *Catalysis* **2018**, *8*, 322. [CrossRef]

38. Sagar Vijay Kumar, P.; Suresh, L.; Vinodkumar, T.; Reddy, B.M.; Chandramouli, G.V.P. Zirconium Doped Ceria Nanoparticles: An Efficient and Reusable Catalyst for a Green Multicomponent Synthesis of Novel Phenyldiazenyl-Chromene Derivatives Using Aqueous Medium. *ACS Sustain. Chem. Eng.* **2016**, *4*, 2376–2386. [CrossRef]

39. Fedorova, E.A.; Kardash, T.Y.; Kibis, L.S.; Stonkus, O.A.; Slavinskaya, E.M.; Svetlichnyi, V.A.; Pollastri, S.; Boronin, A.I. Unraveling the low-temperature activity of Rh-CeO_2 catalysts in CO oxidation: Probing the local structure and Red-Ox transformation of Rh^{3+} species. *Phys. Chem. Chem. Phys.* **2022**, *25*, 2862–2874. [CrossRef]

40. Landa, L.; Remiro, A.; Valecillos, J.; Valle, B.; Bilbao, J.; Gayubo, A.G. Performance of $NiAl_2O_4$ spinel derived catalyst + dolomite in the sorption enhanced steam reforming (SESR) of raw bio-oil in cyclic operation. *Int. J. Hydrogen Energy* **2024**, *58*, 1526–1540. [CrossRef]

41. Lee, J.; Kang, S.; Lee, E.; Kang, M.; Sung, J.; Kim, T.J.; Christopher, P.; Park, J.; Kim, D.H. Aggregation of CeO_2 particles with aligned grains drives sintering of Pt single atoms in Pt/CeO_2 catalysts. *J. Mater. Chem. A* **2022**, *10*, 7029–7035. [CrossRef]

42. Li, J.; Du, C.; Feng, Q.; Zhao, Y.; Liu, S.; Xu, J.; Hu, M.; Zeng, Z.; Zhang, Z.; Shen, H.; et al. Evolution and performances of Ni single atoms trapped by mesoporous ceria in Dry Reforming of Methane. *Appl. Catal. B Environ.* **2024**, *354*, 124069. [CrossRef]

43. Song, S.; Wang, X.; Zhang, H. CeO_2-encapsulated noble metal nanocatalysts: Enhanced activity and stability for catalytic application. *NPG Asia Mater.* **2015**, *7*, e179. [CrossRef]

44. Wei, J.; Iglesia, E. Structural requirements and reaction pathways in methane activation and chemical conversion catalyzed by rhodium. *J. Catal.* **2004**, *225*, 116–127. [CrossRef]

45. Li, M.; Amari, H.; van Veen, A.C. Metal-oxide interaction enhanced CO_2 activation in methanation over ceria supported nickel nanocrystallites. *Appl. Catal. B Environ.* **2018**, *239*, 27–35. [CrossRef]

46. Landa, L.; Remiro, A.; Valecillos, J.; Valle, B.; Bilbao, J.; Gayubo, A.G. Unveiling the deactivation by coke of $NiAl_2O_4$ spinel derived catalysts in the bio-oil steam reforming: Role of individual oxygenates. *Fuel* **2022**, *321*, 124009. [CrossRef]

47. Remiro, A.; Valle, B.; Aguayo, A.T.; Bilbao, J.; Gayubo, A.G. Steam reforming of raw bio-oil in a fluidized bed reactor with prior separation of pyrolytic lignin. *Energy Fuels* **2013**, *27*, 7549–7559. [CrossRef]
48. Landa, L.; Valecillos, J.; Remiro, A.; Valle, B.; Bilbao, J.; Gayubo, A.G. Comparison of the $NiAl_2O_4$ derived catalyst deactivation in the steam reforming and sorption enhanced steam reforming of raw bio-oil in packed and fluidized-bed reactors. *Chem. Eng. J.* **2023**, *458*, 141494. [CrossRef]

Article

Ni-Ag Catalysts for Hydrogen Production through Dry Reforming of Methane: Characterization and Performance Evaluation

Hayat Henni [1,2], Rafik Benrabaa [3,4], Pascal Roussel [5] and Axel Löfberg [5,*]

1   Institut Algérien du Pétrole/Sonatrach, Avenue 1er Novembre, Boumerdes 35000, Algeria; hayat.henni@sonatrach.dz
2   Département de Génie des Procédés, Faculté de Technologie, Université 20 août 1955, Route d'El Hadaiek, Skikda 21000, Algeria
3   Laboratoire de Physico-Chimie des Matériaux, Faculté des Sciences et de la Technologie, Université Chadli Bendjedid, El-Tarf 36000, Algeria; r.benrabaa@univ-eltarf.dz
4   Laboratoire de Chimie des Matériaux, Catalyse et Environnement, Faculté de Chimie, Université des Sciences et de la Technologie Houari Boumediene (USTHB), El-Alia, Bab Ezzouar 16111, Algeria
5   Univ. Lille, CNRS, Centrale Lille, Univ. Artois, UMR 8181—UCCS—Unité de Catalyse et Chimie du Solide, F-59000 Lille, France; pascal.roussel@univ-lille.fr
*   Correspondence: axel.lofberg@univ-lille.fr; Tel.: +33-03-20-43-45-27

Abstract: To investigate the influence of Ag and the loading of Ni species, Ni-Ag type catalysts were synthesized with varying Ni/Ag ratios (1, 1.5 and 2) using the coprecipitation method. The catalysts were extensively characterized using various techniques such as TG-DSC-SM, XRD, ICP, BET, SEM-EDX and TPR and subsequently tested in the CH_4/CO_2 reaction without any pretreatment. Regardless of the ratio employed, a phase mixture containing NiO and Ag was observed after calcination under air between 600 °C and 1200 °C. SEM analysis confirmed the presence of a close interface between Ag and NiO. The specific surface area was found to be significantly higher for the catalyst with lower Ni content (R = 1). TPR analysis demonstrated that the inclusion of Ag facilitated the reduction of Ni at lower temperatures. XRD analyses of the spent catalyst confirmed catalyst reduction during the reaction. Among the samples, a catalyst with Ni/Ag = 1 exhibited superior catalytic activity without any pretreatment under a reduction atmosphere, in which case the conversions of methane and CO_2 at 650 °C amounted to 38 and 45 mol%, respectively, with H_2/CO = 0.7 and 71 mol% of H_2. The presence of Ag species enhances the stability of the Ni catalyst and improves catalytic performance in the dry reforming of methane.

Keywords: H_2 production; CO_2 reforming; CH_4; Ni-Ag catalyst; nickel loading

Citation: Henni, H.; Benrabaa, R.; Roussel, P.; Löfberg, A. Ni-Ag Catalysts for Hydrogen Production through Dry Reforming of Methane: Characterization and Performance Evaluation. *Catalysts* **2024**, *14*, 400. https://doi.org/10.3390/catal14070400

Academic Editors: Georgios Bampos, Paraskevi Panagiotopoulou, Guido Busca and Eleni A. Kyriakidou

Received: 5 April 2024
Revised: 15 June 2024
Accepted: 19 June 2024
Published: 25 June 2024

1. Introduction

Recently, there has been significant attention given to the issue of global warming caused by the greenhouse effect. This has led to increased attention on reducing and utilizing greenhouse gases like carbon dioxide and methane. One promising approach is the catalytic reforming of methane with carbon dioxide, called DRM (dry reforming of methane), to produce synthesis gas which is traditionally obtained through the steam reforming of methane [1]. DRM, a reaction first proposed by Fischer and Tropsch in 1928, is a suggested alternative to steam reforming [2] and its relevance will only grow over time [3]. The reforming of CH_4 with CO_2 offers several advantages, including the production of synthesis gas [4–7] with a lower hydrogen-to-carbon monoxide ratio and higher energy efficiency for hydrocarbon conversion [8–10]. Indeed, compared with commercial methane steam reforming (MSR) process that produces syngas with H_2/CO ratio of 3 [11]. In practice, for MSR processes, it is common to use an O/C (oxygen/carbon) ratio higher than 3 to prevent coke deposition. However, this approach comes with an energy penalty due to the need to heat an excessive amount of steam to approximately 800 °C. Additionally, the process of MSR encounters challenges related to corrosion and necessitates the inclusion of a desulfurization unit, as highlighted by Djinović et al. [12]. Furthermore, the DRM process offers cost advantages compared to other methods due to its ability to bypass the complex gas separation of products [13]. Despite the attractive incentives of carbon dioxide reforming of methane, there is currently no established industrial technology for this process. The formation of carbon deposits quickly deactivates conventional reforming catalysts, due to its endothermic nature [8,14]. Effectively, in comparison to steam reforming, the dry reforming of methane is recognized as the most highly endothermic reaction. This can be attributed to the fundamental difference in the oxidizing agents employed: CO_2, used in DRM, is inherently more stable compared to steam, which is utilized in SRM [15–19].

It is well known that the C-H bond in CH_4 is difficult to activate [20,21], while CO_2 is the utmost oxidized state of carbon, which is also very stable [22,23]. The co-activation of both the C-H bond in CH_4 and the C-O bond in CO_2 faced challenging difficulties. Furthermore, because of the thermodynamics limitation, DRM reaction was usually performed at high temperatures [24–28].

The reactions involved in DRM are listed in Table 1. Reaction (1) represents the DRM reactions involved in the formation of syngas with an H_2/CO ratio of 1:1. The main DRM reaction is favored at temperatures above 727 °C. However, if the reverse water–gas shift (RWGS) (reaction (2)) occurs simultaneously with the DRM reaction, it can lead to a decrease in the H_2/CO ratio. Instead of the RWGS reaction, other side reactions can occur in DRM, depending on the reaction conditions and feed ratio used. These side reactions include oxidative coupling of methane, dehydrogenation of ethane, hydrogenation of CO and CO_2 to ethanol, CO and CO_2 methanation and more. Carbon formation can occur through reactions 3, 4, 5 and 6. In summary, reaction 3 is dominant at higher temperatures (>600 °C), while the other three reactions (4, 5 and 6) tend to be more dominant at lower temperatures (<600 °C) as they are exothermic reactions [8,29,30]. The high temperatures typically result in rapid catalyst deactivation. The main factors contributing to this deactivation are the sintering of the active metal and the accumulation of stable carbon deposits [31]. The different pathways of coke formation mentioned can be understood through CHO ternary diagrams, showing that graphitic carbon formation is thermodynamically favored under complete dry reforming conditions. However, alternative carbons such as amorphous and filamentous carbon may also form, although they are less thermodynamically preferred [32].

Table 1. Reactions in DRM [29].

N°	Reaction Name	Reaction Equation	ΔH_{298K} (kJ/mol)
Main reaction			
1	Dry reforming of CH_4	$CH_4 + CO_2 \leftrightarrow 2CO + 2H_2$	+247
Side reactions			
2	Reverse water–gas shift	$CO_2 + H_2 \leftrightarrow CO + H_2O$	+41
3	Decomposition of CH_4	$CH_4 \leftrightarrow C + 2H_2$	+75
4	Disproportionation of CO	$2CO \leftrightarrow C + CO_2$	−172
5	Hydrogenation of CO_2	$CO_2 + 2H_2 \leftrightarrow C + 2H_2O$	−90
6	Hydrogenation of CO	$CO + H_2 \leftrightarrow C + H_2O$	−131

Considering the unavoidable thermodynamic carbon formation at high temperatures, the most promising approach to enhance catalyst stability might involve implementing a kinetic control strategy, such as designing a catalyst that minimizes carbon formation [8]. Catalysts based on noble metals such as Ru and Rh have demonstrated the highest catalytic activity and resistance to carbon deposition [8,33,34]. Nickel catalysts offer a promising alternative to noble metals, serving as effective catalysts for carbon dioxide reforming [35]. They exhibit a favorable combination of performance and cost, making them a focus of extensive research for the development of new and enhanced catalytic systems at a relatively lower cost. Coke formation and sintering of catalyst are the primary causes of catalyst deactivation that could lead to the low conversion of reactant [36]. Various strategies have been employed to inhibit sintering and coke formation. These include the deposition of nickel on a support (SiO_2 [37], Al_2O_3 [38] or MgO [39]), incorporating the active phase into well-defined structures, such as spinel [40–42], perovskite [43,44], pyrochlore [45] or hydrotalcite [46], and using a second metal as a promoter [47,48] to enhance the dispersion of metal particles on the catalyst surface. The incorporation of promoters enhances the system's robustness [3]. Among these metal promoters, silver has been found to exhibit desirable properties for resisting coke formation in alkane reforming reactions, as reported in references [49,50].

The utilization of Ag as a promoter has been proposed by Van et al. [49], demonstrating its ability to enhance long-term stability by forming surface alloys. Ag tends to nucleate on the step sites of the Ni surface, leading to the formation of Ag islands [51]. However, when heated above 800 K, the mobility of Ag can be increased, resulting in the complete blockage of steps [49,51]. In addition to its role in inhibiting coke deposition, Ag also accelerates the combustion of deposited coke. The promoting effect on coke gasification can be attributed to the fact that Ag destabilizes the formed coke by inhibiting its nucleation and growth, resulting in a reduced formation of graphitic or whisker carbons on Ag-promoted Ni surfaces.

Experimental observations conducted by Vang et al. [49] indicate that Ag preferentially occupies step sites on the Ni surface, leading to a reduced rate of C–C bond cleavage in ethylene hydrogenesis. Subsequently, Parizotto et al. [50] demonstrated that Ag, similar to gold (Au), effectively regulates coke formation on Ni/Al_2O_3 catalysts for methane steam reforming. Similar findings have been reported for Ag-promoted Ni/YSZ catalysts used in internal reforming for fuel cells, although gradual deactivation over time-on-stream has been observed [52]. The promoting effect of Ag in $Ni/MgAl_2O_4$ catalysts has also been explored by Jeong and Kang [53] for butane steam reforming and by Rovik et al. [54] for ethane steam reforming. Zhu et al. [55,56] conducted Density Functional Theory (DFT) calculations to gain insights into Ag's role in dry reforming of methane, suggesting that Ag can act as a promoter by inhibiting coke deposition.

According to Yu et al. [11], silver exhibited excellent stability of the catalyst, as it hindered the formation and growth of coke in a whisker or graphitic form, while promoting coke gasification on the catalyst surface. Indeed, nickel and silver are extensively employed in heterogeneous catalysis due to their favorable catalytic properties. Combining nickel and silver in a bimetallic catalyst, such as Ni-Ag, can offer advantages over monometallic catalysts. In fact, the literature reports that combining two metals to form bimetallic systems enhances catalytic performance, particularly in inhibiting carbon deposition during hydrogen production by reforming [40]. Several effects are attributed to bimetallic systems: (1) increased concentration of active sites; (2) electronic effects resulting from metal–metal contact, leading to better tolerance to carbon formation; (3) a sacrificial role played by one of the elements within the bimetallic system, allowing the second metal to remain available and not poisoned [57–61].

Therefore, it is important to investigate the catalytic performance of Ni-Ag catalysts to understand their impact on reaction kinetics.

The objective of this study was to develop an efficient catalyst for the dry reforming of methane. To achieve this, Ni-Ag catalysts were synthesized using the coprecipitation method, with the Ni/Ag ratio being systematically varied. The obtained compounds were subjected to comprehensive analysis to evaluate their textural, structural and surface properties through various techniques. The reactivity of the catalysts during the reduction process was assessed using hydrogen temperature-programmed reduction for the fresh catalysts, while X-ray diffraction was employed to analyze the reduced catalysts. Furthermore, the catalytic performance of the catalysts in the dry reforming of methane was thoroughly investigated.

2. Results

2.1. Characterization of Fresh Catalysts

2.1.1. Thermal Study (TG-DSC-MS and XRD) of Precursors

The thermal stability of the various precursors was assessed using a combination of techniques including thermogravimetry differential thermal analysis (TG-DSC) coupled with mass spectrometry (MS) and X-ray diffraction (XRD) on compounds annealed at various temperatures (600, 700, 1000 and 1200 °C). TG-DSC was carried out on fresh catalysts in the temperature range from ambient to 1000 °C. The TG-DSC curves (Figure 1) mainly exhibit three weight loss steps up to 600 °C. Peaks in heat flow were observed at 256 °C (endothermic phenomenon) and at 950 °C (unattributed heat flow peak) for the 2AN100 sample. Water and carbon dioxide are identified as the species responsible for weight loss. Mass spectrometry analysis confirms the presence of water ($m/z = 18$) and CO_2 ($m/z = 44$). The first step corresponds to the removal of physisorbed water on the material's surface, while the second step involves the release of interlayer water and carbon dioxide from the precursor's structure. No additional weight loss or significant changes are observed as the temperature continues to increase, suggesting that the material structure has reached a state of relative stability.

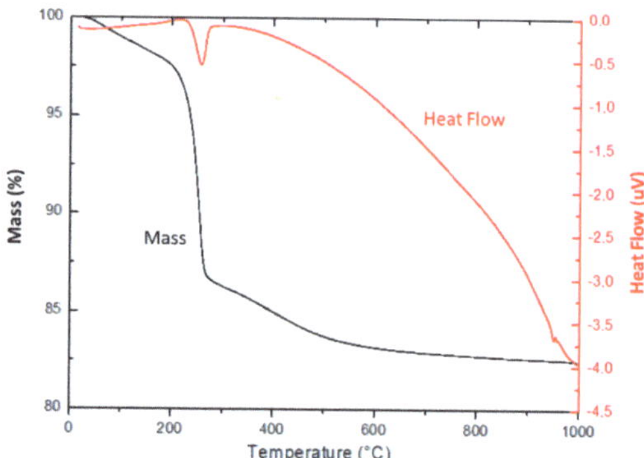

Figure 1. TG-DSC curves of 2NA100 precursors.

To analyze the structural changes in the materials with the temperature, various heat treatments were performed. This approach enabled the assessment of the impact of calcination temperature on the phase composition. The samples were calcined at temperatures of 600, 700, 1000 and 1200 °C (with a rate of 5 °C/min) under a constant airflow for 4 h and then analyzed by XRD. The XRD patterns (Figure 2) obtained for these samples reveal that whatever the annealing temperature, the main phases obtained after calcination are metallic silver (PDF file number 89-3722) and nickel oxide (PDF file number 47-1049). The catalyst's composition remains stable after calcination at different high temperatures.

From TG-TDA-MS, no weight loss is observed above 600 °C. Apart from the peaks corresponding to silver and nickel oxide, XRD analysis reveals that no other diffraction peaks are observed even up to 1200 °C. These findings suggest that the catalyst's composition remains stable at high temperatures. Considering that the DRM reaction occurs within a temperature range of 600 to 850 °C, a calcination temperature of 600 °C has been selected to preserve the structure and texture of the catalyst.

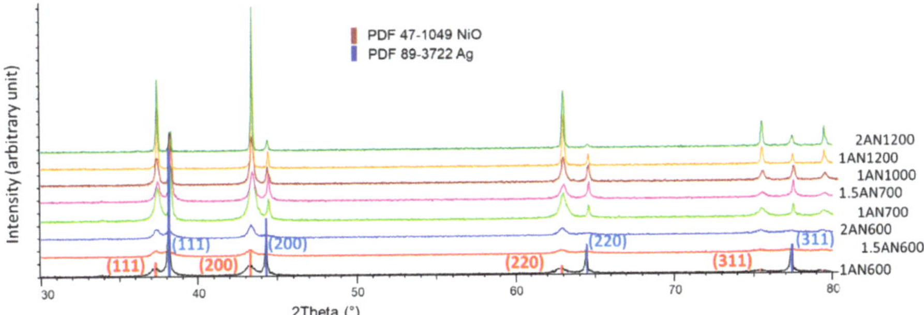

Figure 2. X-ray diffraction profiles of precursors calcined at high temperatures (600, 700, 1000 and 1200 °C).

2.1.2. Study of Stoichiometry of Fresh Catalysts (ICP-OES and EDX)

A study of the stoichiometry of fresh catalysts was conducted using ICP-OES and EDX spectroscopies. The results obtained are summarized in Table 2. As expected, the EDX analysis confirms the presence of silver (Ag) and nickel (Ni) elements. ICP-OES spectroscopy showed that the measured amounts of all analyzed species are slightly lower than the nominal ones used in the preparations (Table 2). The low amount of Ni species in all cases can be interpreted by the low temperature of calcination (600 °C/4 h) or by incomplete precipitation of $Ni(OH)_2$ during preparation synthesis. In contrast, for all three catalysts, the overall EDX analysis reveals a Ni/Ag mass ratio higher than the theoretical value, indicating a relatively higher concentration of nickel in the upper layers of the catalyst. Traces of sodium (Na) and silicon (Si) are detected by EDX in the 1NA600 and 1.5NA600 samples. Additionally, traces of aluminum (Al) and potassium (K) are only observed in the 1.5NA600 sample. The presence of aluminum is attributed to the aluminum sample holder used during the analysis. These impurities are typically present in small amounts and may not be detectable by X-ray diffraction (XRD). In the case of the 2NA600 sample, no signal corresponding to Na-Si-K is observed in the EDX spectra, suggesting that the analyzed regions are not contaminated by Na-Si-K.

Table 2. Composition, refined structural parameters and specific surface area of synthesized catalysts.

Catalyst	B.E.T.		XRD Data				EDX	ICP	XPS
	SSA (m^2/g)	Phase Ratio	Cs [a] (n m)	Cs [b] (nm)	Cs [c] (nm)	a [d] (Å)	Ni/Ag	Ni/Ag	Ni/Ag
1NA600	20	NiO (69) Ag (31)	8.7 (6) 70 (4)	Ni: 12.6 (2) Ag: 21.3 (2) NiO: 3.2 (1)	Ni: 30 (1) Ag: 20.8 (4)	4.177 (2) 4.0875 (5)	1.3	0.73	6.7
1.5NA600	3	NiO (71) Ag (29)	12.6 (5) 6.5 (3)	Ni: 13.1 (1) Ag: 12.7 (1) NiO: 7.4 (1)	Ni: 39 (3) Ag:21 (1)	4.177 (2) 4.092 (2)	1.9	0.96	6.5
2NA600	5	NiO (80) Ag (20)	16.8 (4) 7.5 (3)	Ni: 15.1 (1) Ag: 12.2 (1) NiO: 7.6 (2)	Ni: 35 (2) Ag: 16.3 (1)	4.1801 (7) 4.0917 (9)	3	1.51	6.6

[a] Crystallite size after calcination at 600 °C. [b] Crystallite size after reduction. [c] Crystallite size after the reaction. [d] Cell parameter, between brackets is a mass phase (%).

2.1.3. Structural Characterization of Fresh Catalysts (XRD)

As already pointed out, the diffractograms of the three compositions of the fresh catalysts (after heating at 600 °C) (Figure 2) show the presence of both NiO and Ag materials, with peaks at $2\theta = 37.24°$, $43.27°$, $62.87°$, $75.41°$ and $79.40°$ corresponding to the (111), (200), (220), (311) and (222) planes of the cubic NiO phase (PDF 47-1049) [62], while peaks at 38.12°, 44.307°, 64.45° and 77.41° belong to the (111), (200), (220) and (311) planes of metallic, cubic, Ag (PDF 89-3722). These results are in good agreement with the studies conducted by [63–66]. As shown in Figure 2, XRD patterns of the prepared catalysts exhibit similar diffractograms with slight differences. In order to obtain more information on the relative proportions of the phases present in the catalyst, a refinement using the Rietveld method was carried out on the three preparations annealed at 600 °C.

The Rietveld refinement was conducted using the JANA 2020 program to determine the lattice parameters, the crystallite size (using the fundamental parameter approach) and the relative amount of each crystalline phase. These parameters are given in Table 1, and the results on the Ni/Ag = 1 are illustrated in Figure 3. The crystallite size of NiO varies between 8.7 and 16.8 nm, indicating a difference among the samples. The results also indicate that as the Ni/Ag ratio increases, the crystallite size of NiO also increases. For all the catalysts, NiO is identified as the dominant phase, which aligns with the findings from the EDX analysis. The lattice parameters (a) of detected phases NiO and Ag are similar for all three catalysts. The relatively small crystallite size of NiO observed in the 1NA600 sample corresponds to its large surface area (20 m^2/g, Table 2).

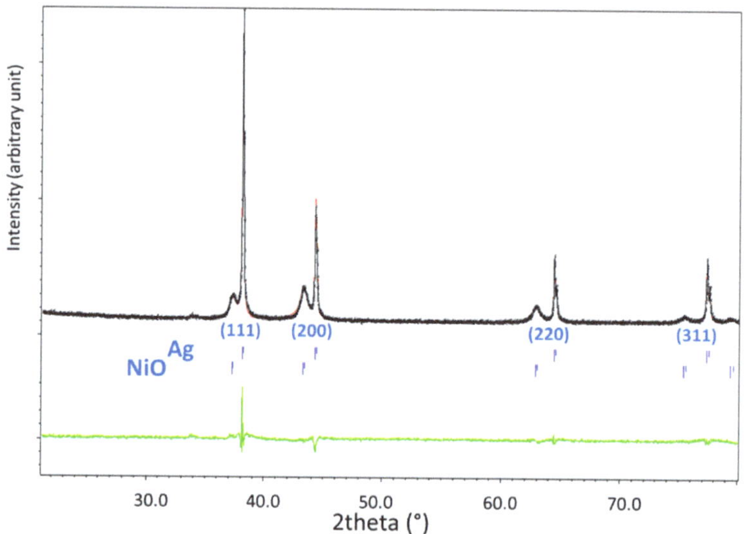

Figure 3. Rietveld refinement of the catalyst calcined at 600 °C (Ni/Ag = 1).

2.1.4. Textural Characterization of Fresh Catalysts (BET, SEM and XPS)

The specific surface area of the catalyst powder was determined through the application of the Brunauer–Emmett–Teller (B.E.T) method, which involves N$_2$ adsorption-desorption analysis, and the results are presented in Table 1. The morphology of the synthesized materials was assessed using scanning electron microscopy. The data S$_{BET}$ indicate a decrease in the specific surface area from 20 m^2/g for the 1NA600 sample to 3 and 5 m^2/g for the 2NA600 and 1.5NA600 samples, respectively. The results agree with the results of the Rietveld refinement analysis. As the surface area increases, the dimensions of the NiO crystallites decrease proportionally. This reduction in the BET surface area corresponds to an increase in the Ni/Ag ratio. This observation is consistent with the findings from the scanning electron microscopy (SEM) analysis and is supported by reference [67].

SEM images of the catalysts are presented in Figure 4, revealing noticeable variations in particle shapes. The contrast in the SEM image is due to the presence of both NiO and Ag phases. For the 1NA600 sample, the micrograph in Figure 4a displays the surface composed of Ag and NiO nanoparticles aggregated to form clusters of different sizes dispersed randomly. The formation of relatively larger spheres can be attributed to the tendency of NiO nanoparticles, which are antiferromagnetic in nature, to aggregate [68]. Regarding the 1.5NA600 sample, SEM micrographs in Figure 4b demonstrate that the powder consists of aggregates of nanoscale particles coexisting with larger-grain aggregates. Pronounced differences in particle shapes can be observed, and agglomeration is more prominent compared to the 1NA600 sample. For the 2NA600 sample, the surface morphology of NiO exhibits the random growth of nanocubes with varying dimensions, as shown in Figure 4c. Small and larger nanocubes are formed, consistent with Ajeet Singh's findings [69]. Figure 4c displays more agglomeration than the 1NA600 sample but less than the 1.5NA600 sample. In summary, it is evident that the Ni loading significantly influences the morphology of the catalysts.

Figure 4. SEM pictures of (**a**) 1NA600, (**b**) 1.5NA600 and (**c**) 2NA600.

The surface chemical properties of the 1NA600, 1.5NA600 and 2NA600 catalysts after calcination at 600 °C for 4 h under airflow were examined by XPS measurements. The binding energy for both species (Ag and Ni) does not vary practically (Figures 5 and 6), whatever the content used during the synthesis. The decomposition of XPS spectra (Figures 5 and 6) does not show any components, and the recorded binding energy values show clearly that Ag and Ni are in a unique oxidation state (only 1 oxidation state for each species), thus explaining the formation of the unique phases corresponding to each species (formation of a single phase for each species) as observed by XRD analysis. The binding energies of Ag 3d photopeaks are at 368.8 eV (1NA600), 368.7 eV (1.5NA600) and 368.5 eV(2NA600) and those of Ni 2p3/2 are at 854 eV (1NA600), 854.1 eV (1.5NA600) and 853.7 eV (2NA600). In the literature, the Ag 3d5/2 and Ag 3d3/2 core level binding energies appear at 368.1 and 374.1 eV, respectively, in good agreement with bulk silver metallic values [70]. In our case, the metallic bulk binding energy values for the Ag 3d core levels were obtained as expected. These results are in good agreement with the presence of fcc metallic silver structure obtained by XRD. For nickel species, the binding energy of Ni(II) was found at close values of 854.9 eV for NiO accompanied with a shoulder at ~1.2 eV [71,72]. This perfectly confirms that Ni(II+) is from the NiO phase and not from another phase that can form during the synthesis containing Ni species (in accordance with XRD data). As observed by EDX analysis, the Ni/Ag ratios (Table 2) were also higher for the three samples suggesting that the surface of 1NA600, 1.5NA600 and 2NA600 samples was enriched by Ni(II) species. This surface behavior is in good accordance with the large amounts of NiO phase estimated by Rietveld Refinement (Table 2).

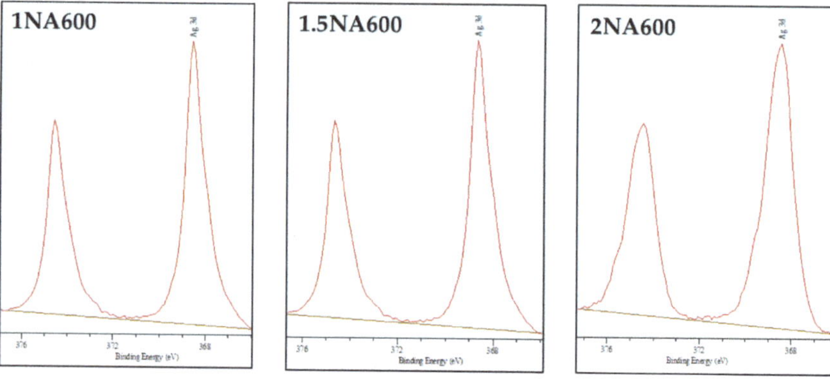

Figure 5. XPS spectra of Ag species of 1NA600, 1.5NA600 and 2NA600 catalysts.

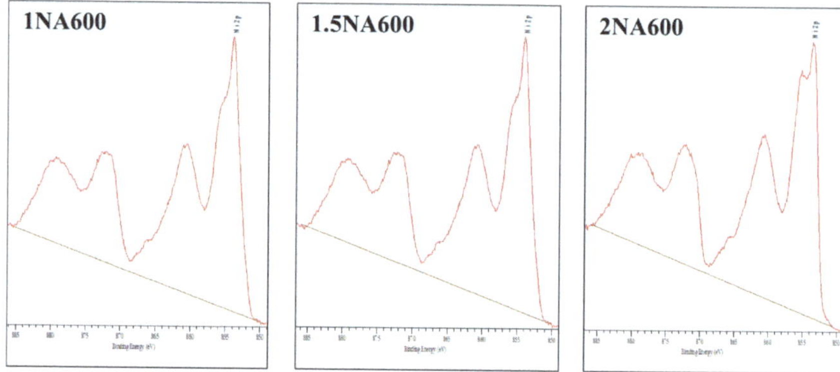

Figure 6. XPS spectra of Ni species of 1NA600, 1.5NA600 and 2NA600 catalysts.

2.1.5. Reducibility Studies of Fresh Catalysts (TPR and XRD after Reduction)

Reducibility studies were conducted using H_2 temperature-programmed reduction (TPR) of the fresh catalysts followed by X-ray diffraction (XRD) analysis for reduced catalysts. These experiments aimed to investigate the influence of Ag on the reducibility of Ni^{2+} ions in the catalysts. Figure 7 illustrates the hydrogen consumption as a function of temperature for the different systems studied. For all three samples, a slight difference in H_2 consumption was observed (~5–9 mmol/g). The TPR curves of the three catalysts exhibit also a similar pattern, showing two main peaks of hydrogen consumption between 100 and 400 °C. The first peak can be attributed to the reduction of nickel species located on the surface, as species on the surface tend to be reduced at lower temperatures compared to those in the volume. The second peak is related to the reduction of surface and/or surface-bulk NiO, representing Ni-reduction mechanisms [73–76]. Previous literature suggests that large NiO particles with lower interaction with other phases can be reduced at low temperatures [77–79]. The peak at 350–500 °C can be attributed to the reduction of NiO [80]. For the 1.5NA600 sample, a small peak observed at very high temperatures (~1000 °C) may be associated with residual species that were not completely eliminated during calcination [81,82]. This different behavior of 1.5NA by the appearance of a peak at high temperature compared to the other two solids, in particular to 1NA600, could be assigned to the reduction of sodium carbonates detected by EDX analysis as mentioned above; this phenomenon is also observed in our previous works [83]. The TPR curves indicate a two-step reduction of nickel, with the second peak showing higher intensity compared to the first peak. As Ni/Ag content increases, especially for the 1NA600 and

2NA600 samples, there is a slight shift towards higher temperatures in TPR peak positions, as well as an increase in peak area. This suggests that the presence of silver lowers the reduction temperature of nickel.

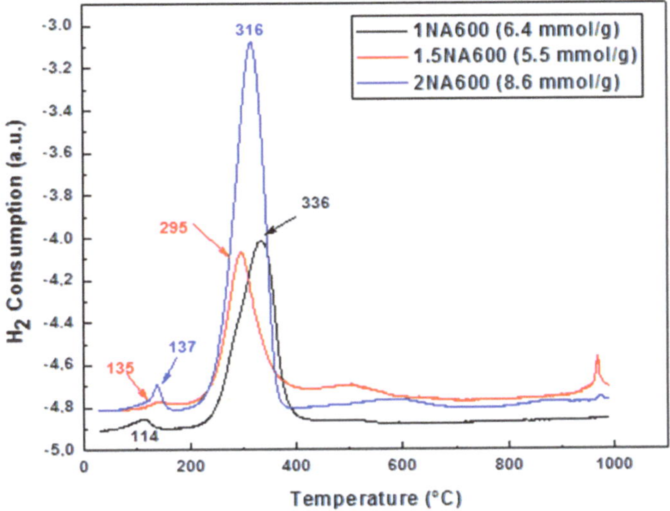

Figure 7. H$_2$-TPR profiles of fresh catalysts.

The catalysts were subjected to an in situ reduction at 400 °C for 1 h after being mounted into the tubular reactor. XRD analyses were then performed on the reduced catalysts. XRD patterns (Figure 8) displayed two or three distinct phases, depending on the concentration. The different peaks were identified as metallic Ag (PDF 89-3722) with peaks at 2θ = 38.12° (111), 44.307° (200), 64.45° (220) and 77.41° (311); metallic Ni (PDF 04-0850) with peaks at 2θ = 44.5° (111), 51.8° (200) and 76.4° (220) and NiO (PDF 47-1049) with peaks at 2θ = 37.24° (111), 43.27° (200), 62.87° (220) and 75.41° (311) for both 1.5NA600 and 2NA600 samples. In the case of the 1NA600 sample, no NiO peaks were observed, indicating that NiO had been completely reduced to metallic Ni during the reduction process. Conversely, the presence of NiO peaks in the diffractogram of the 1.5NA600 and 2NA600 samples suggests that NiO was not fully reduced to Ni during the H$_2$ treatment probably due to the high concentration of nickel oxide present in the sample as observed by EDX analysis (Table 2). The crystallite size of metallic species increases after reduction (Table 2); this increase is very remarkable for the low ratio formulation (Ni/Ag = 1).

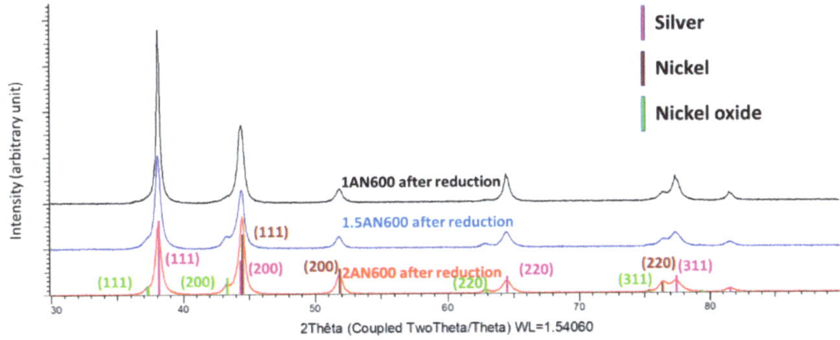

Figure 8. XRD patterns of reduced catalysts after reduction at 400 °C for 1 h.

2.2. Catalytic Performance in CO_2 Reforming of Methane

2.2.1. The Catalytic Activity versus Temperature Test

The catalysts obtained after calcination at 600 °C were tested for DRM. We initially established a standard reaction condition to evaluate the performance of the catalysts. In the first series of experiments, the activities of the catalysts were tested at temperatures ranging from 600 to 850 °C. Figures 9 and 10 illustrate the plots of activity, including CH_4 and CO_2 conversions, H_2 selectivity and H_2/CO ratio, as a function of temperature. The CH_4 and CO_2 conversions exhibit similar trends for the three catalysts. The highest CH_4 and CO_2 conversions are achieved at 650 °C for 1NA600. Catalysts 1.5NA600 and 2NA600 exhibit comparable conversion behaviors. CO_2 activation initiates at 650 °C for catalyst 1NA600, 700 °C for 1.5NA600 and 750 °C for 2NA600. There is a variation in the H_2/CO ratio among the three catalysts. At 650 °C, the H_2/CO ratio reaches its maximum value of 0.7 for 1NA600. In contrast, 1.5NA600 and 2NA600 yield values of 0.1 and 0.4, respectively. Regarding selectivity, 2NA600 demonstrates the highest selectivity. The maximum selectivity of 85% is achieved at 600 °C for 2NA600. At 650 °C and 850 °C, the H_2 selectivity reaches its highest values of 71% and 73%, respectively, for 1NA600. The contribution of side reactions, such as (i) $CO + 3H_2 \rightarrow CH_4 + H_2O$, (ii) $CO_2 + 4H_2 \rightarrow CH_4 + 2H_2O$ and/or (iii) RWGS ($CO_2 + H_2 \leftrightarrow CO + H_2O$), is probable and conducted mainly to lose the activity. The participation of $CO + 3H_2 \rightarrow CH_4 + H_2O$ and $CO_2 + 4H_2 \rightarrow CH_4 + 2H_2O$ reactions can be excluded for thermodynamic reasons ($\Delta G < 0$ if $T < 500$ °C for both reactions, at 800 °C, $\Delta G(i)\sim+55.41$ Kj/mol and $\Delta G(ii)\sim+55.98$ Kj/mol). The contribution of RWGS reaction which tends to reduce H_2 production seems thus the most likely ($\Delta G\sim0.5$ Kj/mol), leading to a reoxidation of the active sites ($Ni°$) by H_2O vapor formed during the catalytic test.

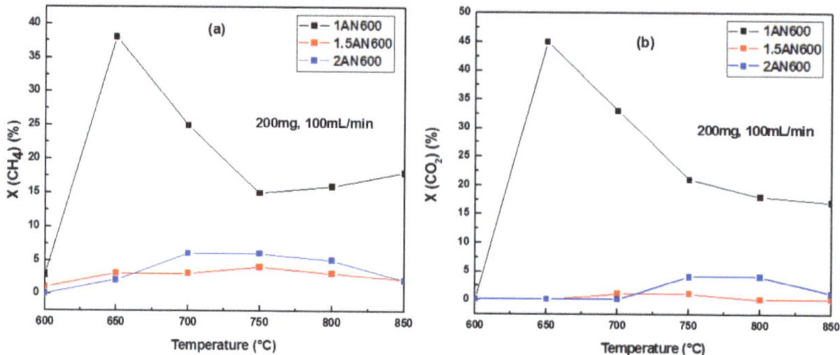

Figure 9. CH_4 (**a**) and CO_2 (**b**) conversions obtained on the fresh catalysts performed between 600 and 850 °C.

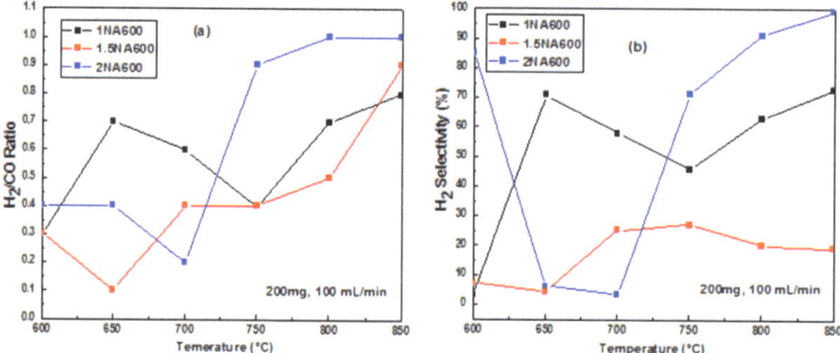

Figure 10. H_2/CO ratio (**a**) and H_2 selectivity (**b**) obtained on the fresh catalysts performed between 600 and 850 °C.

The textural and structural properties of 1NA600 may explain its higher activity compared to 1.5NA600 and 2NA600 catalysts. The higher surface area (20 m^2/g vs. 3–5 m^2/g) and smaller crystallite size of NiO (Cs = 8.7 Å vs. 12.6–16.8 Å) in 1NA600 likely contribute to the increased conversion of reactants. The reducibility patterns (Figure 7) observed for 2NA600 differ from those of 1NA600 and 1.5NA600, which could explain its higher selectivity. The unexpectedly low activity of 1.5NA600 and 2NA600 is quite surprising considering the significant amount of Ni species used (Ni/Ag = 1.5 and 2), which plays a crucial role in the DRM reaction. Above 750 °C, the curves of CH$_4$ and CO$_2$ conversion no longer exhibit significant changes, indicating progressive deactivation of the catalysts, possibly due to sintering [84]. According to previous literature [84–86], coke deposition is attributed to metallic nickel particles. Effectively, catalysts may undergo significant modifications during the reaction. Most probably, the NiO species are reduced to form metallic nickel, which is then could be responsible for the high carbon deposition [84]. It is worth noting that 1NA600 and 2NA600 exhibit low carbon loss, while 1.5NA600 shows no loss. Despite the reduction of NiO, the low carbon deposit observed may be attributed to the introduction of Ag, which minimizes carbon deposition in DRM [11]. Additionally, the disappearance of the active Ni-metallic phase is probably due to the formation of Ni-Ag alloy. Ni-Ag alloy was not detected by XRD analysis (after the isothermal tests). Indeed, we cannot totally exclude the possibility of Ni-Ag alloy formation for many reasons. Ni-Ag alloy could be present in extremely low amounts or maybe well-dispersed, which makes it difficult to detect by the XRD technique. Also, as suggested, Ni-Ag alloy cannot be detected probably due to its amorphous phase.

Based on the research findings presented in references [49,51], it is evident that both Ag and Ni exhibit a significant energetic driving force that favors the formation of a surface alloy instead of a bulk alloy. Specifically, Ag tends to concentrate on the step sites, which are specific crystal defects found on the material's surface. Benrabaa et al. [85] suggest that the stabilization of Ni metallic species in this alloy could prevent surface sintering and, consequently, the formation of coke. In contrast, the absence of coke deposition in 1.5NA600 can be attributed to its low activity, which is further diminished by the presence of impurities and its specific physicochemical characteristics. In conclusion, 1NA600 demonstrates the best catalytic activity indicating that a Ni/Ag ratio of 1 provides the optimal performance among these formulations.

2.2.2. The Time-on-Stream Test

Afterwards, a series of 4 h time-on-stream tests were conducted at 650 °C to evaluate the stability of the synthesized catalysts over time, as illustrated in Figures 11 and 12. Isothermal tests were performed using a new catalyst sample heated to the reaction temperature (650 °C) in inert gas and exposed to reactants for approximately 4 h under the same reaction conditions. Among the catalysts, 1NA600 exhibited higher conversion compared to 1.5NA600 and 2NA600, but the conversion of reactants gradually decreased with time. On the other hand, the conversion of reactants for 1.5NA600 and 2NA600 remained relatively low and stable throughout the studied period (up to 240 min). The CH$_4$ conversion was similar for both 1.5NA600 and 2NA600 catalysts. The sample 1NA600 catalyst showed signs of deactivation starting from the 12th minute, which could be attributed to sintering. However, some residual activity was observed after 150 min. The isothermal reactivity test at 650 °C is aligned with the results of the temperature rise test at the same temperature.

In contrast to the temperature rise test at 650 °C, which showed no conversion of CO$_2$, the isothermal reactivity test of 2NA600 revealed a low conversion of CO$_2$ at the 50th minute. The H$_2$/CO ratio for 1NA600 initially reached its maximum value of 0.5 but gradually decreased and stabilized around 0.3 after the 100th minute. For 2NA600, after an unexpected drop, the H$_2$/CO ratio progressively increased, approaching the value of 0.6. In the case of 1.5NA600, it reached its maximum value at the 30th minute and remained stable. While the H$_2$ selectivity tended to decrease with time for 1NA600, it started to increase again from the 130th minute. On the other hand, it continued to increase over time for 2NA600 and 1.5NA600.

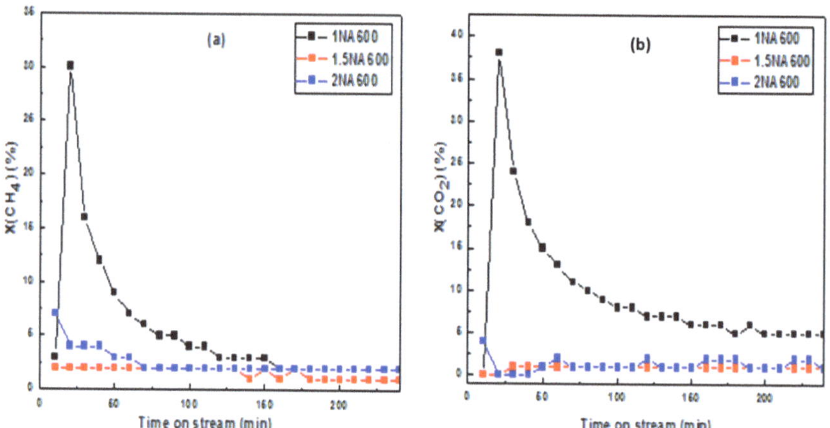

Figure 11. CH_4 (**a**) and CO_2 (**b**) conversions obtained on the fresh catalysts performed at 650 °C.

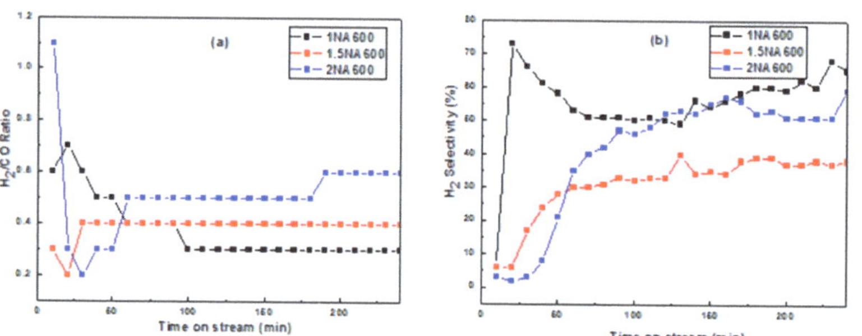

Figure 12. H_2/CO ratio (**a**) and H_2 selectivity (**b**) obtained on the fresh catalysts performed at 650 °C.

It is worth noting that the CH_4 conversion was higher than the CO_2 conversion for the 1.5NA600 and 2NA600 catalysts, which could be attributed to the combustion of methane given the low H_2/CO. This combustion is linked to the reduction of the catalyst, which releases oxygen responsible for the oxidation of methane. The results obtained at the 20 min mark for the three catalysts are consistent with those obtained during the temperature rise catalytic test at the temperature of 650 °C.

XRD (Figure 13) and Raman spectroscopy (Figure 14) analyses were conducted on the catalysts after the isothermal tests. XRD reveals lines corresponding to the cubic structure of silver (PDF 89-3722) with peaks at $2\theta = 38.12°$ (111), 44.307° (200) and 64.45° (220) and the metallic nickel phase (PDF 04-0850) with peaks at $2\theta = 44.5°$ (111), 51.8° (200) and 76.4° (220). Furthermore, despite the prior sieving of the catalysts, characteristic peaks of SiC are observed. Raman spectra show mainly the line of SiC at 760, 780 and 960 cm^{-1}, with no lines at 1500 cm^{-1} suggesting that there is no deposited carbon. The analyses after the reaction indicate that deactivation might be due mainly to sintering phenomena. After the reaction, a volcano change in crystallite size was observed (Table 2); the 1NA600 sample shows the lowest crystallite size value of Ni metallic species (30 nm).

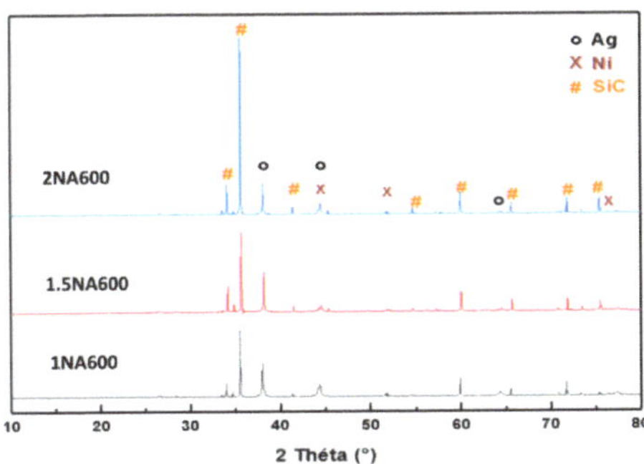

Figure 13. XRD analysis obtained on the spent catalysts performed at 650 °C.

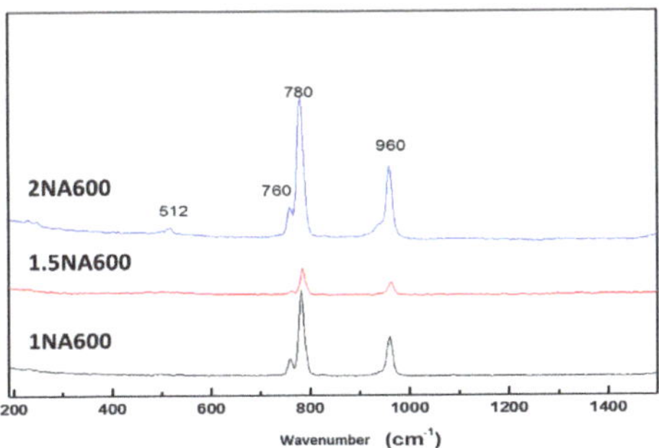

Figure 14. Raman spectra obtained on the spent catalysts performed at 650 °C.

3. Materials and Methods

3.1. Chemicals

Nickel (II) nitrate hexahydrate ($Ni(NO_3)_2,6H_2O$, ≥97%, Rectapur), silver nitrate ($AgNO_3$, ≥98%, Alfa Aesar), sodium hydroxide (NaOH, ≥99%, AnalaR NORMAPUR) and ethanol absolute (C_2H_5OH, ≥99.8%, AnalaR NORMAPUR) were used in catalysts preparation. All reagents were analytical grade and used without any further purification. Distilled water was used in the synthesis and washing processes.

3.2. Catalyst Preparation

Catalysts based on silver and nickel were synthesized using the coprecipitation method by varying the molar ratio of Ni and Ag (Ni/Ag = 1, 1.5 and 2). The starting salts, silver nitrate and nickel nitrate, are dissolved separately in distilled water. Then the two solutions are mixed and placed under moderate stirring. The precipitating agent NaOH (2M) was added drop by drop until the pH of the mixture reached the value of 10. The product obtained was washed with distilled water and ethanol, dried at 100 °C, then ground and calcined at 600 °C. The obtained powders are afterward denoted 1NA600, 1.5NA600 and 2NA600 in this manuscript.

3.3. Catalysts Characterization

Several physicochemical methods were used for the characterization of the catalysts before and after heating treatment and also after catalytic tests.

Thermogravimetry differential thermal analysis (TG-DSC) was performed on a TA instrument SDT Q600. The sample was heated at 5 °C/min from 25 to 1000 °C. The analyses were carried out with a total flow rate of 100 mL/min (20% O_2 in helium). The released gases that evolved during the analysis were monitored by a Pfeiffer Omnistar mass spectrometer. X-ray powder diffraction (XRD) was performed using a Bruker AXS D8 Advance diffractometer working in Bragg–Brentano geometry using Cu Kα radiation (λ = 1.54 Å), equipped with a LynxEye detector. Patterns were collected at room temperature, in the 2θ = 10–80° range, with a 0.02° step and 96s counting time per step. The EVA (version 6.1.0.4 Bruker AXS GmbH, Karlsruhe, Germany) software was used for phase identification, and the unit cell parameters were optimized by Rietveld refinement using the Jana2020 crystallographic software. The elemental analysis was performed by inductively coupled plasma-optic emission spectroscopy (ICP-OES) 720-ES ICP-OES (Agilent, Santa Clara, CA, USA) with axially viewing and simultaneous CCD detection. The quantitative determination of metal content in the catalysts was based on the analysis of certificated standard solutions. The ICP ExpertTM software (version 2.0.4) provides the concentration of metal in samples allowing estimating the weight percentage of Ag and Ni. All the analyses were performed 40 min after the spectrometer was turned on to achieve a stable plasma as well as constant and reproducible sample introduction. The sample preparation was made by dissolving 10 mg of dried and ground samples catalyst in concentrated 4 mL of HNO$_3$. All the sample solutions were heated to 80 °C in a sonicator for 2 h. Then, the solution was diluted up to 20 mL with ultrapure water before being analyzed by ICP-OES. The surface areas were calculated from N$_2$ adsorption–desorption isotherms measured on a FlowSorb III Micromeritics (Norcross, GA, USA) analyzer by Brunauer–Emmett–Teller (B.E.T). The morphology of catalysts was examined by scanning electron microscopy (SEM) (S-3400N, Hitachi, Tokyo, Japan). The energy-dispersive X-ray spectra (EDX) (Thermo, Waltham, MA, USA) SCIENTIFIC UltraDry) of samples were also acquired during their conventional scanning electron microscopic investigations. X-ray photoelectron spectroscopy (XPS) was carried out on Escalab 220 XL (Vacuum Generators, Seoul, Republic of Korea) spectrometer. A monochromatic Al Kα X-ray source was used, and electron energies were measured in the constant analyzer energy mode. The pass energy was 100 eV for the survey of spectra and 40 eV for the single-element spectra. All XPS binding energies were referred to C$1s$ core level at 285 eV. The angle between the incident X-rays and the analyzer was 58°, with photoelectrons being collected perpendicularly to the sample surface. Spectra were analyzed with the CasaXPS software (version 2.3.25). The reducible species that existed in the catalysts were profiled by temperature-programmed reduction. Hydrogen temperature-programmed reduction (H$_2$-TPR) was measured on a Micromeritics AutoChem II 2920 apparatus with a thermal conductivity detector (TCD) to monitor the H$_2$ consumption. After calibration of H$_2$ on the TCD, samples were sealed in a U-shaped quartz tube reactor and pre-treated in an argon atmosphere to remove surface impurities. Then, the temperature was raised from 25 to 1000 °C at 5 °C/min in a stream of 5% v/v H$_2$/ Ar. Laser-Raman spectra were recorded from 200 to 1500 cm^{-1} at room temperature using an FT-Raman spectrometer (Dilor XY Raman, Horiba France, Palaiseau, France) at an excitation wavelength of 647.1 nm, laser power of 3 mW and spectral resolution of 0.5 cm^{-1}.

3.4. Catalytic Reforming Experiments

The catalytic performance test was conducted under atmospheric pressure in a fixed-bed quartz flow reactor, which was placed inside a programmable oven. Prior to loading into the reactor, the catalyst (200 mg) was thoroughly mixed with silicon carbide (SiC) powder (1000 mg). A mixture of methane, carbon dioxide and argon (CH$_4$:CO$_2$:Ar = 20:20:60) with a total flow of 100 mL/min was introduced into the reactor, and the reaction was initiated by gradually increasing the temperature from room temperature up to 850 °C at a heating rate of 5 °C/min. Online mass spectrometry (OmniStar, Pfeiffer Vacuum, Asslar, Germany) was used to analyze all effluents during the reaction.

In the initial phase of the experiment, the catalysts were tested for their activities at temperatures ranging from 600 to 850 °C. Subsequently, isothermal reactivity tests were performed using a fresh sample of the catalyst heated to a reaction temperature of 650 °C. The samples were exposed to the same reaction conditions for approximately 4 h. After the evaluation through isothermal tests, each catalyst was characterized using XRD analysis.

4. Conclusions

In summary, the findings from this study on CO_2 reforming of methane lead to several significant conclusions. Firstly, the catalysts exhibited relative stability in composition and structure at high calcination temperatures. XRD analysis confirmed the presence of metallic silver and nickel oxide as the main phases in the catalysts after calcination at different temperatures. EDX analysis further detected the presence of Ag and Ni elements, with a higher concentration of nickel observed in the upper layers of the catalysts. The morphological and surface properties of the catalysts were also investigated. BET analysis revealed a decrease in specific surface area with an increase in the Ni/Ag ratio, indicating the agglomeration of silver particles on the catalyst surface. SEM images showed variations in particle shapes and agglomeration patterns influenced by the loading of nickel in the catalysts. Regarding catalytic performance, catalyst 1NA600 demonstrated the highest activity, with the highest CH_4, CO_2 conversions and H_2/CO ratio achieved at 650 °C. 2NA600 catalyst exhibited the highest selectivity, reaching a maximum of 85% at 600 °C. 1NA600 catalyst showed higher activity attributed to its higher specific surface area and smaller crystallite size of NiO. Both 1NA600 and 2NA600 demonstrated low carbon loss, while 1.5NA600 showed no loss. This study opens the path for further investigation through adequate process conditions (e.g., by adding another metal) in order to slow the reduction of the material or through further investigation of material design and synthesis to better stabilize the active species.

Author Contributions: Conceptualization and methodology: A.L. and R.B.; catalysts preparation: H.H.; characterizations: R.B., P.R. and A.L.; writing—original draft preparation: H.H., R.B., P.R. and A.L.; DRM experiments: H.H.; supervision: A.L. and R.B. All authors have read and agreed to the published version of the manuscript.

Funding: This work was partially supported by the Algerian Petroleum Institute/Sonatrach and UCCS-Lille.

Data Availability Statement: Data available within the article.

Acknowledgments: H.H. gratefully acknowledges UCCS in providing the necessary facilities to conduct the work. Additionally, H.H. extends her heartfelt appreciation to Pardis Simon, Olivier Gardol, Laurence Burylo and Martine Trentesaux for the technical assistance and fruitful discussion. Chevreul Institute (CNRS FR2638) is acknowledged for funding X-ray and XPS facilities. H.H. is grateful for the support from the IAP Institute.

Conflicts of Interest: The authors declare that they have no known competing financial interests or personal relationships that could have appeared to influence the work reported in this paper.

References

1. Le Saché, E.; Reina, T.R. Analysis of Dry Reforming as direct route for gas phase CO_2 conversion. The past, the present and future of catalytic DRM technologies. *Prog. Energy Combust. Sci.* **2022**, *89*, 100970. [CrossRef]
2. Fischer, F.; Tropsch, H. The Preparation of Synthetic Oil Mixture from Carbon Monoxide and Hydrogen. *Brennstoff-Chem* **1928**, *9*, 539.
3. Tungatarova, S.; Xanthopoulou, G. Ni-Al Self-Propagating High-Temperature Synthesis Catalysts in Dry Reforming of Methane to Hydrogen-Enriched Fuel Mixtures. *Catalysts* **2022**, *12*, 1270. [CrossRef]
4. Ashcroft, A.; Cheetham, A.K. Partial oxidation of methane to synthesis gas using carbon dioxide. *Nature* **1991**, *352*, 225–226. [CrossRef]
5. Liu, C.J.; Ye, J. Progresses in the preparation of coke resistant Ni-based catalyst for steam and CO_2 reforming of methane. *ChemCatChem* **2011**, *3*, 529–541. [CrossRef]
6. Hu, Y.H. Advances in catalysts for CO_2 reforming of methane. Advances in CO_2 conversion and utilization. *ACS Publ.* **2010**, *1056*, 155–174. [CrossRef]
7. Chen, W.; Zhao, G. High carbon-resistance Ni/CeAlO$_3$-Al$_2$O$_3$ catalyst for CH_4/CO_2 reforming. *Appl. Catal. B Environ.* **2013**, *136*, 260–268. [CrossRef]
8. Aziz, M.; Setiabudi, H. A review of heterogeneous catalysts for syngas production via dry reforming. *J. Taiwan Inst. Chem. Eng.* **2019**, *101*, 139–158. [CrossRef]
9. Rostrup-Nielsen, J.R. *Catalytic Steam Reforming*; Springer: Berlin/Heidelberg, Germany, 1984. [CrossRef]

10. Lercher, J.A.; Bitter, J.H.; Steghuis, A.G.; Van Ommen, J.G.; Seshan, K. Methane utilisation via synthesis gas generation-catalytic chemistry and technology. In *Environmental Catalysis*; World Scientific Publishing Co Pte Ltd.: Singapore, 1999. [CrossRef]

11. Yu, M.; Zhu, Y.A.; Lu, Y.; Tong, G.; Zhu, K.; Zhou, X. The promoting role of Ag in Ni-CeO$_2$ catalyzed CH$_4$-CO$_2$ dry reforming reaction. *Appl. Catal. B Environ.* **2015**, *165*, 43–56. [CrossRef]

12. Djinović, P.; Črnivec, I.G.O. Influence of active metal loading and oxygen mobility on coke-free dry reforming of Ni–Co bimetallic catalysts. *Appl. Catal. B Environ.* **2012**, *125*, 259–270. [CrossRef]

13. San-José-Alonso, D.; Juan-Juan, J. Ni, Co and bimetallic Ni–Co catalysts for the dry reforming of methane. *Appl. Catal. A Gen.* **2009**, *371*, 54–59. [CrossRef]

14. Gaur, S.; Haynes, D.J. Rh, Ni, and Ca substituted pyrochlore catalysts for dry reforming of methane. *Appl. Catal. A Gen.* **2011**, *403*, 142–151. [CrossRef]

15. Lavoie, J.M. Review on dry reforming of methane, a potentially more environmentally-friendly approach to the increasing natural gas exploitation. *Front. Chem.* **2014**, *2*, 81. [CrossRef] [PubMed]

16. Ginsburg, J.M.; Piña, J.; El Solh, T.; De Lasa, H.I. Coke formation over a nickel catalyst under methane dry reforming conditions: Thermodynamic and kinetic models. *Ind. Eng. Chem. Res.* **2005**, *44*, 4846–4854. [CrossRef]

17. Bradford, M.C.; Vannice, M.A. Catalytic reforming of methane with carbon dioxide over nickel catalysts II. Reaction kinetics. *Appl. Catal. A Gen.* **1996**, *142*, 97–122. [CrossRef]

18. Pakhare, D.; Spivey, J. A review of dry (CO$_2$) reforming of methane over noble metal catalysts. *Chem. Soc. Rev.* **2014**, *43*, 7813–7837. [CrossRef]

19. James, O.O.; Maity, S. Towards reforming technologies for production of hydrogen exclusively from renewable resources. *Green Chem.* **2011**, *13*, 2272–2284. [CrossRef]

20. Labinger, J.A.; Bercaw, J.E. Understanding and exploiting C–H bond activation. *Nature* **2002**, *417*, 507–514. [CrossRef] [PubMed]

21. Yao, L.; Shi, J.; Xu, H.; Shen, W.; Hu, C. Low-temperature CO$_2$ reforming of methane on Zr-promoted Ni/SiO$_2$ catalyst. *Fuel Process. Technol.* **2016**, *144*, 1–7. [CrossRef]

22. Tomiyama, S.; Takahashi, R.; Sato, S.; Sodesawa, T.; Yoshida, S. Preparation of Ni/SiO$_2$ catalyst with high thermal stability for CO$_2$-reforming of CH$_4$. *Appl. Catal. A Gen.* **2003**, *241*, 349–361. [CrossRef]

23. Uchida, T.; Ikeda, I.Y.; Takeya, S.; Kamata, Y.; Ohmura, R.; Nagao, J.; Buffett, B.A. Kinetics and stability of CH$_4$–CO$_2$ mixed gas hydrates during formation and long-term storage. *ChemPhysChem* **2005**, *6*, 646–654. [CrossRef]

24. Al-Fatesh, A.S.; Amin, A.; Ibrahim, A.A.; Khan, W.U.; Soliman, M.A.; AL-Otaibi, R.L.; Fakeeha, A.H. Effect of Ce and Co addition to Fe/Al$_2$O$_3$ for catalytic methane decomposition. *Catalysts* **2016**, *6*, 40. [CrossRef]

25. Mette, K.; Kühl, S.; Tarasov, A.; Düdder, H.; Kähler, K.; Muhler, M.; Behrens, M. Redox dynamics of Ni catalysts in CO$_2$ reforming of methane. *Catal. Today* **2015**, *242*, 101–110. [CrossRef]

26. Angeli, S.D.; Turchetti, L.; Monteleone, G.; Lemonidou, A.A. Catalyst development for steam reforming of methane and model biogas at low temperature. *Appl. Catal. B Environ.* **2016**, *181*, 34–46. [CrossRef]

27. Cao, C.; Bourane, A.; Schlup, J.R.; Hohn, K.L. In situ IR investigation of activation and catalytic ignition of methane over Rh/Al$_2$O$_3$ catalysts. *Appl. Catal. A Gen.* **2008**, *344*, 78–87. [CrossRef]

28. Bian, Z.; Kawi, S. Highly carbon-resistant Ni–Co/SiO$_2$ catalysts derived from phyllosilicates for dry reforming of methane. *J. CO2 Util.* **2017**, *18*, 345–352. [CrossRef]

29. Nikoo, M.K.; Amin, N. Thermodynamic analysis of carbon dioxide reforming of methane in view of solid carbon formation. *Fuel Process. Technol.* **2011**, *92*, 678–691. [CrossRef]

30. Challiwala, M.S.; Ibrahim, G.; Choudhury, H.A.; Elbashir, N.O. Scaling Up the Advanced Dry Reforming of Methane (DRM) Reactor System for Multi-Walled Carbon Nanotubes and Syngas Production: An Experimental and Modeling Study. *Chem. Eng. Process. Process Intensif.* **2024**, *197*, 109693. [CrossRef]

31. Pinto, D.; Hu, L.; Urakawa, A. Enabling complete conversion of CH$_4$ and CO$_2$ in dynamic coke-mediated dry reforming (DC-DRM) on Ni catalysts. *Chem. Eng. J.* **2023**, *474*, 145641. [CrossRef]

32. Gamal, A.; Eid, K.; Abdullah, A.M. Engineering of Pt-based nanostructures for efficient dry (CO$_2$) reforming: Strategy and mechanism for rich-hydrogen production. *Int. J. Hydrogen Energy* **2022**, *47*, 5901–5928. [CrossRef]

33. Faroldi, B.; Múnera, J. Well-dispersed Rh nanoparticles with high activity for the dry reforming of methane. *Int. J. Hydrogen Energy* **2017**, *42*, 16127–16138. [CrossRef]

34. Luisetto, I.; Sarno, C. Packed and monolithic reactors for the dry reforming of methane: Ni supported on γ-Al$_2$O$_3$ promoted by Ru. *Adv. Sci. Lett.* **2017**, *23*, 5977–5979. [CrossRef]

35. Xanthopoulou, G.; Varitis, S.; Zhumabek, M.; Karanasios, K.; Vekinis, G.; Tungatarova, S.A.; Baizhumanova, T.S. Direct Reduction in Greenhouse Gases by Continuous Dry (CO$_2$) Reforming of Methane over Ni-Containing SHS Catalysts. *Energies* **2021**, *14*, 6078. [CrossRef]

36. Abdullah, B.; Abd Ghani, N.A. Recent advances in dry reforming of methane over Ni-based catalysts. *J. Clean. Prod.* **2017**, *162*, 170–185. [CrossRef]

37. Chen, L.W.; Ng, K.H. Hydrothermally synthesized tremella-like monometallic Ni/SiO$_2$ for effective and stable dry reforming of methane (DRM) to syngas. *Process Saf. Environ. Prot.* **2024**, *183*, 244–259. [CrossRef]

38. Zhao, R.; Cao, K.; Ye, R.; Tang, Y.; Du, C.; Liu, F.; Shan, B. Deciphering the stability mechanism of Pt-Ni/Al$_2$O$_3$ catalysts in syngas production via DRM. *Chem. Eng. J.* **2024**, *491*, 151966. [CrossRef]

39. Gutiérrez, M.C.; Hernández, P.S.; Anzures, F.M.; Martínez, A.G.; Galicia, G.M.; García, M.E.F.; Pérez-Hernández, R. MgO impregnation to Al_2O_3 supported Ni catalyst for SYNGAS production using greenhouse gases: Some aspects of chemical state of Ni species. *Int. J. Hydrogen Energy* **2024**, *52*, 1131–1140. [CrossRef]

40. Alabi, W.O.; Adesanmi, B.M.; Wang, H.; Patzig, C. Correlation of MgO loading to spinel inversion, octahedral site occupancy, site generation and performance of bimetal Co–Ni catalyst for dry reforming of CH_4. *Int. J. Hydrogen Energy* **2024**, *51*, 1087–1098. [CrossRef]

41. Wen, F.; Xu, C.; Huang, N.; Wang, T.; Sun, X.; Li, H.; Xia, G. Exceptional stability of spinel Ni–$MgAl_2O_4$ catalyst with ordered mesoporous structure for dry reforming of methane. *Int. J. Hydrogen Energy* **2024**, *69*, 1481–1491. [CrossRef]

42. Zhang, Z.; Bi, G.; Zhong, J.; Xie, J. Key features of nano NixMg1-xAl_2O_4 spinel for biogas bi-reforming: CO_2 activation and coke elimination. *Int. J. Hydrogen Energy* **2024**, *53*, 383–393. [CrossRef]

43. Yang, T.; Zhang, J.; Rao, Q.; Gai, Z.; Li, Y.; Li, P.; Jin, H. Reversible exsolution of iron from perovskites for highly selective syngas production via chemical looping dry reforming of methane. *Fuel* **2024**, *366*, 131386. [CrossRef]

44. Cao, D.; Luo, C.; Luo, T.; Shi, Z.; Wu, F.; Li, X.; Zhang, L. Dry reforming of methane by La_2NiO_4 perovskite oxide: B-site substitution improving reactivity and stability. *Chem. Eng. J.* **2024**, *482*, 148701. [CrossRef]

45. Jagadeesh, P.; Varun, Y.; Reddy, B.H.; Sreedhar, I.; Singh, S.A. A short review on recent advancements of dry reforming of methane (DRM) over pyrochlores. *Mater. Today Proc.* **2024**, *72*, 361–369. [CrossRef]

46. Cheng, H.; Su, P.; Yuan, S.W.; Wang, Z.; Liu, M. Preparation of a green and highly active vermiculite-derived hydrotalcite and its application in methane dry reforming. *Int. J. Hydrogen Energy* **2024**, *50*, 726–733. [CrossRef]

47. Maziviero, F.V.; Melo, D.M.; Medeiros, R.L.; Silva, J.C.; Araújo, T.R.; Oliveira, Â.A.; Melo, M.A. Influence of Mn, Mg, Ce and P promoters on Ni-X/Al_2O_3 catalysts for dry reforming of methane. *J. Energy Inst.* **2024**, *113*, 101523. [CrossRef]

48. Chaudhary, K.J.; Al-Fatesh, A.S.; Ibrahim, A.A.; Osman, A.I.; Fakeeha, A.H.; Alhoshan, M.; Kumar, R. Enhanced hydrogen production through methane dry reforming: Evaluating the effects of promoter-induced variations in reducibility, basicity, and crystallinity on Ni/ZSM-5 catalyst performance. *Energy Convers. Manag.* **2024**, *23*, 100631. [CrossRef]

49. Vang, R.T.; Honkala, K. Controlling the catalytic bond-breaking selectivity of Ni surfaces by step blocking. *Nat. Mater.* **2005**, *4*, 160–162. [CrossRef] [PubMed]

50. Parizotto, N.; Rocha, K. Alumina-supported Ni catalysts modified with silver for the steam reforming of methane: Effect of Ag on the control of coke formation. *Appl. Catal. A Gen.* **2007**, *330*, 12–22. [CrossRef]

51. Lauritsen, J.V.; Vang, R.T. From atom-resolved scanning tunneling microscopy (STM) studies to the design of new catalysts. *Catal. Today* **2006**, *111*, 34–43. [CrossRef]

52. Gavrielatos, I.; Montinaro, D. Thermogravimetric and Electrocatalytic Study of Carbon Deposition of Ag-doped Ni/YSZ Electrodes under Internal CH_4 Steam Reforming Conditions. *Fuel Cells* **2009**, *9*, 883–890. [CrossRef]

53. Jeong, H.; Kang, M. Hydrogen production from butane steam reforming over Ni/Ag loaded $MgAl_2O_4$ catalyst. *Appl. Catal. B Environ.* **2010**, *95*, 446–455. [CrossRef]

54. Rovik, A.K.; Klitgaard, S.K. Effect of alloying on carbon formation during ethane dehydrogenation. *Appl. Catal. A Gen.* **2009**, *358*, 269–278. [CrossRef]

55. Xu, Y.; Fan, C. Effect of Ag on the control of Ni-catalyzed carbon formation: A density functional theory study. *Catal. Today* **2012**, *186*, 54–62. [CrossRef]

56. Zhu, Y.A.; Chen, D. First-principles calculations of C diffusion through the surface and subsurface of Ag/Ni (1 0 0) and reconstructed Ag/Ni (1 0 0). *Surf. Sci.* **2010**, *604*, 186–195. [CrossRef]

57. Boldrin, P.; Ruiz-Trejo, E.; Mermelstein, J.; Bermúdez Men´endez, J.M.; Ramírez Reina, T.; Brandon, N.P. Strategies for Carbon and Sulfur Tolerant Solid Oxide Fuel Cell Materials, Incorporating Lessons from Heterogeneous Catalysis. *Chem. Rev.* **2016**, *116*, 13633–13684. [CrossRef] [PubMed]

58. Carrasco-Ruiz, S.; Zhang, Q.; Gándara-Loe, J.; Pastor-Pérez, L.; Odriozola, J.A.; Reina, T.R.; Bobadilla, L.F. H_2-rich syngas production from biogas reforming: Overcoming coking and sintering using bimetallic Ni-based catalysts. *Int. J. Hydrogen Energy* **2023**, *48*, 27907–27917. [CrossRef]

59. Fang, X.; Mao, L.; Xu, L.; Shen, J.; Xu, J.; Xu, X.; Wang, X. Ni-Fe/La_2O_3 bimetallic catalysts for methane dry reforming: Elucidating the role of Fe for improving coke resistance. *Fuel* **2024**, *357*, 129950. [CrossRef]

60. Armengol-Profitós, M.; Braga, A.; Pascua-Solé, L.; Lucentini, I.; Garcia, X.; Soler, L.; Llorca, J. Enhancing the performance of a novel CoRu/CeO_2 bimetallic catalyst for the dry reforming of methane via a mechanochemical process. *Appl. Catal. B Environ.* **2024**, *345*, 123624. [CrossRef]

61. Diao, J.F.; Zhang, T.; Xu, Z.N.; Guo, G.C. The atomic-level adjacent NiFe bimetallic catalyst significantly improves the activity and stability for plasma-involved dry reforming reaction of CH_4 and CO_2. *Chem. Eng. J.* **2023**, *467*, 143271. [CrossRef]

62. Pandey, S.K.; Tripathi, M.K.; Ramanathan, V.; Mishra, P.K.; Tiwary, D. Highly facile Ag/NiO nanocomposite synthesized by sol-gel method for mineralization of rhodamine B. *J. Phys. Chem. Solids* **2021**, *159*, 110287. [CrossRef]

63. Zhou, L.X.; Yang, Y.Y. In situ synthesis of Ag/NiO derived from hetero-metallic MOF for supercapacitor application. *Chem. Pap.* **2021**, *75*, 1795–1807. [CrossRef]

64. Nagamuthu, S.; Ryu, K.S. Synthesis of Ag/NiO honeycomb structured nanoarrays as the electrode material for high performance asymmetric supercapacitor devices. *Sci. Rep.* **2019**, *9*, 4864. [CrossRef]

65. Aydoghmish, S.M.; Hassanzadeh-Tabrizi, S. Facile synthesis and investigation of NiO–ZnO–Ag nanocomposites as efficient photocatalysts for degradation of methylene blue dye. *Ceram. Int.* **2019**, *45*, 14934–14942. [CrossRef]

66. Khedkar, C.V.; Vedpathak, A.S. Synthesis, characterization, electrochemical and catalytic performance of NiO nanostructures and Ag-NiO nanocomposite. *Chem. Phys. Impact* **2023**, *6*, 100153. [CrossRef]

67. Zhang, X.W.; Su, Y.X. Effect of Ag on deNOx performance of SCR-C_3H_6 over Fe/Al-PILC catalysts. *J. Fuel Chem. Technol.* **2019**, *47*, 1368–1378. [CrossRef]

68. Predoi, D.; Jitianu, A. Study of FexOy–SiO_2 nanoparticles obtained by sol-gel synthesis. *Dig. J. Nanomater. Biostruct* **2006**, *1*, 93–97.

69. Singh, A.; Yadav, B.C. Green synthesized ZnO/NiO heterostructures based quick responsive LPG sensor for the detection of below LEL with DFT calculations. *Results Surf. Interfaces* **2023**, *11*, 100103. [CrossRef]

70. Prieto, P.; Nistor, V.; Nouneh, K.; Oyama, M.; Abd-Lefdil, M.; Díaz, R. XPS study of silver, nickel and bimetallic silver–nickel nanoparticles prepared by seed-mediated growth. *Appl. Surf. Sci.* **2012**, *258*, 8807–8813. [CrossRef]

71. X-ray Photoelectron Spectroscopy (XPS) reference pages. Available online: http://www.xpsfitting.com/ (accessed on 1 March 2024).

72. Biesinger, M.C.; Payne, B.P.; Grosvenor, A.P.; Lau, L.W.; Gerson, A.R.; Smart, R.S.C. Resolving surface chemical states in XPS analysis of first row transition metals, oxides and hydroxides: Cr, Mn, Fe, Co and Ni. *Appl. Surf. Sci.* **2011**, *257*, 2717–2730. [CrossRef]

73. Wandekar, R.; Ali, M. Physicochemical studies of NiO–GDC composites. *Mater. Chem. Phys.* **2006**, *99*, 289–294. [CrossRef]

74. Augusto, B.L.; Costa, L.O. Ethanol reforming over Ni/CeGd catalysts with low Ni content. *Int. J. Hydrogen Energy* **2012**, *37*, 12258–12270. [CrossRef]

75. Pérez-Hernández, R.; Galicia, G.M. Synthesis and characterization of bimetallic Cu–Ni/ZrO_2 nanocatalysts: H_2 production by oxidative steam reforming of methanol. *Int. J. Hydrogen Energy* **2008**, *33*, 4569–4576. [CrossRef]

76. Ashok, J.; Subrahmanyam, M. Hydrotalcite structure derived Ni–Cu–Al catalysts for the production of H_2 by CH_4 decomposition. *Int. J. Hydrogen Energy* **2008**, *33*, 2704–2713. [CrossRef]

77. Montoya, J.A.; Romero-Pascual, E. Methane reforming with CO_2 over Ni/ZrO_2–CeO_2 catalysts prepared by sol–gel. *Catal. Today* **2000**, *63*, 71–85. [CrossRef]

78. Shan, W.; Luo, M. Reduction property and catalytic activity of $Ce_{1-x}Ni_xO_2$ mixed oxide catalysts for CH_4 oxidation. *Appl. Catal. A Gen.* **2003**, *246*, 1–9. [CrossRef]

79. Kirumakki, S.R.; Shpeizer, B.G. Hydrogenation of naphthalene over NiO/SiO_2–Al_2O_3 catalysts: Structure–activity correlation. *J. Catal.* **2006**, *242*, 319–331. [CrossRef]

80. Joo, O.S.; Jung, K.D. CH_4 dry reforming on alumina-supported nickel catalyst. *Bull. Korean Chem. Soc.* **2002**, *23*, 1149–1153. [CrossRef]

81. Casenave, S.; Martinez, H. Acid–base properties of Mg–Ni–Al mixed oxides using LDH as precursors. *Thermochim. Acta* **2001**, *379*, 85–93. [CrossRef]

82. Casenave, S.; Martinez, H. Acid-Base properties of MgCuAl mixed oxides. *J. Therm. Anal. Calorim.* **2003**, *72*, 191–198. [CrossRef]

83. Benrabaa, R.; Löfberg, A.; Rubbens, A.; Bordes-Richard, E.; Vannier, R.N.; Barama, A. Structure, reactivity and catalytic properties of nanoparticles of nickel ferrite in the dry reforming of methane. *Catal. Today* **2013**, *203*, 188–195. [CrossRef]

84. Hallassi, M.; Benrabaa, R. Characterization and Syngas Production at Low Temperature via Dry Reforming of Methane over Ni-M (M= Fe, Cr) Catalysts Tailored from LDH Structure. *Catalysts* **2022**, *12*, 1507. [CrossRef]

85. Benrabaa, R.; Boukhlouf, H.; Löfberg, A.; Rubbens, A.; Vannier, R.N.; Bordes-Richard, E.; Barama, A. Nickel ferrite spinel as catalyst precursor in the dry reforming of methane: Synthesis, characterization and catalytic properties. *J. Nat. Gas Chem.* **2012**, *21*, 595–604. [CrossRef]

86. Slagtern, Å.; Olsbye, U. Partial oxidation of methane to synthesis gas using La-MO catalysts. *Appl. Catal. A Gen.* **1994**, *110*, 99–108. [CrossRef]

MDPI AG
Grosspeteranlage 5
4052 Basel
Switzerland
Tel.: +41 61 683 77 34

Catalysts Editorial Office
E-mail: catalysts@mdpi.com
www.mdpi.com/journal/catalysts